INTRODUCTION TO MICROWAVE TECHNOLOGY

INTRODUCTION TO MICROWAVE TECHNOLOGY

FRED MONACO
Los Angeles Trade-Technical College

Merrill Publishing Company
A Bell & Howell Information Company
Columbus Toronto London Melbourne

Published by Merrill Publishing Company
A Bell & Howell Information Company
Columbus, Ohio 43216

This book was set in Times Roman.
Administrative Editor: Stephen Helba
Production Coordinator: Anne Daly
Art Coordinator: Patrick Welch
Cover Designer: Brian Deep

Credits. All photos copyrighted by individuals or companies listed. Page 209, Epsco; p. 211, Frequency Sources/a subsidiary of Loral Corp.; pp. 52, 186, Hewlett-Packard; pp. 148, 154, 158, Russell Illig; p. 144, Maury Microwave Corp.; pp. 166, 178, Gail Meese; p. 141, Microtech Inc.; p. 31, Fred Monaco; pp. 174, 184, 235, Raytheon. Figure 5–27, p. 184, from Hewlett-Packard Application Note No. 12., Fig. 1, "How A Helix Backward-Wave Tube Works", Oct. 15, 1965. Concepts for Figures 3.14, p. 87, and 7.2, p. 229, by Roberta Monaco.

Library of Congress Catalog Card Number: 88–062356
International Standard Book Number: 0–675–21030–5
Printed in the United States of America
1 2 3 4 5 6 7 8 9—93 92 91 90 89

This book is dedicated to my wife, Roberta, and my two children, Petrouchka and Lonnie, whose love served as an inspiration throughout this book.

PREFACE

Introduction to Microwave Technology is intended primarily for students in advanced technician-level programs of vocational-technical schools, community colleges, and technical institutes. The textbook may be used either as the principal resource in a semester-length microwave course, or as a supplementary reference in a general one-year communications program. Emphasis is on introductory concepts in microwave technology with applications to telecommunications and radar. Although a basic understanding of algebra and simple trigonometry is the only mathematics requirement, more advanced students will enjoy working through the few calculus derivations of the first chapter. Other students may skip these with no loss of comprehension.

The text begins with an in-depth discussion of transmission line theory basic to a thorough understanding of all microwave phenomena. Subsequent chapters take the student from an elementary analysis of the wave propagation model to a concluding discussion of radar principles. The sequence of chapters follows a logical flow from rudimentary ideas to advanced topics; however, since each chapter is essentially self-contained, the order of presentation (after the first chapter) may be varied to suit the needs of individual instructors and programs.

Essential AM and FM topics are summarized in an appendix. The basic concepts covered in this appendix may be applied to microwave topics at the instructor's discretion. Other appendices include an extensive glossary of terms and formulas used throughout the text; microwave safety precautions; and a comprehensive discussion of decibels.

I wish to thank all those people in industry who provided helpful ideas and suggestions concerning the content of this text. I also wish to acknowledge a personal debt to my wife, Roberta, whose advice, understanding, and tireless efforts have been so very crucial to the text's completion. My thanks go also to these reviewers who offered useful comments and suggestions: K.H. Bailey (Wake Technical College, Raleigh,

N. C.), Thomas Bingham (St. Louis Community College at Florrisant Valley, St. Louis, Mo.), Charles Dewater (ITT Technical College, Grand Rapids, Mich.), Jim Georgias (DeVry Institute of Technology, Chicago, Ill.), Charles Higgins (Hawkeye Institute of Technology, Waterloo, Iowa.), Sam Kibler (ITT Technical Institute, La Mesa, Calif.), Ron Moody (Pima Community College, Tuscon, Ariz.), Walter Newlon (Hocking Technical College, Nelsonville, Ohio), Robert Tataronis (Wentworth Institute of Technology, Boston, Mass.), and Paul Young (Arizona State University, Tempe, Ariz.).

NOTICE TO THE READER

CONTENTS

1 Transmission Lines 1

MERRILL'S INTERNATIONAL SERIES IN ELECTRICAL AND ELECTRONICS TECHNOLOGY

ADAMSON	*Applied Pascal for Technology,* 20771–1
	Structured BASIC Applied to Technology, 20772–X
ASSER/ STIGLIANO/ BAHRENBURG	*Microcomputer Servicing,* 20907–2
	Microcomputer Theory and Servicing, 20659–6
	Lab Manual to accompany Microcomputer Theory and Servicing, 21109–3
BATESON	*Introduction to Control System Technology, Third Edition,* 21010–0
BEACH/JUSTICE	*DC/AC Circuit Essentials,* 20193–4
BERLIN	*Experiments in Electronic Devices, Second Edition,* 20881–5
	The Illustrated Electronics Dictionary, 20451–8
BERLIN/GETZ	*Principles of Electronic Instrumentation and Measurement,* 20449–6
	Experiments in Instrumentation and Measurement, 20450-X
BOGART	*Electronic Devices and Circuits,* 20317–1
BOGART/BROWN	*Experiments in Electronic Devices and Circuits,* 20488–7
BOYLESTAD	*DC/AC: The Basics,* 20918–8
	Introductory Circuit Analysis, Fifth Edition, 20631–6
BOYLESTAD/ KOUSOUROU	*Experiments in Circuit Analysis, Fifth Edition,* 20743–6
	Experiments in DC/AC Basics, 21131–X
BREY	*Microprocessors and Peripherals: Hardware, Software, Interfacing, and Applications, Second Edition,* 20884–X
	8086/8088 Microprocessor: Architecture, Programming and Interfacing, 20443–7
BUCHLA	*Experiments in Electric Circuits Fundamentals,* 20836–X
	Experiments in Electronics Fundamentals: Circuits, Devices and Applications, 20736–3
	Digital Experiments: Emphasizing Systems and Design, 20562–X
	Lab Manual to accompany Principles of Electric Circuits, Third Edition, 20657–X
COX	*Digital Experiments: Emphasizing Troubleshooting,* 20518–2
DELKER	*Experiments in 8085 Microprocessor Programming and Interfacing,* 20663–4
FLOYD	*Electric Circuits Fundamentals,* 20756–8
	Electronics Fundamentals: Circuits, Devices and Applications, 20714–2
	Digital Fundamentals, Third Edition, 20517–4
	Electronic Devices, Second Edition, 20883–1
	Essentials of Electronic Devices, 20062–8
	Principles of Electric Circuits, Third Edition, 21062–3
	Electric Circuits, Electron Flow Version, 20037–7
GAONKAR	*Microprocessor Architecture, Programming, and Applications with the 8085/8080A, Second Edition,* 20675–8
	The Z80 Microprocessor: Architecture, Interfacing, Programming, and Design, 20540–9

1

TRANSMISSION LINES

To the student encountering the topic of microwave technology for the first time, it may seem strange that we begin with a discussion as prosaic as transmission lines. After all, one generally thinks of the subject of microwaves in the broader context of parabolic antennas or exotic communication satellites orbiting through space. However, as we shall discover, many of the functional elements of such microwave systems are based, in principle, on transmission line phenomena. For example, waveguides, resonant cavities, and antennas are best understood as an extension of transmission line theory. It is, therefore, essential that we develop our topic from a conceptual viewpoint and treat the transmission line as an elemental microwave component. We will now begin to see how this is possible.

The simple act of completing a circuit between a load and a voltage source produces an effect on the connecting wires which is variously referred to as a "disturbance" or "wave." This wave travels down the wires at roughly the speed of light even though the actual electrons travel much more slowly. If the source frequency happens to be that of an ordinary 60 Hz supply, for example, one might inquire as to how far this "disturbance" travels in the time required for the voltage to complete one cycle. This distance, as we shall learn later, is called the wavelength, and for 60 Hz turns out to be about 3,200 miles, or slightly more than 5,000 km. This is roughly the distance between New York and Los Angeles as the crow flies.

Since ordinary circuits are never quite this long as one continuous run, whatever the voltage happened to be when the switch was closed is also what the load experiences. However, as frequencies get higher and higher, the wavelength gets shorter and shorter until, in the microwave region, wavelengths are measured in centimeters and millimeters. Consequently, the voltage or current at any point along the line may be completely different from that at the source. This situation results in the "disturbance" or "wave" moving down the line section by section, with each section being one wavelength long.

As a result of this type of wave progression, ordinary wires no longer behave as simple conductors, but begin to acquire special characteristics beyond their usual ohmic properties. In other words, at microwave frequencies, wires behave as additional circuit *components,* not merely conductors. This chapter will describe these new components and how they behave as a *transmission line.* Chapter 2 will deal with the nature of the disturbance or wave on the line.

THE NEED FOR IMPEDANCE MATCHING

In your earlier studies of electronics, you were probably told that maximum power transfer from source to load could be achieved by matching the source resistance to the load resistance. In figure 1–1 this condition of maximum power transfer would occur if r (the source resistance) was made equal to R (the load resistance). You may verify this assertion for specific cases of r versus R. For example, if $r = 5$ ohms and $V = 10$ volts (values chosen for simplicity), you might make a table like that in table 1–1 for different values of R.

FIGURE 1–1
Simple DC series circuit.

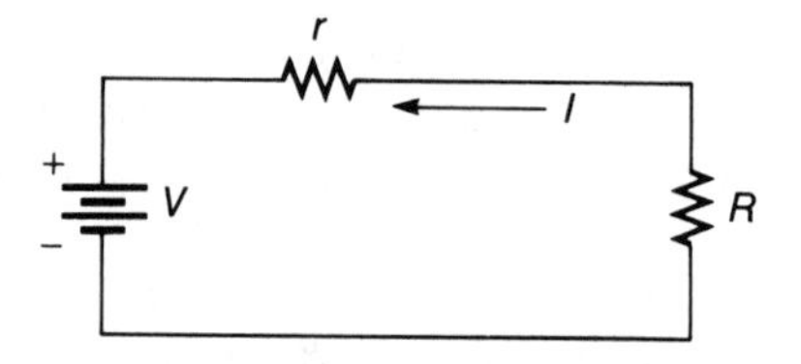

r = internal (source) resistance
R = load resistance
I = total current
V = source voltage

TABLE 1–1
Power transfer as a function of load resistance.

R ohms	I amps	P_{load}	P_{total} watts	η (efficiency) $\eta = R/(r + R) \times 100$*
1.0	1.67	2.78	16.67	16.7%
2.0	1.43	4.08	14.29	28.6%
3.0	1.25	4.69	12.50	37.5%
4.0	1.11	4.94	11.11	44.5%
5.0	**1.00**	**5.00**	**10.00**	**50.0%**
6.0	0.91	4.96	9.09	54.6%
7.0	0.83	4.86	8.33	58.3%
8.0	0.77	4.73	7.69	61.5%
9.0	0.71	4.59	7.14	64.3%
10.0	0.67	4.44	6.67	66.6%

*The formula for efficiency, $\eta = R/(r + R) \times 100$, follows from the basic definition $\eta = P_{out}/P_{in} = P_{load}/P_{total}$. The derivation of this formula is left to the student as an exercise.

Note that values in the third column of table 1–1, P_{load}, increase steadily, reaching a peak of 5 watts when $r = R = 5$ ohms. Thereafter, the values decline. Note, too, that under matched-load conditions, the efficiency of the circuit is exactly 50%. This means that only half the total power is being used by the load.

From table 1–1, it is also obvious that while higher circuit efficiencies may be attained, it is not possible to transfer more than half the total power to the load. A transfer of half the available power is simply the best one can hope for under the best conditions.

It should now be apparent why everything possible must be done to obtain this matched-resistance condition if one wishes to deliver maximum power. And, of course, one does wish to transfer maximum power for reasons of load performance as well as economics. Remember, it takes money to generate power, and any power produced that is not delivered to the load represents an economic loss.

While the above exercise will verify the matched-impedance assertion for the specific case of 5 ohms and 10 volts, a more general proof does exist. If the student has a basic understanding of the calculus, it may be instructive to review the steps that follow. However, the procedure outlined next may be omitted with no loss of understanding.

From figure 1–1, $I = V/(r + R)$ and power in load is given by

$$P = I^2R$$

or

$$P = \frac{V^2}{(r + R)^2} \times R = \frac{V^2R}{r^2 + 2rR + R^2} = \frac{u}{v}$$

Differentiating, dP/dR according to

$$\frac{d}{dx}\left(\frac{u}{v}\right) = \frac{v\left(\frac{du}{dx}\right) - u\left(\frac{dv}{dx}\right)}{v^2}$$

gives

$$\frac{dP}{dR} = \frac{(r^2 + 2rR + R^2)V^2 - V^2R(2r + 2R)}{(r^2 + 2rR + R^2)^2}$$

Simplifying the numerator, we get

$$\frac{V^2(r^2 + 2rR + R^2 - 2rR - 2R^2)}{(r^2 + 2rR + R^2)^2} = \frac{V^2(r^2 - R^2)}{(r^2 + 2rR + R^2)^2}$$

and, $dP/dR = 0$ (a maximum) iff $r = R$.

In the preceding discussion, it was assumed that the resistance of the wires carrying the power from source to load was zero. In reality, such is not the case, and this resistance must be taken into account.

Since any power dissipated in the wires is not transferable to the load, we may simply combine their effect with that of the source and continue to regard the wires as lossless. That is, if the wires have a total resistance of r', then we may add this to the source to obtain $r + r' = r''$ as the total source resistance. This simplifies computation, and the wires are once again considered to be lossless.

In closing this section, it should be mentioned that it may not always be desirable in every application to match source and load impedances for the purpose of achieving maximum power transfer. For example, in small voltage amplifiers one is only concerned with obtaining as much of the signal's voltage across the input impedance of the amplifier as possible (see figure 1–2).

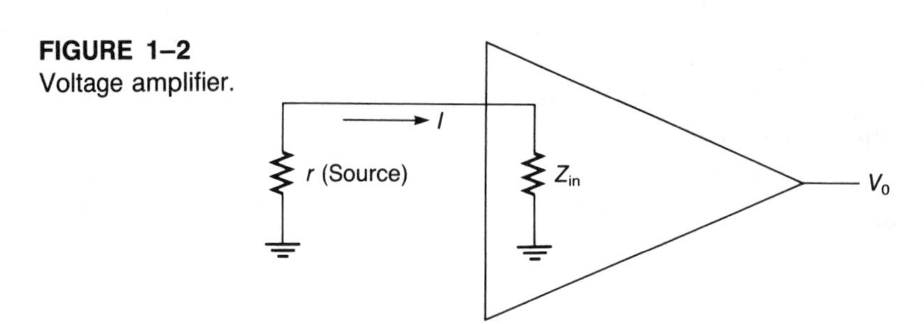

FIGURE 1–2
Voltage amplifier.

It is apparent from figure 1–2 that the source and amplifier input impedances form a series circuit. As such, the maximum voltage drop will occur across the largest impedance. Therefore, if source and load impedances were matched, only half the signal voltage would be available at the input terminals of the amplifier. Ideally, then, this situation calls for zero source impedance and an infinite input (load) impedance.

If we consider a power amplifier, however, we find that the output impedance must, indeed, match the load impedance in order for maximum power transfer to occur (see figure 1–3).

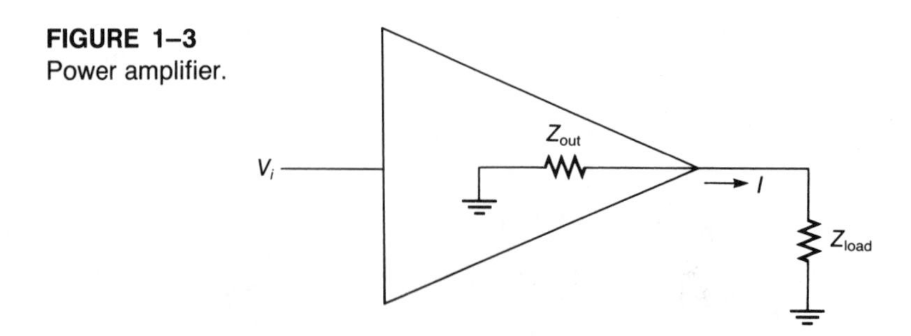

FIGURE 1–3
Power amplifier.

It is evident from figure 1–3 that maximum power transfer occurs only if $Z_{out} = Z_{load}$. In communication work, Z_{out} might be the output impedance of a transmitter, while Z_{load} might be an antenna's input impedance.

THE TRANSMISSION LINE

In the foregoing discussion, we referred to the conductor's linking source to load as merely "wires." (After all, that is basically what they are.) However, in specific instances where the sole purpose of these wires is to efficiently transfer power, it has become common to refer to the physical assembly of two or more such wires as a *transmission line*.

On the surface, this nomenclature may seem like mere window dressing in an attempt to elevate the status of the lowly wire. But such is not the case, and the reason will shortly become evident. In the interim, a formal definition will be suggested.

> A transmission line is a system of conductors having a precise geometry and arrangement which is used to transfer power from source to load with minimum loss.

Geometry will mean the cross-sectional shape, dimensions, and spacing of the individual conductors, and *arrangement* will mean the manner in which the conductors are arrayed relative to one another.

In communication work, there are many types of transmission lines. However, two basic types are quite common and so will be presented here for analysis. The first type of line is the *parallel-wire* line, so-called because its two conductors, embedded in an insulating medium, run parallel to one another. A parallel-wire line is shown in figure 1–4. A parallel-wire transmission line is also referred to as a *balanced* line since each wire has the same electrical capacitance relative to ground. In actual practice, however, this ideal balance may be difficult to achieve and maintain.

A second type of transmission line is known as the *coaxial cable*. In this line, the two conductors are arranged concentrically and share a common axis; hence, the description *co-axial* is aptly applied. The inner conductor is insulated from the outer conductor (called the shield) and is precisely centered by means of a dielectric material. A coaxial transmission line is shown in figure 1–5.

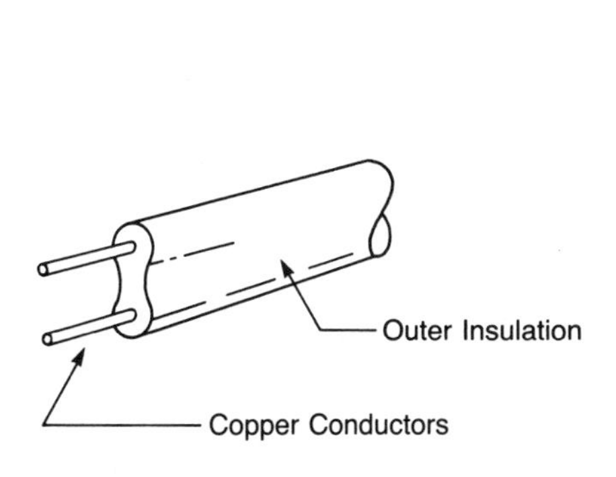

FIGURE 1–4
A parallel-wire transmission line.

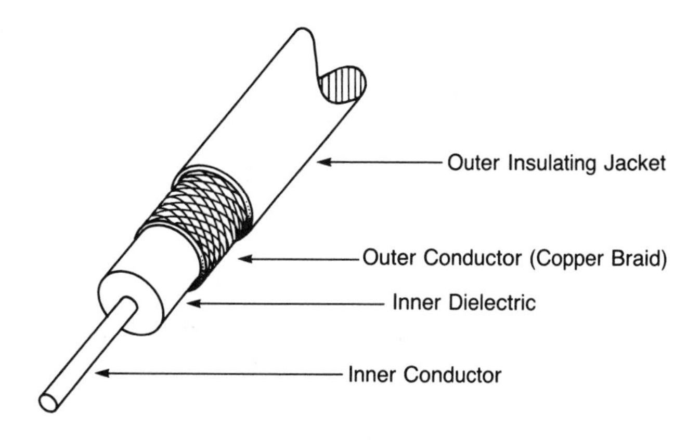

FIGURE 1–5
A coaxial transmission line.

The coaxial line is sometimes referred to as an *unbalanced* line since the inner conductor does not have the same electrical capacitance to ground as the outer conductor. Indeed, if the outer conductor is grounded, the inner conductor will not be influenced by the external environment at all. It is said to be ''shielded'' from external conditions.

While parallel-wire line is comparatively cheaper than coaxial cable, it has higher radiation loss. This makes its application unsuitable for use in the UHF/SHF ranges. Coaxial cable, on the other hand, has the signal confined entirely between the inner conductor and outer shield. Consequently, radiation is quite negligible.

Practically, coaxial cable may only be used up to about 15 GHz. Beyond that, signal attenuation due to dielectric losses makes coax impractical, and one must turn to waveguides for efficient transfer of microwave power. (Waveguides will be discussed in chapter 4.)

FUNDAMENTAL IDEAS

Energy that travels along a transmission line does so in the form of a transverse electromagnetic (TEM) wave. This wave consists of an electric field E (V/m) and a magnetic field H (A/m), which are at right angles to one another and also mutually perpendicular to their direction of propagation (see figure 1–6).

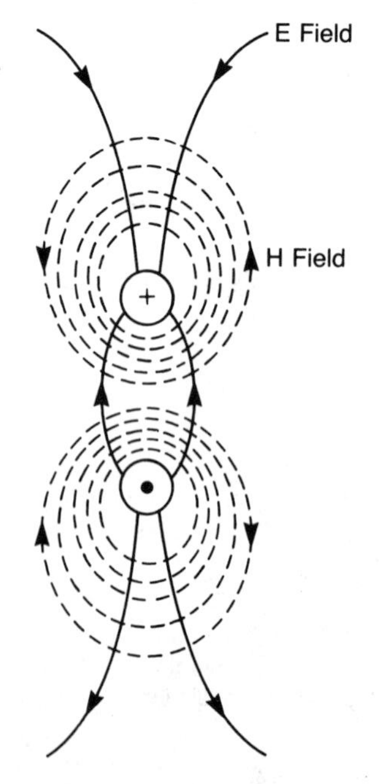

FIGURE 1–6
Electromagnetic field around a parallel-wire transmission line.

In chapter 2, we will discuss in detail how such TEM waves are formed and how they propagate in space. In the interim, however, we shall accept that TEM waves may also travel along a transmission line and that the following fundamental ideas have been verified experimentally.

Phase Velocity

Phase velocity (v_p) may be defined as the velocity with which a point of constant phase (on a progressive periodic TEM wave) is propagated (see figure 1–7). The point referred to will have both an electric and a magnetic field moving with it. Note that the voltage at a fixed point A on the line varies sinusoidally with time, while P maintains a constant magnitude and polarity (phase) as it advances along the line.

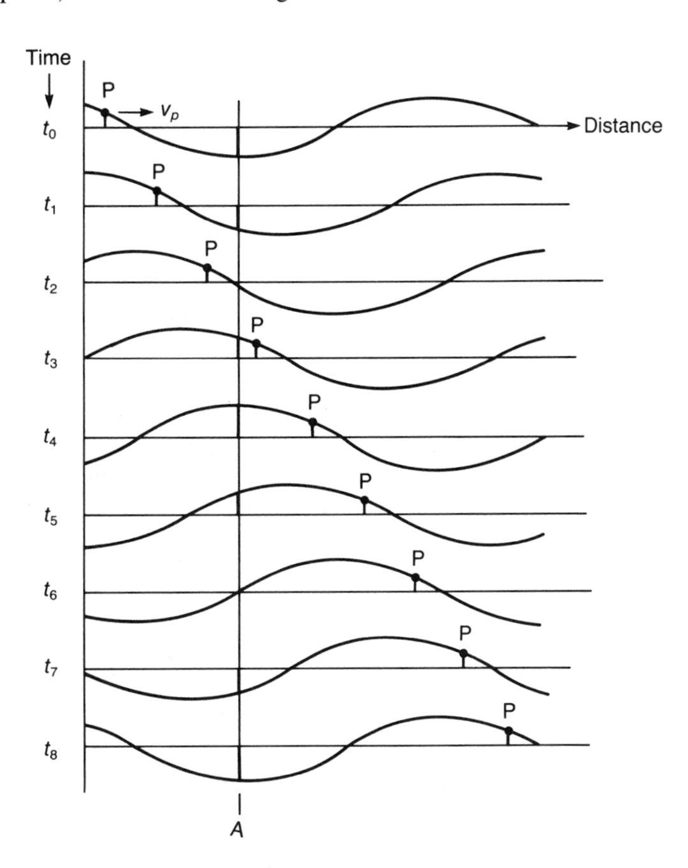

FIGURE 1–7
Point P propagates to the right at v_p.

For any TEM wave, the electric field (E) has an electric flux density (D) given by:

$$D = \epsilon E \tag{1–1}$$

where ϵ = the absolute permittivity of the medium. For free space (to be explained in chapter 2), its value is 8.854×10^{-12} F/m (farads/meter). The unit of D is coulombs/ m^2 (C/m^2).

The magnetic field (H) has a magnetic flux density (B) given by

$$B = \mu H \tag{1–2}$$

where μ = the absolute permeability of the medium. In free space, this value is 1.26 $\times 10^{-6}$ H/m (henrys/meter). The unit of B is teslas.

It has been determined empirically that an electric field density (D) moving at a phase velocity (v_p) will always induce a magnetic field (H) proportional to v_p. That is,

$$\mathrm{H} = Dv_p = \epsilon \mathrm{E} v_p \qquad \textbf{(1–3)}$$

Also, a magnetic field density (B) moving at v_p will similarly induce an electric field E according to

$$\mathrm{E} = Bv_p = \mu \mathrm{H} v_p \qquad \textbf{(1–4)}$$

In other words, each type of *moving* field generates the other. This explains (as will be seen later) why TEM waves are self-sustaining as they leave an antenna.

Taking this last expression, $\mathrm{E} = \mu \mathrm{H} v_p$, and substituting for H, we obtain:

$$\mathrm{E} = \mu(\epsilon \mathrm{E} v_p)v_p$$

On solving for v_p, we get

$$v_p = \frac{1}{\sqrt{\mu\epsilon}} \text{ m/S} \qquad \textbf{(1–5)}$$

where v_p is called the *phase velocity*. Note that if the numerical values of free-space permittivity and permeability are substituted into this expression, one obtains the velocity of light (c) in free space (3×10^8 m/S). That is, $v_p = c$.

Now, the absolute permittivity of a medium is given by

$$\epsilon = \epsilon_r \epsilon_0 \qquad \textbf{(1–6)}$$

where ϵ_r is the relative permittivity (dielectric constant) which may vary between 1 to 5, and ϵ_0 is the permittivity of free space (8.854×10^{-12} F/m). Upon substituting this expression into the equation for v_p (along with the values of μ and ϵ_0), one obtains:

$$v_p = \frac{c}{\sqrt{\epsilon_r}} \text{ m/S} \qquad \textbf{(1–7)}$$

The factor $1/(\sqrt{\epsilon_r})$ is called the *velocity factor*. Equation (1–7) shows that the phase velocity of a TEM wave on a transmission line is always less than it would be in free space.

Wavelength

Wavelength (λ) may be defined as the distance traveled by a point on a periodic TEM wave in the time required to complete one cycle. That is,

$$\lambda = \frac{v_p}{f} \text{ meters} \qquad \textbf{(1–8)}$$

where v_p is the phase velocity and f is the frequency in hertz.

If $v_p = c$, the resultant wavelength (λ_0) is called the free-space wavelength. Substituting for v_p from equation (1–7)

$$\lambda = \frac{c}{\sqrt{\epsilon_r}} \times \frac{1}{f} = \frac{\lambda_0}{\sqrt{\epsilon_r}} \qquad \textbf{(1–9)}$$

The wavelength given by equation (1–9) is the value that must always be used for transmission-line calculations since $v_p < c$.

Average Power Density

From the basic circuit relationship, $P = V^2/R$, we may write

$$P_d = \frac{E^2}{Z_0} \text{ W/m}^2 \qquad \textbf{(1–10)}$$

where P_d is the power density in the TEM wave. Since $Z_0 = E/H$, then $E = H \times Z_0$. Therefore, we may write equation (1–10) as

$$P_d = \frac{(H \times Z_0)^2}{Z_0} = H^2 Z_0 \text{ W/m}^2 \qquad \textbf{(1–11)}$$

WAVE IMPEDANCE (Z_w)

It turns out that the ratio of E/H is a constant value which has units in ohms. That is, if E (V/m) is divided by H (A/m) the quotient is in units of volts/amperes, which may be interpreted as ohms. In other words,

$$\frac{E}{H} = \frac{\mu H v_p}{\epsilon E v_p}, \text{ or } \frac{E^2}{H^2} = \frac{\mu}{\epsilon}$$

On solving for the ratio E/H, we obtain

$$\frac{E}{H} = \sqrt{\frac{\mu}{\epsilon}}\ \Omega \qquad \textbf{(1–12)}$$

If we now substitute the numerical values of μ and ϵ as given above for free space, we obtain

$$Z_w = \sqrt{\frac{\mu}{\epsilon}} = 377\ \Omega \qquad \textbf{(1–13)}$$

This value of 377 ohms (a constant) is an intrinsic property of free space, and is called the wave impedance (Z_w) or characteristic impedance of free space. We shall make use of the relationship $\sqrt{\mu/\epsilon}$ later when we compute the characteristic impedance (Z_0) of a transmission line.

CHARACTERISTIC IMPEDANCE

Many of us may already be familiar with the parallel-wire hook-up line that connects our television receiver with the antenna on the roof. Indeed, we may even know that this television transmission line is referred to as a 300-ohm twin lead, and that its value of 300 ohms is somehow a constant. Let us speculate on the nature of this value.

At first, we might be tempted to think of this 300 ohms as a DC resistance. But if the cable is made of 24 gage copper wire, as is common practice, it will require a little over a mile to obtain 300 ohms of DC resistance![1] But since much shorter lengths are common, we must conclude that not only would each length have a different ohmic value, but the value would be considerably less than 300 ohms. Such conditions hardly qualify the twin lead as having a constant resistance. Alas, our first attempt to explain this 300-ohm value has led us astray.

Perhaps we could take a more inventive approach and regard this transmission line as being made up of multiple sections of resistance, as shown in figure 1–8. This view is justified since twin lead does have some series resistance as well as shunt conductance since no insulator is perfect. In figure 1–8, R_s represents series resistance and R_p is the parallel-leakage resistance. For simplicity, all series resistance per section is shown as lumped in one conductor.

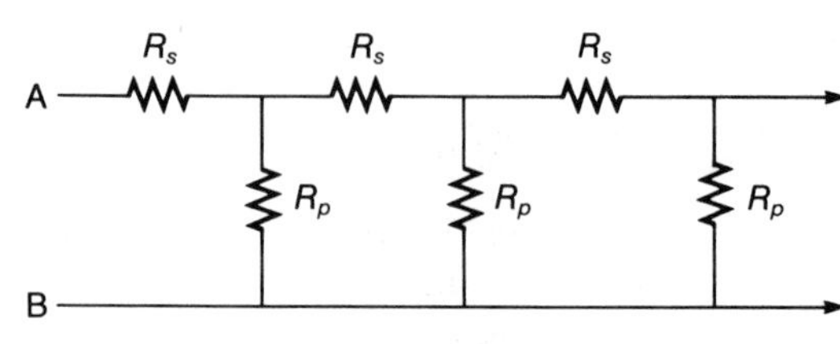

FIGURE 1–8
A DC transmission line showing three sections.

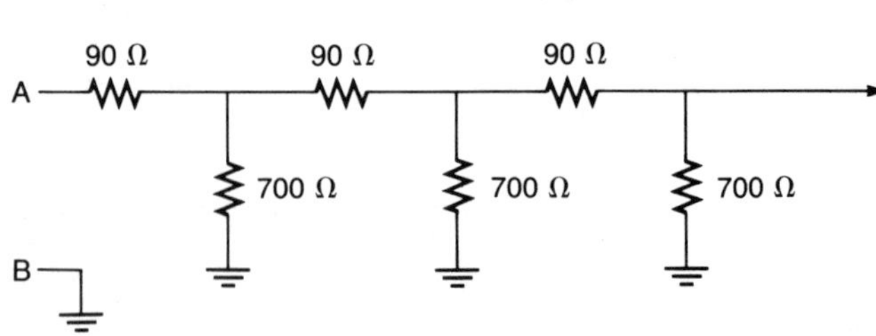

FIGURE 1–9
A 300-ohm DC transmission line.

We will now show that for one of many other possible combinations of R_s and R_p values, the DC resistance measured at the input terminals A–B of figure 1–8 will be 300 ohms *no matter how long the line is!* Let us assume that the value of R_s per section is 90 ohms and the value of R_p per section is 700 ohms.[2] We redraw figure 1–8 and insert the specific values of $R_s = 90$ and $R_p = 700$ (see figure 1–9). In figure 1–9, the common conductor symbol is used in lieu of showing the return path, and all of R_s per unit section has been lumped in the 90-ohm value.

---

[1] #24 AWG copper wire has 26.17 ohms/1,000 feet. In order to measure 300 ohms across the input terminals of a shorted length of twin lead, the cable would need to be 1.09 miles long.

[2] Other values of R_s and R_p will also work, and it is left to the student as an exercise to demonstrate this.

If only one section is used for our DC transmission line, the DC resistance read at terminals A–B would obviously be the simple sum of 90 and 700 ohms, or 790 ohms. As more and more sections are added, the value read at A–B will change as shown in table 1–2. As may be seen from the table, the DC transmission line rapidly approaches a limiting value of 300 ohms. This limit is called the characteristic resistance (R_0). It would be instructive to verify the values in this table by means of a simple program written in BASIC. One such program showing the new DC value obtained as additional sections are added is:

```
10 LET R0=790
20 PRINT"Enter number of additional sections S"
30 INPUT S
40 FOR N=1 TO S
50 LET R0 = 1/(1/R0 + 1/700) + 90
60 NEXT N
70 PRINT"The result is";R0
80 END
```

Even after 1,000,000 sections have been added (for which the computer requires 3.5 hours to calculate in BASIC), the limiting value of R_0 remains 300 ohms.

TABLE 1–2
Resistance measured at terminals A–B for increasing number of sections.

Section Number	Resistance at A–B
2	461.1
3	368.0
4	331.2
5	314.8
6	307.2
7	303.5
8	301.7
9	300.8
10	300.4
11	300.2
12	300.1
13	300.0
14	300.0
15	300.0
16	300.0
.	.
.	.
.	.
(any other)	300.0

Perhaps even more astonishing than the idea that a so-called DC transmission line thus conceived can have a characteristic resistance of 300 ohms is the idea that this line, no matter how short or long, will show the same 300-ohm value measured at its input if it is terminated in a 300-ohm resistor! For example, figure 1–10 shows a single-section line terminated in a 300-ohm resistive load.

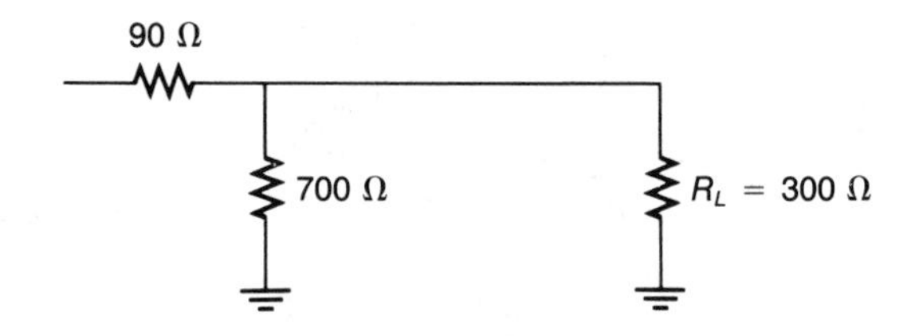

FIGURE 1–10
DC transmission line terminated in 300-ohm load resistance.

A simple calculation reveals that the total input resistance is, indeed, 300 ohms. Moreover, no matter how many sections comprise the line, if it is terminated in 300 ohms, the input resistance will *always* be 300 ohms.[3] This is an impressive as well as useful result, and will be referred to later.

While the above analysis does, in fact, give us the expected constant 300-ohm DC value, we must remember that the intended purpose of twin lead is to transfer AC power, not DC. However, the idea of "distributed" resistance sections that yield a characteristic (constant) resistance seems to be very appealing. Therefore, we shall pursue this idea as an AC phenomenon and see what happens.

The first question one might reasonably ask is, "What are the distributed components for an AC transmission line?" To answer this question, first consider that any time we have two conductors separated by an insulator, there is a certain amount of capacitance present. This is exactly the case for twin lead. Furthermore, any straight section of wire has some inductance, no matter how diminutive. And, since some series resistance and shunt leakage is always present, the picture of our AC line emerges as shown in figure 1–11.

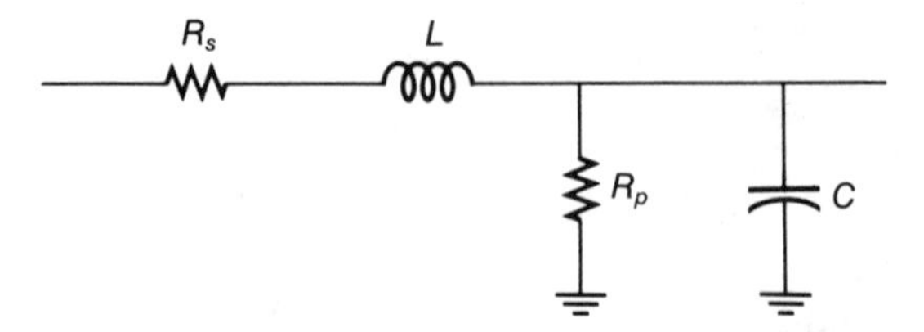

FIGURE 1–11
A section of AC transmission line.

In figure 1–11, all of the series resistance (R_s) and all of the series inductance (L) per section has been shown in one conductor for the sake of simplicity. It should be remembered that figure 1–11 shows only a *single* section of the transmission line as seen by the passage of a *single* wave.

We may further simplify our AC line model by noting that at radio frequencies, the inductive reactance is considerably more than the series resistance. Moreover, the

[3]The student should verify this fact for a line of various sections terminated in 300 ohms.

capacitive reactance is much less than any shunt conductance. Therefore, we may omit R_s and R_p from our unit section entirely and regard the transmission line as ''lossless.'' That is, very little power (I^2R) loss occurs in the conductors. In actual practice, this omission of the resistance yields quite a good approximation of a real line. Our final simplified transmission line is shown in figure 1–12.

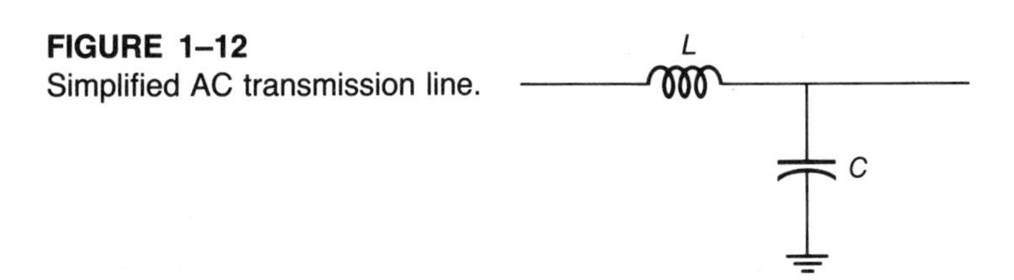

FIGURE 1–12
Simplified AC transmission line.

We are now ready to show that the opposition measured at the input to this AC transmission line has a constant characteristic value that is quite *independent* of frequency. For students with a basic understanding of the calculus, the following explanation will prove to be enlightening and satisfying. A non-calculus demonstration of the concept will follow the calculus derivation.

The voltage drop (v_L) across an inductor depends on the rate at which the current is flowing through it. This is shown by the equation

$$v_L = L\frac{di}{dt} \tag{1–14}$$

Also, the current through a capacitor (i_c) depends on the rate at which the voltage is changing across it. This may also be written as the equation

$$i_c = C\frac{dv}{dt} \tag{1–15}$$

We may rewrite equations (1–14) and (1–15) as

$$v_L dt = L di \text{ and } i_c dt = C dv$$

If we then solve these equations for the appropriate variable, we obtain

$$v_L t = Li \text{ and } i_c t = Cv$$

or, dropping the subscripts, we obtain the more general expression:

$$vt = Li \text{ and } it = Cv$$

If we now divide the former by the latter, we get:

$$\frac{vt}{it} = \frac{Li}{Cv}$$

Rearranging, we have:

$$\frac{v^2}{i^2} = \frac{L}{C} = Z^2$$

Therefore,

$$Z_0 = \sqrt{\frac{L}{C}} \tag{1–16}$$

This value of impedance is called the *characteristic impedance* (Z_0) or surge impedance of the transmission line. Note that it is *not* frequency dependent since L and C depend entirely on physical properties; that is, the values of L and C depend only upon the cross-sectional geometry of the transmission line.

As yet another derivation of Z_0, we begin with an assertion from iterative network theory which states that, for sinusoidal voltages, the ratio of series impedance to shunt admittance is given by,

$$Z_0^2 = \frac{Z}{Y} = \frac{R + j\omega L}{G + j\omega C}$$

where ω = angular velocity = $2\pi f$. Upon taking the square root of each side we obtain

$$Z_0 = \sqrt{\frac{R + j\omega L}{G + j\omega C}}$$

If $R \ll j\omega L$ and $G \ll j\omega C$ (which is quite an accurate approximation at higher frequencies), then the above expression reduces to

$$Z_0 = \sqrt{\frac{j\omega L}{j\omega C}} = \sqrt{\frac{L}{C}}$$

which is the same result obtained in equation (1–16).

An important consequence of Z_0's frequency independence is that it acts as if it were entirely *resistive*. Moreover, as was the case with our DC transmission line, any length of AC transmission line terminated in a purely resistive load equal to its characteristic impedance (Z_0) will have Z_0 measured as its input impedance.

One question that might be asked is, "Since each section has this value of Z_0, and since there are countless numbers of such sections, why isn't the input impedance their sum?" The answer to this question is that since the wavelength is so short at RF, the wave propagates down the line section by section, seeing only one section at a time. Remember that our model of the transmission lines shown in figures 1–11 and 1–12 was based on what a *single* wave would experience as it passed through a *single* section of line. It was upon this assumption that our expression for Z_0 was derived.

This explains why L and C are *distributed* parameters, always measured and given per unit length of line. It is meaningless to lump these values.

Now, since $Z_0 = \sqrt{L/C}$, we may choose values of L and C such that the resulting Z_0 will be 300 ohms.[4] However, the important result here is not the numerical value,[5] but rather that Z_0 is frequency *independent* and its value depends entirely upon the cross-sectional geometry of the line and *not* how long the line is.

CALCULATING Z_0

As was mentioned in a previous section, the characteristic impedance (Z_0) of a transmission line is a function of its distributed inductance and capacitance per unit length of conductor. Moreover, the actual values of L and C depend on the cross-sectional geometry of the line. The implication of this statement is that one should be capable of determining the value of Z_0 from a knowledge of the line's physical dimensions.

Moreover, just as the medium of free space had a permittivity (ϵ) and a permeability (μ), so, too, does the insulating material used in the fabrication of a transmission line. Therefore, we ought to be able to find the characteristic impedance of a transmission line by solving the simple equation

$$Z_0 = \sqrt{\frac{\mu}{\epsilon}} \tag{1–17}$$

where μ would be the actual inductance (L) per unit length in H/m and ϵ the capacitance (C) per unit length in F/m. That is, $Z_0 = \sqrt{L/C}$. Note that this formula is the same we had obtained earlier as equation (1–16) through a somewhat different analysis of the distributed properties of a transmission line.

We will now develop a formula for both parallel-wire and coaxial transmission line characteristic impedance (Z_0) in terms of μ, ϵ, and line cross-sectional geometry. For parallel-wire line, it may be shown that for conductors embedded in a medium of permittivity ϵ (F/m), permeability, μ (H/m), and with dimensions in meters, the distributed L and C are given approximately by

$$L = \frac{\mu}{\pi} \ln \frac{2D}{d} \text{ H/m} \tag{1–18}$$

$$C = \frac{\pi\epsilon}{\ln \frac{2D}{d}} \text{ F/m} \tag{1–19}$$

where D = the center-to-center distance of the conductors
d = their diameter

---

[4]For example, let C = 98 pF/m and L = 8.82 μH/m. Then Z_0 = 300 ohms.

[5]Typically, parallel transmission lines have values of Z_0 ranging from 150 to 600 ohms.

Upon substituting the expressions for L and C above into the equation

$$Z_0 = \sqrt{\frac{L}{C}}$$

we obtain

$$Z_0 = \frac{1}{\pi}\sqrt{\frac{\mu}{\epsilon}}\ \ln\frac{2D}{d}\ \Omega$$

Using the natural-to-common log conversion formula,[6] we may rewrite this expression as:

$$Z_0 = \frac{1}{\pi}\sqrt{\frac{\mu}{\epsilon}}\left(\log_{10}\frac{2D}{d}\right) \times \frac{1}{0.434} \qquad \textbf{(1–20)}$$

Now, the permeability (μ) of most dielectrics used in practice is the same as that of free space: $\mu = 1.26 \times 10^{-6}$ H/m. Moreover, the permittivity will be $\epsilon = \epsilon_r \times \epsilon_0$ where ϵ_0 (the permittivity of free space) $= 8.854 \times 10^{-12}$ F/m, and ϵ_r is the relative permittivity (the "dielectric constant") whose range is typically $1 < \epsilon_r < 5$.

Substituting these specific values into equation (1–20) gives:

$$Z_0 = \frac{276}{\sqrt{\epsilon_r}}\log_{10}\frac{2D}{d}\ \Omega \qquad \textbf{(1–21)}$$

For coaxial lines,

$$L = \frac{\mu}{2\pi}\ln\frac{D}{d}\ \text{H/m} \qquad \textbf{(1–22)}$$

and

$$C = \frac{2\pi\epsilon}{\ln\dfrac{D}{d}}\ \text{F/m} \qquad \textbf{(1–23)}$$

where D = outside diameter of outer conductor
d = diameter of inner conductor

---

[6]In what follows, $\ln x = \log_\epsilon x$, where ϵ = base of natural logs (about 2.718). $\log_\epsilon 2D/d = k$. So, $\epsilon^k = 2D/d$ or, $\epsilon^{(\log_\epsilon 2D/d)} = 2D/d$. Taking $\log_{10}$ of both sides, $\log_{10} \epsilon^{(\log_\epsilon 2D/d)} = \log_{10} 2D/d$, which may be written as $\log_\epsilon 2D/d\ \log_{10}\epsilon = \log_{10} 2D/d$. But since $\log_{10}\epsilon = 0.434$, then, $\log_\epsilon 2D/d = (\log_{10} 2D/d)/0.434$.

Therefore,

$$Z_0 = \sqrt{\frac{L}{C}} = \frac{1}{2\pi}\sqrt{\frac{\mu}{\epsilon}}\ \ln\frac{D}{d}\ \Omega$$

Using the same log conversion as above, and the same values of μ and ϵ_0, we may write

$$Z_0 = \frac{138}{\sqrt{\epsilon_r}}\log_{10}\frac{D}{d}\ \Omega \qquad \textbf{(1–24)}^7$$

RF PERFORMANCE OF LOSSLESS LINES

In a future section, we will learn how to use a transmission line calculating device called the Smith chart. This chart will allow us to quickly solve problems concerning line-load impedance matching without recourse to complicated mathematics. In this section, we will develop some ideas concerning transmission line behaviors that will aid our understanding of the Smith chart and its many applications.

Standing Waves

From a conceptual viewpoint, most microwave application systems may be reduced to three fundamental building blocks: *source, line,* and *load.* The source might be anything from a simple Gunn diode oscillator circuit to an entire X-band radar transmitter. The line might be a few meters of coaxial cable or a complex waveguide network stretching out hundreds of feet. The load may range from a simple resistive termination to some sort of deep-space antenna array, active repeater, or similar devices. Whatever the actual components involved, we may represent these three fundamental building blocks as shown in figure 1–13.

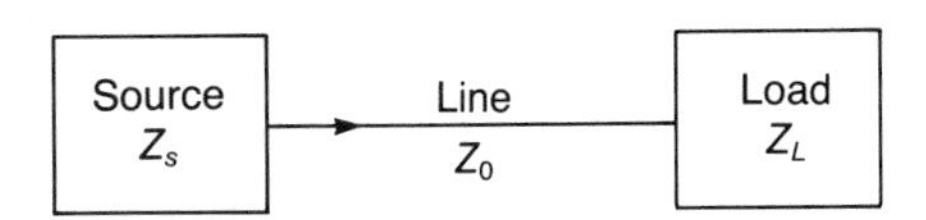

FIGURE 1–13
Basic building blocks of a microwave system.

Note that each block exhibits a certain impedance: Z_s is the output impedance of the source; Z_0 is the characteristic impedance of the transmission line; and Z_L is the load impedance.

---

[7] It has become common practice to express the formula for Z_0 in terms of base-ten logarithms. However, the use of the natural log will yield equally correct results.

For the purpose of our analysis, we shall consider the transmission line as having negligible conduction (I^2R) loss. We will also consider radiation and dielectric losses minimal. These are reasonable assumptions for most lines encountered in practice, and they formed the basis of our earlier development of the equation for Z_0. Our transmission line is now considered to be free of all significant losses.

The energy produced by the source is carried down such a lossless transmission line in the form of TEM waves and is delivered to the load. If the load impedance is either a pure resistance (or can be made to look like a pure resistance) and has a value equal to Z_0, then we have a perfect impedance match between the line and load. Moreover, if the source impedance is now matched to Z_0 (hence Z_L), then we have also achieved a match between the line and the source. Under these conditions, we have attained the ideal situation mentioned at the beginning of this chapter: there will be a maximum transfer of power down the line to the load. This means that we have made the most efficient use of the power delivered by the source. In practice, it is usually a simpler matter to match the source to the line rather than match the line to the load. Consequently, we must develop a strategy for dealing with mismatched loads.

The result of any impedance mismatch manifests itself in the form of TEM waves (hence, power) being reflected from its intended destination. For example, suppose $Z_L \neq Z_0$. In this situation, part or all of the waves sent down the line to the load are reflected back to where they originated. The amount of the reflected power (reflected waves) depends on the degree of mismatch.

An interesting and important consequence of this impedance-mismatch situation is that any time a *reflected* wave meets a *traveling* wave (i.e., a wave coming the other way), a new wave, called a *standing* wave, is produced. A standing wave is formed as a result of the two waves combining in such a way as to form a new wave. The idea is similar to the result obtained by throwing two small stones into a pool of water. As the ripples spread out from each stone, they will eventually meet at some point and produce a new wave where they interact.

For TEM waves, the resulting composite is called a standing wave because its nodes (points of minimum value) are stationary on the transmission line, as shown in figure 1–14.

There are several important features of figure 1–14 that deserve comment. First, note that the waves shown are standing waves of *voltage*. The reason for this is that voltage variations along the line are easiest to detect with the simplest instrumentation. Note, too, that all the waves are shown on one side of the line even though they actually undergo polarity reversals just as the periodic source voltage does. However, detection

FIGURE 1–14
Standing waves of voltage.

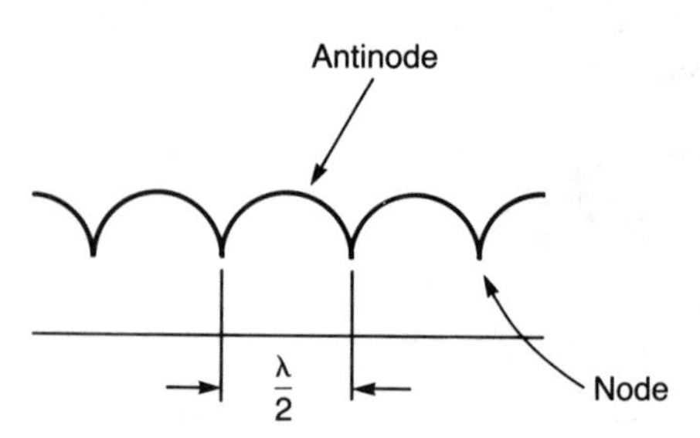

methods pay no attention to polarity, and so it has become common to show standing waves in the positive sense.

Furthermore, it should be mentioned that the nodes are stationary in the sense that they represent positions on the line where the voltage is permanently zero. The antinodes are also stationary in position (always midway between two adjacent nodes), but they vary in amplitude as the source voltage.

Finally, the distance between nodes is one-half wavelength. Of course, there is also $\lambda/2$ between antinodes, but the nodes provide a sharper indication of position for conventional detecting apparatus. Hence, we usually refer to the nodes rather than to antinodes.

With this picture of a standing wave and an understanding of how and why they are formed, we may now derive a quantitative means of expressing the degree of mismatch for a given line-load impedance condition.

Propagation Constant

For any actual transmission line, the variation of voltage and current with distance along the line is determined by a complex quantity called the *propagation constant* (γ) as expressed by

$$I = I_s \epsilon^{-\gamma l} \qquad \textbf{(1–25)}$$

and

$$V = V_s \epsilon^{-\gamma l} \qquad \textbf{(1–26)}$$

where the subscript s indicates a source condition; ϵ is the base of the natural logarithms; γ is the propagation constant; and l is the distance along the line as measured from the source.

Although beyond the scope of this treatment, the propagation constant (γ) may be shown to depend on the primary line constants (L, C, R, and G) and the angular velocity of the signal (ω).

The propagation constant may be written as

$$\gamma = \alpha + j\beta \qquad \textbf{(1–27)}$$

where α is the attenuation coefficient and determines how voltage and current decrease with distance, and β is the phase-shift coefficient and determines the phase shift of voltage or current with distance along the line. That is

$$\beta = \frac{2\pi}{\lambda} \text{ r/m}$$

Since our line is considered to be lossless, we may regard the propagation constant as

$$\gamma = j\beta \qquad \textbf{(1–27a)}$$

Voltage Reflection Coefficient

With the equation (1–27a) as our revised constant, we may now write the equation of any reflected voltage (V_r) at a distance (l) from the load end of a lossless line as

$$V_r = V_R \epsilon^{-j\beta l} \quad \textbf{(1–28)}$$

where V_R is the value of the reflected voltage *at the load,* and may be complex.

The incident voltage (V_i) at the same distance from the load end of the line may similarly be written as

$$V_i = V_I \epsilon^{j\beta l} \quad \textbf{(1–29)}$$

where V_I is the value of the incident voltage *at the load,* and may be complex. Note that V_r and V_i may also be complex. Furthermore, the sign of the complex exponent $j\beta l$ in equation (1–29) is + since the direction of measurement is opposite that shown in equation (1–28).

Now, at any point along the line

$$V = V_r + V_i \quad \textbf{(1–30)}$$

In particular, at the load

$$V_L = V_I + V_R \quad \textbf{(1–31)}$$

Moreover, the incident current is

$$I_i = \frac{V_i}{Z_0} \quad \textbf{(1–32)}$$

and the reflected current is

$$I_r = -\frac{V_r}{Z_0} \quad \textbf{(1–33)}$$

Note the negative sign in equation (1–33) signifying the phase change (180 degrees) from the incident current. So, at any point along the line

$$I = I_i + I_r \quad \textbf{(1–34)}$$

Or, by substituting from equations (1–32) and (1–33),

$$I = \frac{V_i - V_r}{Z_0} \quad \textbf{(1–35)}$$

In particular, at the load,

$$I_L = \frac{V_I - V_R}{Z_0} \quad \textbf{(1–36)}$$

Now, define *voltage reflection coefficient* Γ_L at the load as

$$\Gamma_L = \frac{V_R}{V_I} \tag{1–37}$$

where Γ may be complex, with magnitude ρ and phase angle θ. That is, $\Gamma_L = \rho \angle\theta$. The magnitude ρ (where ρ = absolute value of Γ_L) indicates the fraction of the incident voltage reflected from the load. The range of values ρ may assume is $0 < \rho < 1$. That is, 0 to 100% of the incident voltage wave may be reflected. The phase angle (θ) represents the angle between the incident and the reflected voltage waves. The range of values θ may assume is $-180 < \theta < +180$ degrees.

We may derive a more useful expression for Γ_L in terms of Z_L and Z_0, which are easier to determine in practice since actual voltage measurements are not required.

Since $Z_L = V_L/I_L$, we may substitute from equations (1–31) and (1–36) and write

$$Z_L = \frac{V_L}{I_L} = \frac{V_I + V_R}{1} \frac{Z_0}{V_I - V_R}$$

or,

$$Z_L(V_I - V_R) = Z_0(V_I + V_R)$$

or

$$V_I(Z_L - Z_0) = V_R(Z_L + Z_0)$$

and finally,

$$\frac{V_R}{V_I} = \Gamma_L = \frac{Z_L - Z_0}{Z_L + Z_0} \tag{1–38}$$

Voltage Standing-Wave Ratio

We have at last arrived at a point where it is possible to derive a numerical expression that will serve as a quantitative assessment of the degree of any line-load impedance mismatch condition. If we define *voltage standing-wave ratio (VSWR)* as the ratio of maximum to minimum standing wave voltages present at the load, then the larger the numerical value of this ratio, the bigger the mismatch. That is,

$$\text{VSWR} = \frac{V_{\text{max}}}{V_{\text{min}}} \tag{1–39}$$

where, since the line is lossless, all the values of V_{max} are the same and all values of V_{min} are the same. That is,

$$\text{VSWR} = \frac{V_{\text{max}}}{V_{\text{min}}} = \frac{V_I + V_R}{V_I - V_R}$$

Dividing numerator and denominator by V_I results in

$$\text{VSWR} = \frac{1 + \dfrac{V_R}{V_I}}{1 - \dfrac{V_R}{V_I}}$$

or,

$$\text{VSWR} = \frac{1 + \rho}{1 - \rho} \qquad \textbf{(1–40)}$$

We now have a convenient expression for VSWR in terms of the magnitude (ρ) of the reflection coefficient Γ_L, and it is not actually necessary to measure either V_{max} or V_{min}.

Note that a complex load $Z_L = R + jX$ may assume any value between a short ($Z_L = 0$) and an open ($Z_L = \infty$). Then, from equations (1–38) and (1–40), it may be seen that the range of VSWR is

$$1 < \text{VSWR} < \infty$$

and the ideal situation is VSWR $= 1.00$ since this represents the case where Z_L is a pure resistance equal to Z_0. Note that if the VSWR is known, then

$$\rho = \frac{\text{VSWR} - 1}{\text{VSWR} + 1} \qquad \textbf{(1–41)}$$

Quite often in the literature of the microwave industry, the lower-case Greek letter sigma (σ) is used to denote VSWR. Equation (1–40) is then written as

$$\sigma = \frac{1 + \rho}{1 - \rho} \qquad \textbf{(1–41a)}$$

and equation (1–41) becomes

$$\rho = \frac{\sigma - 1}{\sigma + 1} \qquad \textbf{(1–41b)}$$

It is important that one not confuse Γ_L (voltage reflection coefficient) with VSWR (voltage standing-wave ratio). The former is a *complex* variable (ρ, θ) expressing both the ratio V_R/V_I as well as their phase difference. The latter is simply a *scalar* quantity expressing a voltage ratio.

REACTANCE PROPERTIES OF TRANSMISSION LINES

At microwave frequencies, the wavelength approaches the physical dimensions of many discrete capacitors and inductors. Moreover, small stray values of L and C become highly significant as frequencies climb toward 10 and 12 digits. Therefore, it is quite possible for resistors to behave like capacitors, inductors may start to resonate, and simple L–C circuits might refuse to behave as designed. Consequently, the idea of lumped parameters must be abandoned in favor of more predictable circuit elements.

It may be demonstrated that various short lengths of transmission line with their distributed constants of L and C may be made to function as either inductors or capacitors, impedance-matching transformers, wave traps, resonant circuits, and a variety of other circuit elements and even insulators. The following will demonstrate how transmission lines may perform as circuit elements, and this will lead us directly into applications of the Smith chart.

Equation (1–37) gave us the value of the voltage reflection coefficient Γ_L at the load. Now, at any other point on the line, it follows from equations (1–28), (1–29), and (1–37) that

$$\Gamma = \frac{V_r}{V_i} \tag{1–42}$$

where equation (1–37) is a special case of this. Moreover, from equation (1–38), we may find the impedance at any point on the line in terms of the voltage reflection coefficient (Γ) at that point.

$$Z = Z_0 \frac{1 + \Gamma}{1 - \Gamma} \tag{1–43}$$

This last equation will now allow us to look at three special cases of transmission line: (1) $Z_L = 0$ (short circuit); (2) $Z_L =$ open circuit; and (3) a quarter-wavelength section of line.

Case 1 $Z_L = 0$ (Short Circuit) From equation (1–38), $\Gamma_L = (Z_L - Z_0)/(Z_L + Z_0)$, $\Gamma_L = -1$, and from equation (1–42)

$$\Gamma = \frac{V_r}{V_i} = \frac{V_R\,\epsilon^{-j\beta l}}{V_I\,\epsilon^{j\beta l}} = -1\epsilon^{-j2\beta l}$$

where the -1 comes from the fact that $\Gamma_L = -1 = V_R/V_I$.

If we substitute this into equation (1–43), we obtain,

$$Z = Z_0 \frac{1 + \Gamma}{1 - \Gamma} = Z_0 \frac{1 - \epsilon^{-j2\beta l}}{1 + \epsilon^{-j2\beta l}} = jZ_0 \tan \beta l \tag{1–44}$$

This follows from the trigonometric identity[8]

$$\tan x = -j\,\frac{\epsilon^{jx} - \epsilon^{-jx}}{\epsilon^{jx} + \epsilon^{-jx}}$$

$$= -j\,\frac{1 - \epsilon^{-j2x}}{1 + \epsilon^{-j2x}}$$

If l is expressed in fractional parts of wavelength, then the argument of equation (1–44) reduces to multiples of π. Then, values of Z versus l appear as in figure 1–15.

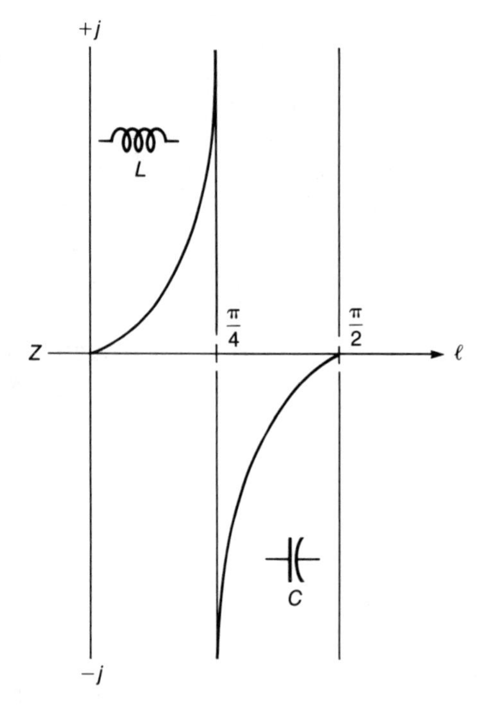

FIGURE 1–15
Reactance versus length of short-circuited line.

Since j is positive for $0 < l < \pi/4$, Z appears inductive. Similarly, for $\pi/4 < l < \pi/2$, Z is capacitive. This pattern repeats ad infinitum.

Case #2 Z_L Open Circuit From a discussion similar to the one above for $Z_L = 0$, we find that

$$Z = -jZ_0 \cot \beta l \qquad \textbf{(1–45)}$$

The graph of Z versus l for the open-circuited line is shown in figure 1–16.

From the figure, it may be seen that for $0 < l < \pi/4$, the line appears capacitive.

---

[8]See Tuma, J. J. *Technology Mathematics Handbook*. New York: McGraw-Hill, 1975.

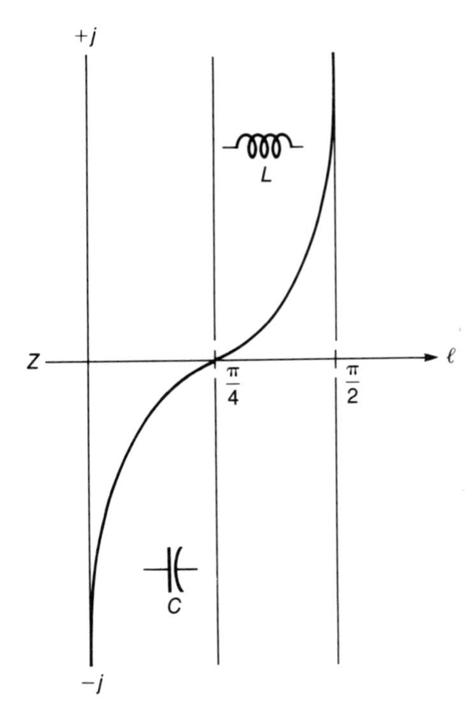

FIGURE 1–16
Reactance versus length of open-circuited line.

Moreover, for $\pi/4 < l < \pi/2$, the line appears to be inductive since j is now positive. This pattern repeats ad infinitum. The equivalent reactance seen by the generator for shorted and open circuit transmission lines of length $0 < l < \lambda/2$ is shown in figure 1–17.

Case #3 Transmission Line λ/4 Long From equation (1–42), $\Gamma = V_r/V_i = (V_R\,\epsilon^{-j\beta l})/(V_I\,\epsilon^{-j\beta l})$. Therefore, $\Gamma = (V_R/V_I)\,\epsilon^{-j2\beta l}$. And, since $\beta = 2\pi/\lambda$ and $l = \lambda/4$, then $2\beta l = \pi$. So, $\Gamma = \Gamma_L\,\epsilon^{-j\pi}$. And, from equation (1–43)

$$Z = Z_0\,\frac{1 + \Gamma}{1 - \Gamma} = Z_0\,\frac{1 + \Gamma_L\,\epsilon^{-j\pi}}{1 - \Gamma_L\,\epsilon^{-j\pi}} \qquad \textbf{(1–46)}$$

Now, since $\epsilon^{-j\pi} = -1$, then equation (1–46) becomes

$$Z = Z_0\,\frac{1 - \Gamma_L}{1 + \Gamma_L} = Z_0\,\frac{1 - \dfrac{Z_L - Z_0}{Z_L + Z_0}}{1 + \dfrac{Z_L - Z_0}{Z_L + Z_0}}$$

which, when simplified, becomes

$$Z = \frac{Z_0^2}{Z_L} \qquad \textbf{(1–47)}$$

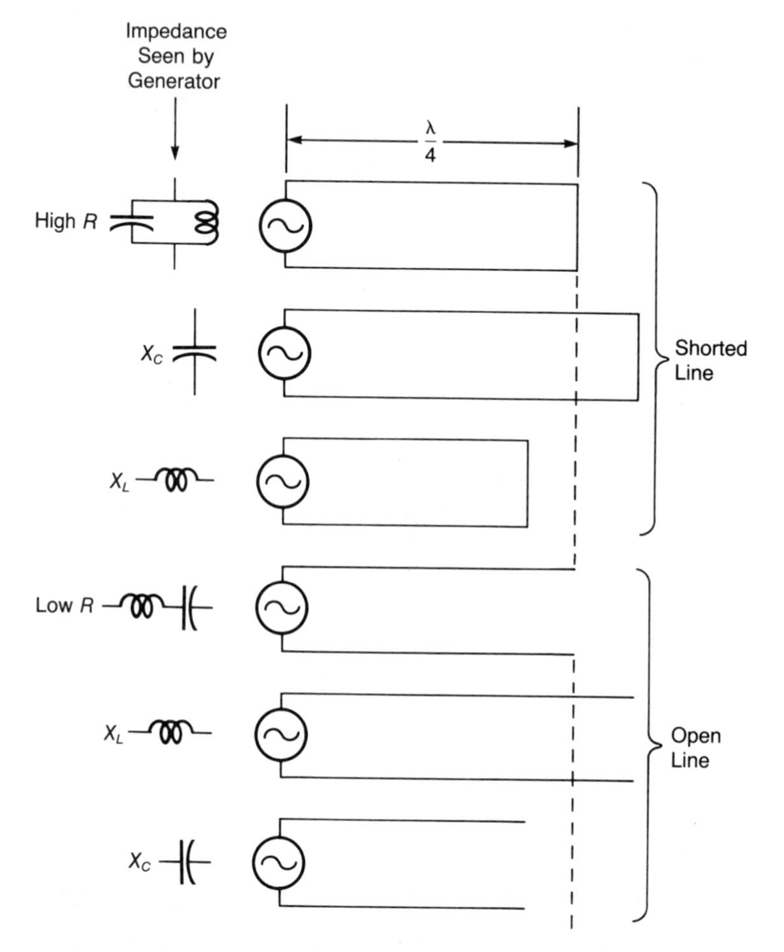

FIGURE 1–17
Reactance properties of shorted and open transmission lines.

Equation (1–47) is an important result since it states that the source impedance Z_s (Z in equation 1–47) to a quarter-wavelength section of transmission line depends on both Z_0 and the load impedance Z_L. The implication of this fact is that a section of transmission line of proper impedance (Z'_0) may be used in exactly the same way as an ordinary matching transformer at lower frequencies. Indeed, a $\lambda/4$ transmission line used in this way at microwave frequencies is called a *quarter-wave transformer*.

For example, suppose $Z_0 = 300$ ohms and Z_L is a pure resistance of 150 ohms, as shown in figure 1–18.

From equation (1–47), $Z_s = Z_0^2/Z_L$, we know that if the input to the $\lambda/4$ transformer (Z_s) is to be 300 ohms, with $Z_L = 150$ ohms, then solving for Z'_0 (the value of Z_0 for the transformer) yields

$$Z'_0 = \sqrt{Z_s \cdot Z_L} = \sqrt{300 \cdot 150} = 212\ \Omega$$

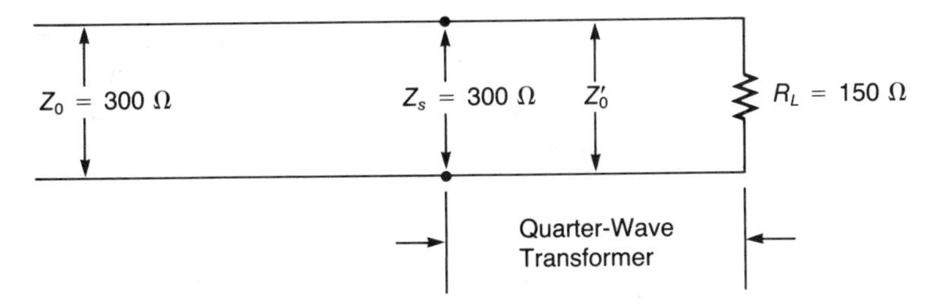

FIGURE 1–18
Use of the quarter-wave transformer to match 300-ohm line to 150-ohm load.

The idea of having to come up with a 212-ohm transmission line on the spot is not a very appealing idea for the practicing microwave technician or engineer. Moreover, the transformer may be applied only in those cases where Z_L is a pure resistance or can be made to appear as if it were a pure resistance. However, the real value of the quarter-wave transformer lies not so much in its applied value, but rather in an even more important theoretical concept that will now be presented. This idea is the basis of the Smith chart.

If we divide both sides of equation (1–47) by Z_0, we get

$$\frac{Z_s}{Z_0} = \frac{Z_0}{Z_L} \tag{1–48}$$

We will define the process of dividing an impedance by Z_0 as *normalization,* and say that any number so divided has a *normalized value,* which we will represent by lower-case letters. In equation (1–48), Z_s has a normalized value z_s. Similarly, any admittance divided by Y_0, where $Y_0 = 1/Z_0$, is said to be normalized and has a normalized value represented by y (lower case). For example, $Y_L/Y_0 = y_L$. As will be seen, these normalized values are simply a numerical strategy that allows us to perform certain types of computations easily. Normalized values are unitless. From equation (1–48) then, $Z_s/Z_0 = z_s$ (as stated before) and $Z_0/Z_L = 1/z_L = y_L$.[9]

From this normalization process, we see that

$$z_s = \frac{1}{z_L} = y_L \tag{1–49}$$

This relationship is important for the following reason. Suppose a certain normalized load impedance is $z_L = r + jx$, where r and x are normalized values, as defined above. Then, from equation (1–49), $y_L = 1/(r + jx) = r' - jx'$. Note carefully the change of sign from + to −. The implication here is that the normalized value of z_L started out as *inductive* ($+jx$), but ended up *capacitive* ($-jx$) one-quarter wavelength

---

[9]The student will need to recall certain terms and the relationship between them: $Z = R + jX$ and $Y = G + jB$ where Z is impedance; R is resistance; X is reactance; Y (admittance) $= 1/Z$; G (conductance) $= 1/R$; B (susceptance) $= 1/X$.

away. The significant conclusion of this is that at some point between these two extremes, the normalized load (z_L) must have looked like a pure, normalized resistance (r).

If we knew precisely where this point of pure resistance was, we could connect a quarter-wave transformer there and create an impedance match *even when the load is complex,* not just a pure resistance as was shown by figure 1–18.

For example, suppose a transmission line having a characteristic impedance of 50 ohms is connected to a complex load of $Z_L = 70 - j55$ ohms. At what point must one connect a quarter-wave transformer to provide a matched-impedance situation (see figure 1–19)?

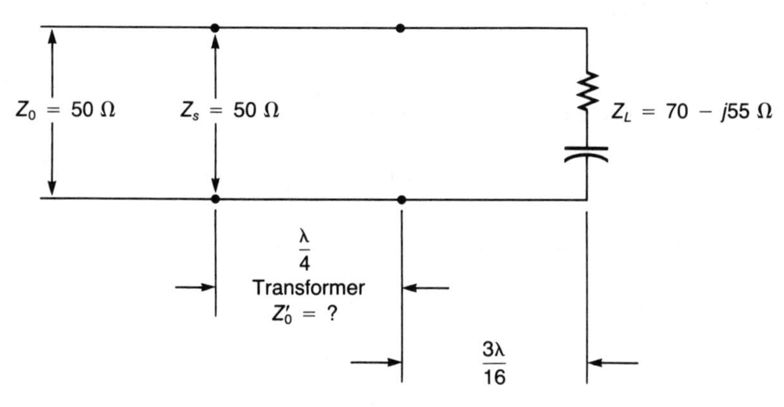

FIGURE 1–19
Matching a complex load using a quarter-wave transformer.

If you knew how to use the Smith chart (which you will shortly), you would know that a quarter-wave transformer connected about 3/16 of the generator wavelength away from the complex load Z_L will see a normalized resistance of $r = 0.39$ (no units). That is, there is a point at which $y_L = r + jb = 0.39 + 0$. The *real* value of this resistance is 19.5 ohms since

$$r = \frac{R}{Z_0} \text{ so } R = rZ_0 = 0.39(50) = 19.5\ \Omega$$

Then, from equation (1–47), $Z'_0 = \sqrt{Z_s Z_L} = \sqrt{50(19.5)} = 31.2$ ohms. Of course, this value of 31.2 ohms is not exactly a common, off-the-shelf transmission line impedance. However, the idea is that there is *some* resistance point at which the transformer may be connected. But there is yet an even more important implication. At some *other* point, the value of the normalized admittance will be $y_L = 1 + jb$ (or $1 - jb$). It is obvious that this point is not a pure resistance as was the previous situation.

However, as mentioned earlier, short lengths of transmission line may exhibit reactance properties (see equations 1–44 and 1–45) which, if connected across the main line at the proper point, will tune out the normalized susceptance $\pm jb$ of the normalized load admittance. Short lines used in this way are called *stubs*.

Therefore, with the stub of proper length in place, the normalized admittance becomes $y_L = 1 + 0$. In other words, $y_L = r = 1$, and since $r = R/Z_0$ (that is, $1 = R/Z_0$), then $R = Z_0$, which is a perfect resistive match to the line! We no longer need the use of a quarter-wave transformer with its curious values of Z'_0. The situation is shown in figure 1–20. Figure 1–20 shows that a stub having the proper length (L) will

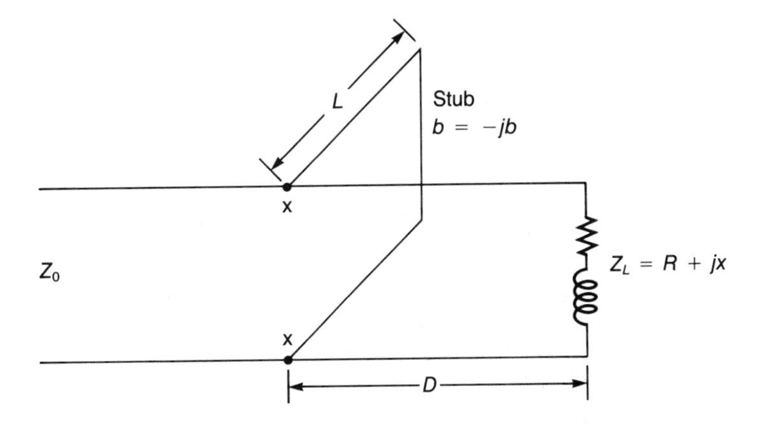

FIGURE 1–20
Stub used to match load with line.

have a susceptance $-jb$. If this stub is placed at the proper distance (D) back from the load at x, the normalized load looks like $y_L = 1 + 0$, which is a pure resistance point. The length (L) and the distance (D) are determined with the use of a Smith chart.

Before leaving this section, we should mention two special situations involving Case 3: the quarter-wave length transmission line. From equation (1–47), $Z = Z_0^2/Z_L$, if the $\lambda/4$ line is terminated in a short circuit, then $Z = \infty$. This property of an infinitely high impedance makes the line appear very much like a parallel-resonant circuit. Conversely, if Z_L = open circuit, $Z = 0$, this makes the $\lambda/4$ line appear much as a series-resonant circuit (see figure 1–17).

THE SMITH CHART

The transmission line calculator (Smith chart[10]), sometimes called a polar impedance diagram, eliminates much of the laborious and time-consuming calculations that would otherwise have to be made in answering line-load matching problems and additional questions concerning line conditions. The universal Smith chart is shown in figure 1–21.

A convenient and durable Smith chart calculator made of plastic is available for easy and rapid solution of many common transmission line problems. The calculator may be purchased from:

Analog Instruments Company
P.O. Box 808
New Providence, NJ 07974
(201) 464–4214

[10]Smith, Phillip H. "An Improved Transmission Line Calculator." *Electronics* 17, (January, 1944): 130.

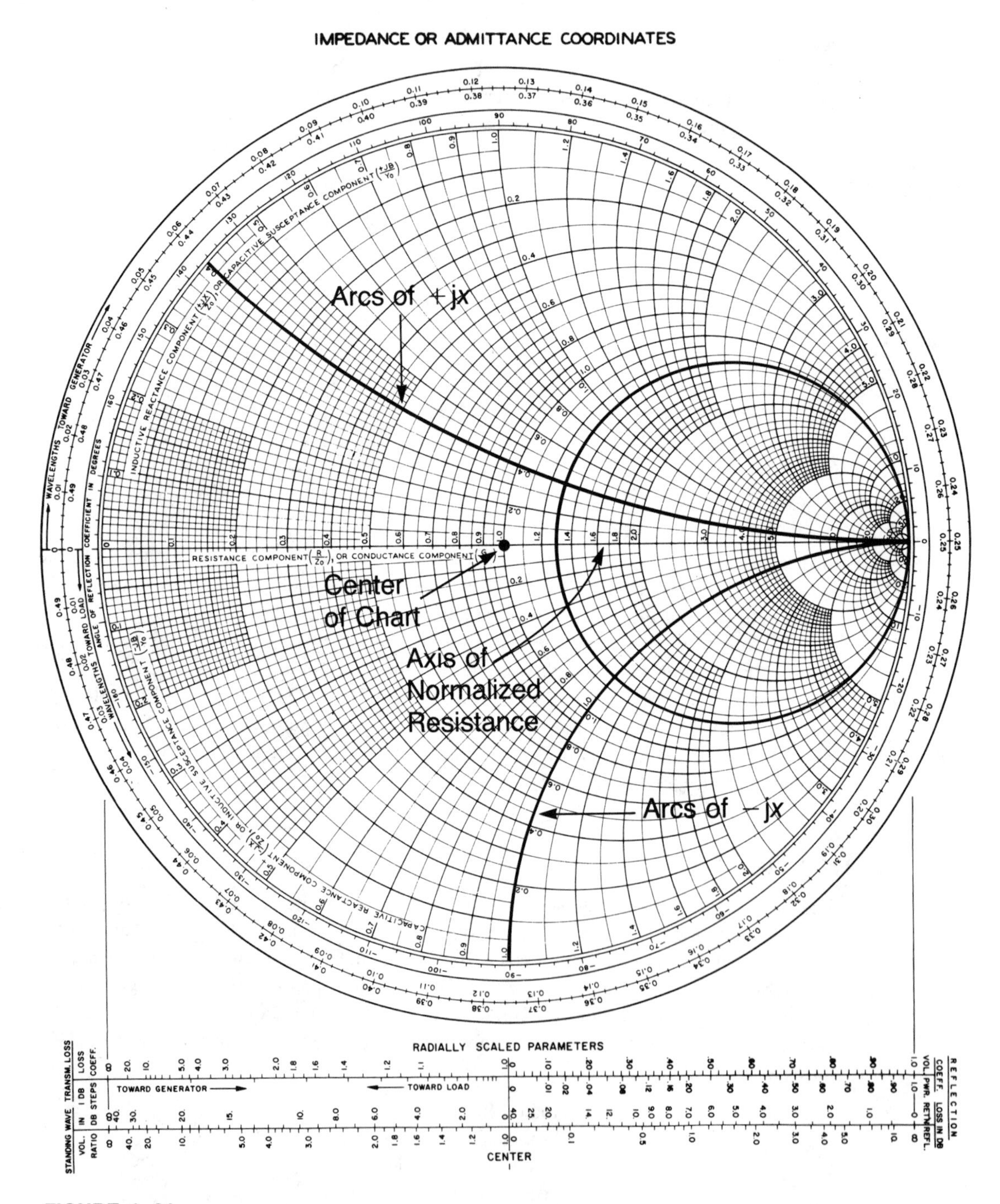

FIGURE 1–21
The universal Smith chart. (Courtesy Anita Smith, Analog Instruments Co., New Providence, NJ 07974.)

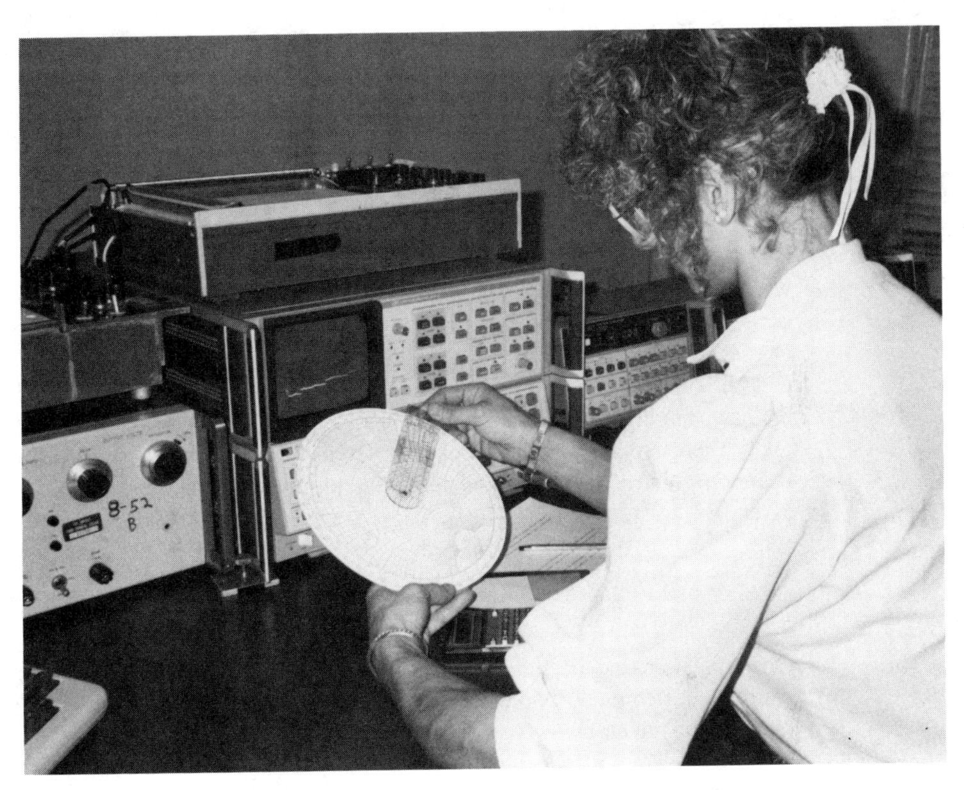

FIGURE 1–22
The Smith chart calculator in use. (Reproduction of calculator in photo by permission of Anita M. Smith, Analog Instruments Co.)

Price as of this writing is $27.50 plus $3.00 shipping and handling. The calculator is shown being used in figure 1–22.

The development of the Smith chart, while interesting, is beyond the scope of this text. Instead, we will concentrate on learning how to use the chart to solve transmission-line problems. Let us begin with a discussion of the chart's anatomy.

To most students, the chart appears formidable at first glance. There seems to be no customary beginning, and the familiar straight lines of other common charts are missing. In fact, there is only one straight line on the entire field of the chart. To add confusion, there is a set of straight scales at the bottom of the chart that seems to relate to something on the circular scales. The first question, then, might well be, "Where do we begin?"

If you look at the very bottom of the chart, you will see the word CENTER printed beneath a small straight vertical line. If you project this line directly upwards a few inches, you will see that it intersects the horizontal line (which cuts the circular chart in half) at a point marked 1.0. We say that this is the start of the chart since this is the center of a circle *the student must draw* in order to use the chart. (We should mention here that the student will require a bow compass and straightedge in order to use the chart.)

The non-linear horizontal scale which divides the chart in half is called the axis of normalized resistance and is numbered, beginning at the left, 0, 0.1, 0.2, . . . 1.0 at the center, . . . 20, 50 and infinity at the right. These numbers represent normalized resistance values as given by

$$r = \frac{R}{Z_0} \tag{1–50}$$

where R is the actual resistive part of the complex load $Z = R + jX$. You may recall our earlier reference to the process of normalization in connection with equation (1–48). The complete circles correspond to the normalized values of resistance. Note that all complete circles are tangent at one point on the far right of the chart. One such arc has been highlighted in figure 1–21.

There are also two sets of arcs on the chart. One set curves upward from its common point of tangency and represents normalized values of inductive reactance as given by

$$x = \frac{jX}{Z_0} \tag{1–51}$$

One typical arc has been highlighted in figure 1–21. The other set of arcs curves downward from a common point of tangency and represents normalized values of capacitive reactance as given by

$$x = -\frac{jX}{Z_0} \tag{1–52}$$

One such arc has been highlighted in figure 1–21. The normalized values of either type of reactance are read from the appropriate curved scale near the periphery of the chart. In figure 1–21, one of the highlighted arcs is that of $+j0.4$, and the other arc is that of $-j1.0$. It is often necessary to interpolate values when reading these scales.

Note that all the complete circles intersect all the arcs. Any given point of intersection is a normalized load impedance value as given by

$$z_L = \frac{Z_L}{Z_0} = r + jx \tag{1–53}$$

As an example, suppose we had a complex load of $Z_L = 35 + j65$ ohms terminating a 50-ohm transmission line. The normalized value of this impedance is

$$\frac{Z_L}{Z_0} = \frac{(35 + j65)}{50} = 0.7 + j1.3$$

This normalized impedance[11] is located as point A in figure 1–23.

[11] All values must be normalized in order to enter the universal Smith chart. However, there are charts for use with specific values of Z_0 (50 ohms, for example) that do not use normalized values.

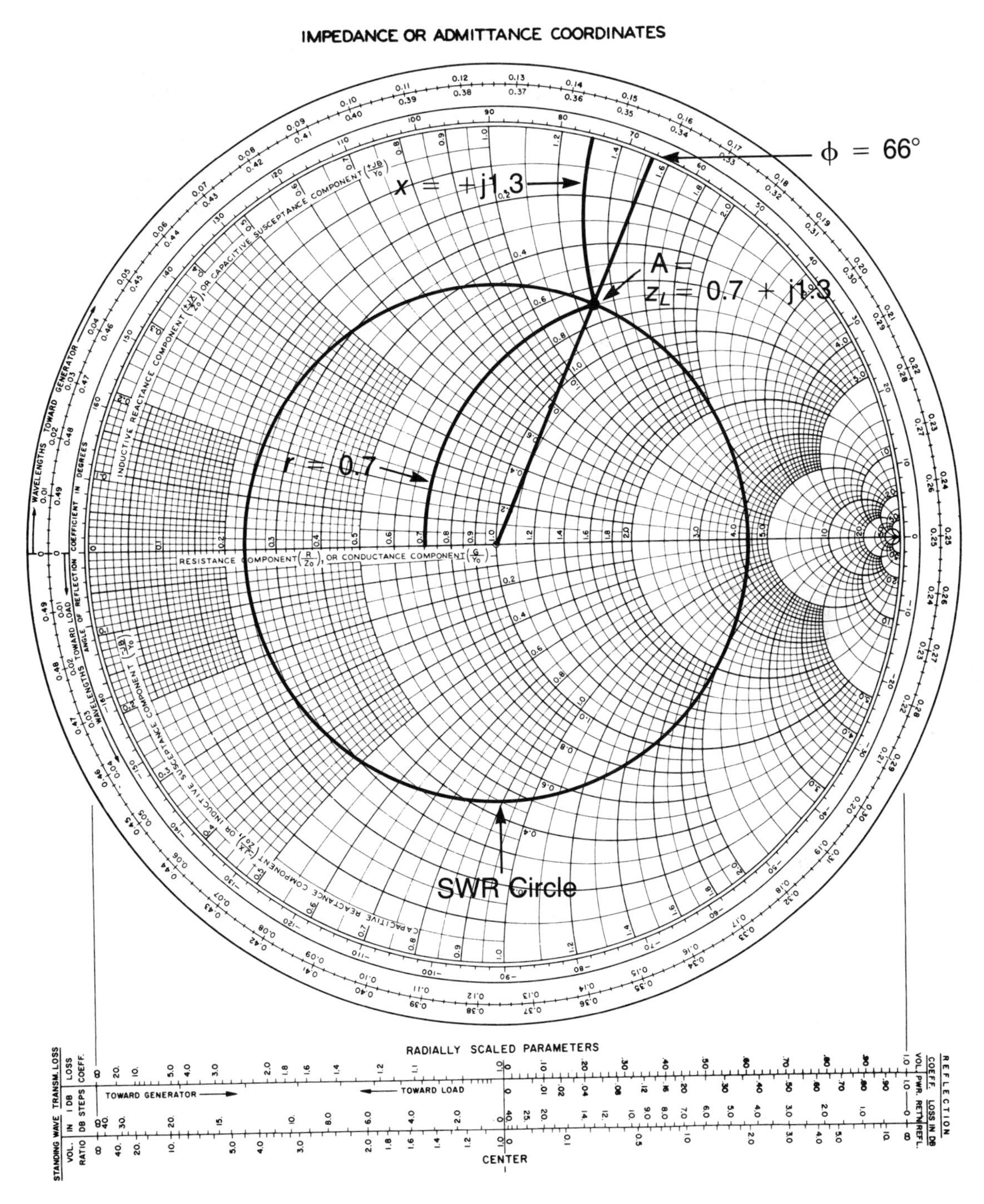

FIGURE 1–23
A normalized complex impedance plotted on the Smith chart.

If one now places the point of a bow compass at the center of the chart (1.0) and extends the other leg to point A, a complete circle may be drawn on the chart, as shown in figure 1–23. This circle is called the SWR circle because it intersects the horizontal scale at the SWR value of 4.3 for this particular load. The SWR value has *no units* since it is the ratio of two voltages.

Note how much more involved it would have been to calculate the SWR value using equations (1–38) and (1–40). From equation (1–38), $\Gamma_L = (Z_L - Z_0)/(Z_L + Z_0)$,

$$\Gamma_L = \frac{(35 + j65) - 50}{(35 + j65) + 50} = \frac{-15 + j65}{85 + j65}$$

$$= \frac{66.7 \angle 103^\circ}{107 \angle 37.4^\circ} = 0.623 \angle 65.6^\circ$$

where $\rho = 0.623$ and $\theta = 65.6$ degrees.

Now, from equation (1–40)

$$\text{VSWR} = \frac{1 + \rho}{1 - \rho} = \frac{1.623}{0.377} = 4.3$$

Note that the phase angle 65.6 degrees of the reflection coefficient Γ_L may be read directly from the chart by drawing a straight line from 1.0 through point A out past the rim of the chart. This line will intersect the phase angle at about +66 degrees, as shown in figure 1–23.

The value of the VSWR as well as the magnitude of the reflection coefficient may also be read directly from the scales located at the bottom of the chart labeled RADIALLY SCALED PARAMETERS. As shown in figure 1–24, lines drawn perpendicular to the horizontal axis where the SWR circle intersects to the bottom of the chart intersect the VSWR value of 4.3 and the magnitude (ρ) of Γ_L at 0.62.

If one extends the line down a little further past $\rho = 0.62$, one will encounter another radially scaled parameter shown as RETURN LOSS in dB. The line is shown in figure 1–24 as intersecting the scale at 4.2 decibels. This new term will now be explained.

If all the power sent to the load fails to be absorbed and returns to the source, we say that no signal has been "lost" to the load. In other words, the returned power suffered no loss. This reasoning gives rise to a new and useful term, *return loss,* which relates VSWR and ρ in a single number.

Recall from an earlier discussion that the only way to prevent a device from absorbing any power whatsoever is to short out the input since a short itself absorbs no energy. Conversely, a perfect load will absorb all the power it receives. Conceptually, then, the return loss (L_r) of a practical load may be taken as

$$L_r = \text{dB(short)} - \text{dB(open)} \tag{1–54}$$

where *open* refers to the non-shorted condition of the load.

Equation (1–54) may also be written as

$$L_r = 20 \log V_I - 20 \log V_R \tag{1–55}$$

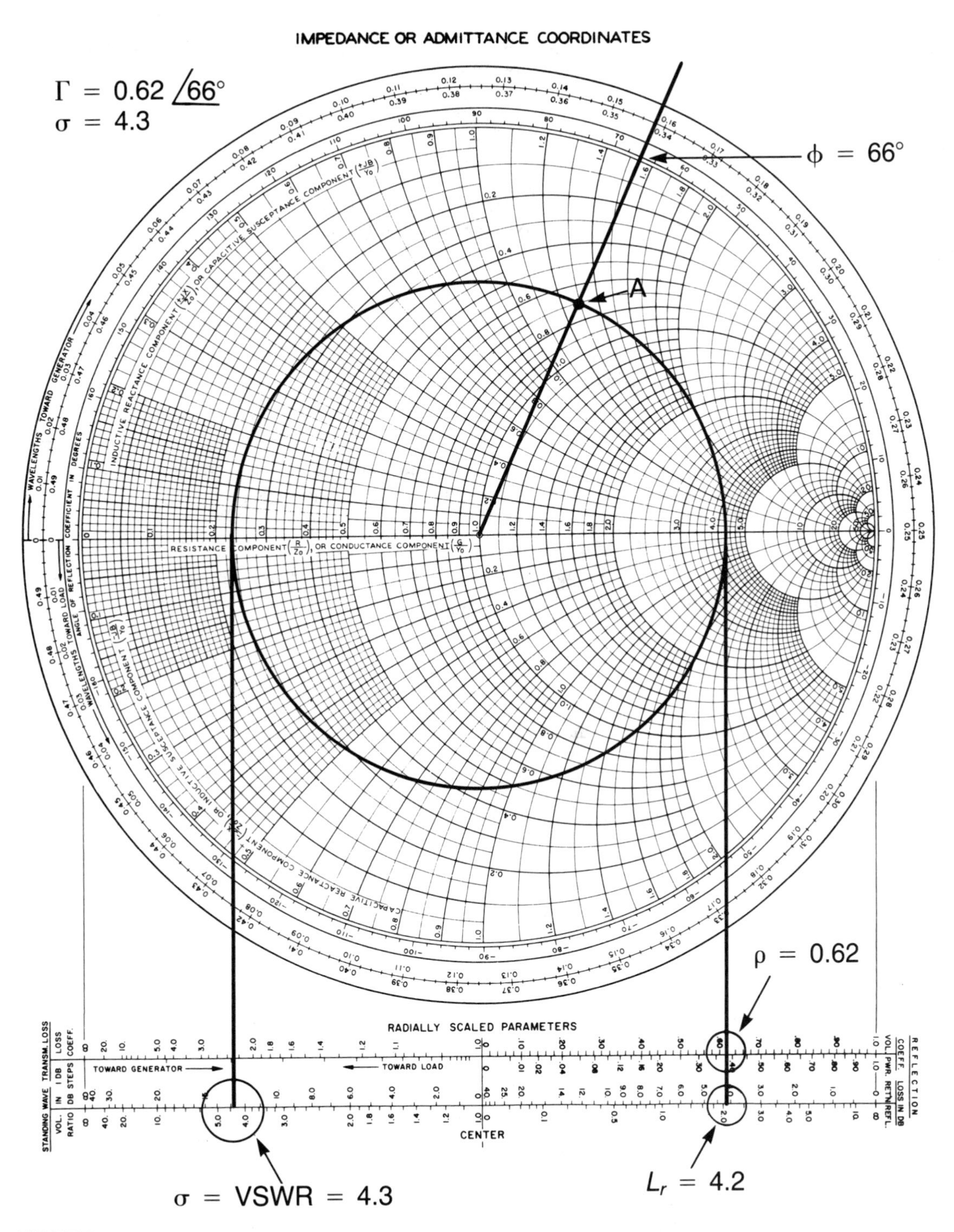

FIGURE 1–24
Use of the radially scaled parameters.

where V_I is the incident signal voltage and V_R is the reflected voltage. Using the laws of logarithms, we may rewrite equation (1–55) as

$$L_r = 20 \log \frac{V_I}{V_R} \tag{1–56}$$

Recall from equation (1–37) that $\rho = V_R/V_I$. Then, from equation (1–56), $V_I/V_R = 1/\rho$. Therefore, we may write equation (1–56) as

$$L_r = 20 \log \left(\frac{1}{\rho}\right) = -20 \log \rho \tag{1–57}$$

In the example shown in figure 1–24, $L_r = -20 \log (0.62) = 4.15$, or approximately 4.2 as read on the scales below the Smith chart.

Before applying our knowledge of the Smith chart to the solution of a particular problem, we shall first look at two additional points on the Smith chart which deserve special attention. At one point, the impedance and the voltage are both a maximum, and at the other point, both the impedance and the voltage are a minimum. These points are shown in figure 1–25.

In order for a voltage maximum to exist, V_r and V_i must both be in phase. That is, the phase angle of Γ must be zero. This condition occurs only at a point of resistance. Now, using equation (1–43)

$$Z = Z_0 \frac{1 + \Gamma}{1 - \Gamma}$$

we normalize this impedance to obtain

$$z_{max} = \frac{1 + \Gamma}{1 - \Gamma} = \text{VSWR}$$

This result shows that the point of maximum impedance is also equal to the VSWR. In figure 1–25, $z_{max} = r = \text{VSWR} = 4.3$. The actual value of the resistance at this point is $rZ_0 = 4.3(50) = 215$ ohms.

At a voltage minimum, V_r and V_i must be 180 degrees out of phase. Then, from equation (1–46), we obtain

$$Z = Z_0 \frac{1 - \Gamma}{1 + \Gamma}$$

Normalizing this, we have

$$z_{min} = \frac{1 - \Gamma}{1 + \Gamma} = \frac{1}{\text{VSWR}}$$

In figure 1–25, the SWR circle cuts the r-axis at 0.23. Note that the reciprocal of the VSWR (4.3) equals 0.23. The actual value of resistance at this point is $rZ_0 = 0.23(50) = 11.5$ ohms.

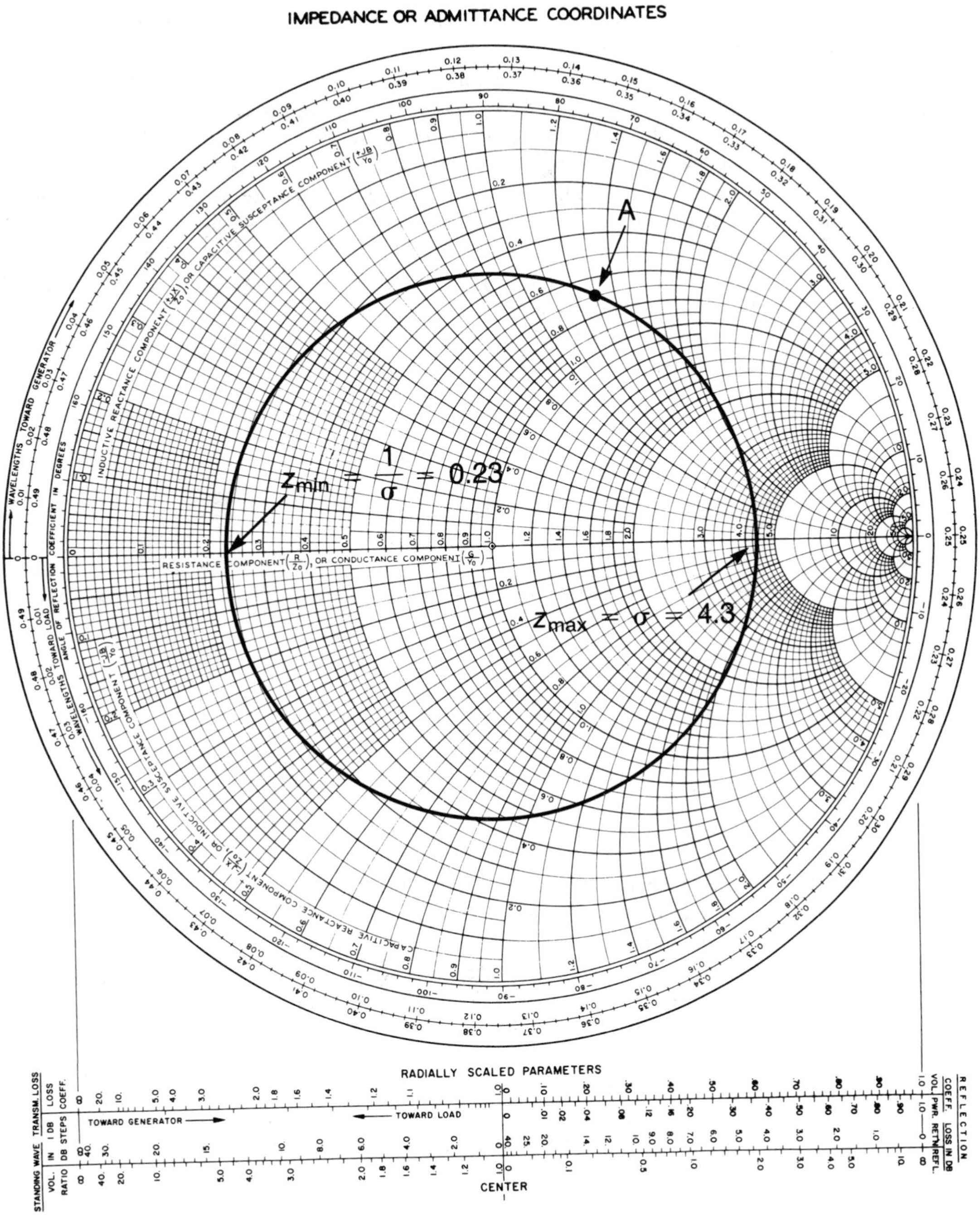

IMPEDANCE OR ADMITTANCE COORDINATES

FIGURE 1–25
Points of z_{min} and z_{max}

Recall our earlier discussion concerning the use of a stub in matching a complex load to the line in order to achieve complete power transfer. In that case, the VSWR will be 1.00. We will now see how this is accomplished using the Smith chart and our example load whose normalized impedance is plotted again as A in figure 1–26.

To begin with, it must be recognized that since the stub is to be placed in parallel with the line and load, we must deal with admittance, not impedance.[12]

From equation (1–49), $z_S = y_L$, we know that the normalized impedance has a normalized admittance value one-quarter wavelength away. In figure 1–26, point A is the normalized impedance $z_L = 0.7 + j1.3$. Diametrically opposite point A is point B, which is the normalized admittance $y_L = 1/z_L = 0.32 - j0.6$. This value may be read directly from the Smith chart at B.

This value of y_L should be confirmed by the student using complex algebra. Note that there are at least three approaches to the problem:

1. $$y_L = \frac{1}{z_L} = \frac{1}{0.7 + j1.3} \cdot \frac{0.7 - j1.3}{0.7 - j1.3} = 0.32 - j0.6$$

2. $$y_L = \frac{1\ \angle 0^\circ}{1.476\ \angle 61.6^\circ} = .7\ \angle -61.7^\circ$$

$$\Rightarrow = 0.32 - j0.6$$

3. $y_L = Y_L/Y_0$ where Y_0 is the characteristic load admittance given by $Y_0 = 1/Z_0$.

$$y_L = \frac{\left(\dfrac{1}{Z_L}\right)}{\left(\dfrac{1}{Z_0}\right)} = \frac{0.014\ \angle -61.7^\circ}{0.02\ \angle 0^\circ}$$

$$= 0.677\ \angle -61.7^\circ$$

$$\Rightarrow 0.32 - j0.6$$

Since the total distance around the Smith chart represents one-half wavelength, then one-half that distance represents one-quarter wavelength. This is why point B is diametrically opposite point A, and is said to be $\lambda/4$ away.

Note that decimal parts of a wavelength are printed along the two outermost scales from 0 to $\lambda/2$. Movement in the clockwise direction corresponds to movement along a lossless line *from* the load *toward* the generator. The opposite is true when one moves counterclockwise. So, the direction one moves is obviously important.

A line is now drawn from A through B and extended out past the rim of the chart. Note that this line intersects 0.409, and is our reference wavelength value. Recall from our earlier discussion that a stub placed at the proper point sees a normalized admittance

---

[12] This situation is similar to resistors in parallel whose conductances add arithmetically.

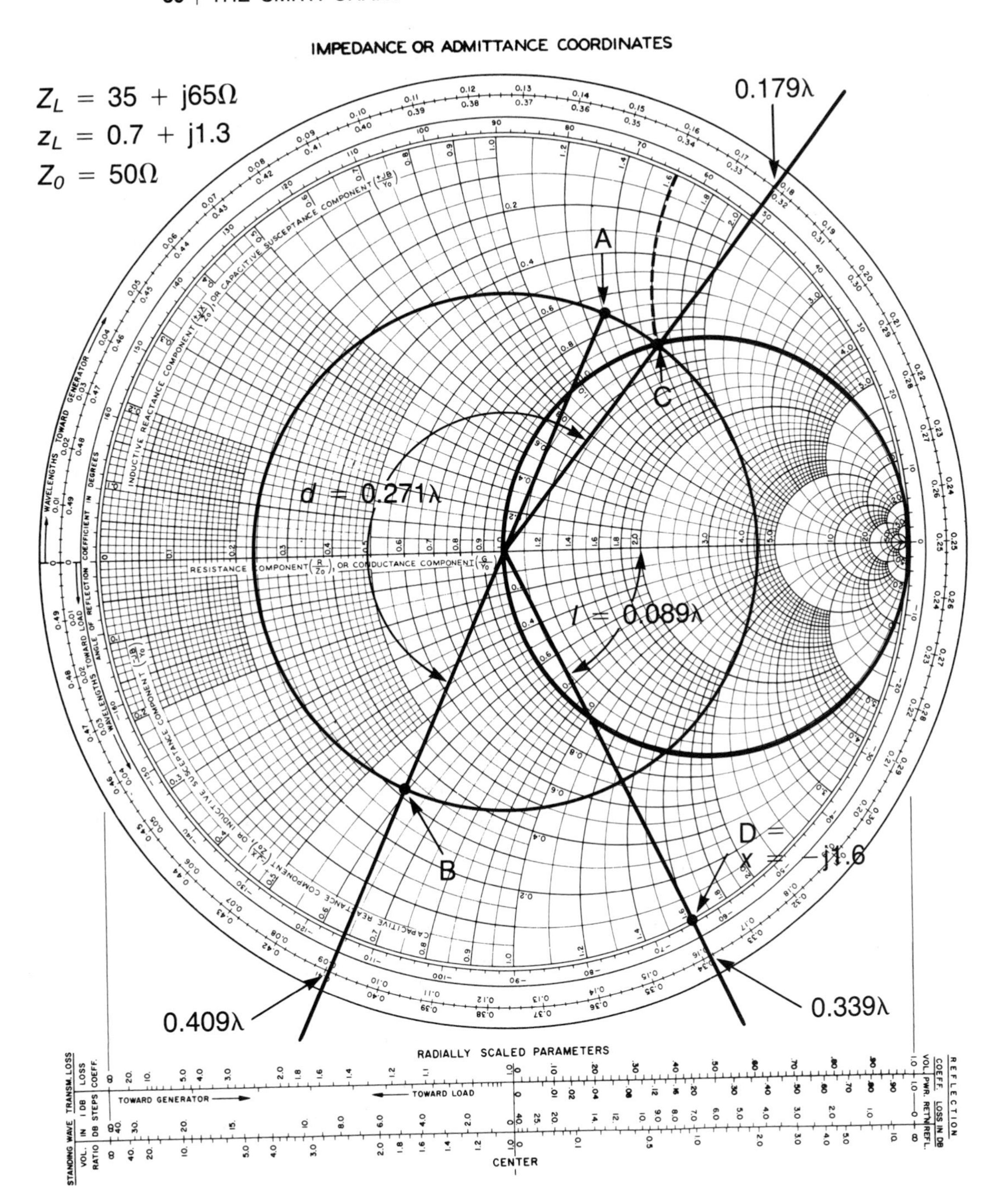

FIGURE 1–26
Smith chart solution for matching $Z_L = 35 + j65$ ohms to $Z_0 = 50$ ohm line.

of 1 + jb. The question is now, "How far must we move along the line from B in order to see this value?"

It should be apparent that as one moves away from the load's normalized admittance at B (CW direction), one will eventually reach the point of intersection of the $r = 1$ circle. This is shown as point C in figure 1–26. Note that the susceptance at this point is +j1.6 as read from the chart (follow arc from C to inner scale. Read 1.6). In other words, at point C, the admittance at this point on the line is 1 + j1.6. The CW distance d we have traveled to reach this point may now be calculated as follows.

The distance from B to the horizontal line is 0.5 − 0.409 = 0.091 wavelength. Moreover, the distance from the horizontal line to C is 0.179 as read from the outermost scale. This point was determined by drawing a line from the center of the chart through C and extending to the scale. Therefore, the total distance moved is 0.091 + 0.179 = 0.27 wavelength. To find the actual distance in feet or meters, one need only multiply this value of 0.27 by the actual value of the wavelength as computed by equation (1–9),

$$\lambda = \frac{\lambda_0}{\sqrt{\epsilon_r}}$$

where λ_0 = freespace wavelength computed as c/f.

Now that we know where to place the stub, the next question is, "How long must the stub be?" We know from the above that at C the line has a normalized admittance of 1 + j1.6. Therefore, a susceptance of −j1.6 will exactly tune out this value, leaving us with a pure resistance value $y_L = 1.0$. Locate point D where the susceptance is −j1.6 as read from the innermost scale of the chart. Draw a line from the center of the chart through this point, and read the reference wavelength 0.339 from the chart's rim. The value 0.339 is *not* the length of the stub.

Since the stub will have a short circuit at its far end, its admittance at that point will be infinity, j-infinity, which is located at the extreme right hand of the r-axis. The wavelength at this point is labeled 0.25. Therefore, the length of the stub will be 0.339 − 0.25 = 0.089, as shown in figure 1–26. One must always start at the righthand edge of the chart at infinity whenever a short-circuited stub is being used.[13]

In conclusion, a short-circuited stub 0.089 wavelength long located on the line 0.27 wavelength away from a complex load of Z_L = 35 + j65 ohms will reduce the VSWR on the line to 1.00, and a maximum transfer of power from source to load will be achieved.

Figure 1–27 shows another example of a Smith chart problem where a load impedance consisting of a 45-ohm resistor in series with a 26.5 pF capacitor is connected across a generator of 0.12 GHz. The stub length (l) and distance (d) required for matching are as shown in the figure. The details of the solution are left to the student as an exercise.

---

[13]While open-circuit stubs may also be used in theory, their high radiation losses make them unsuitable in practice.

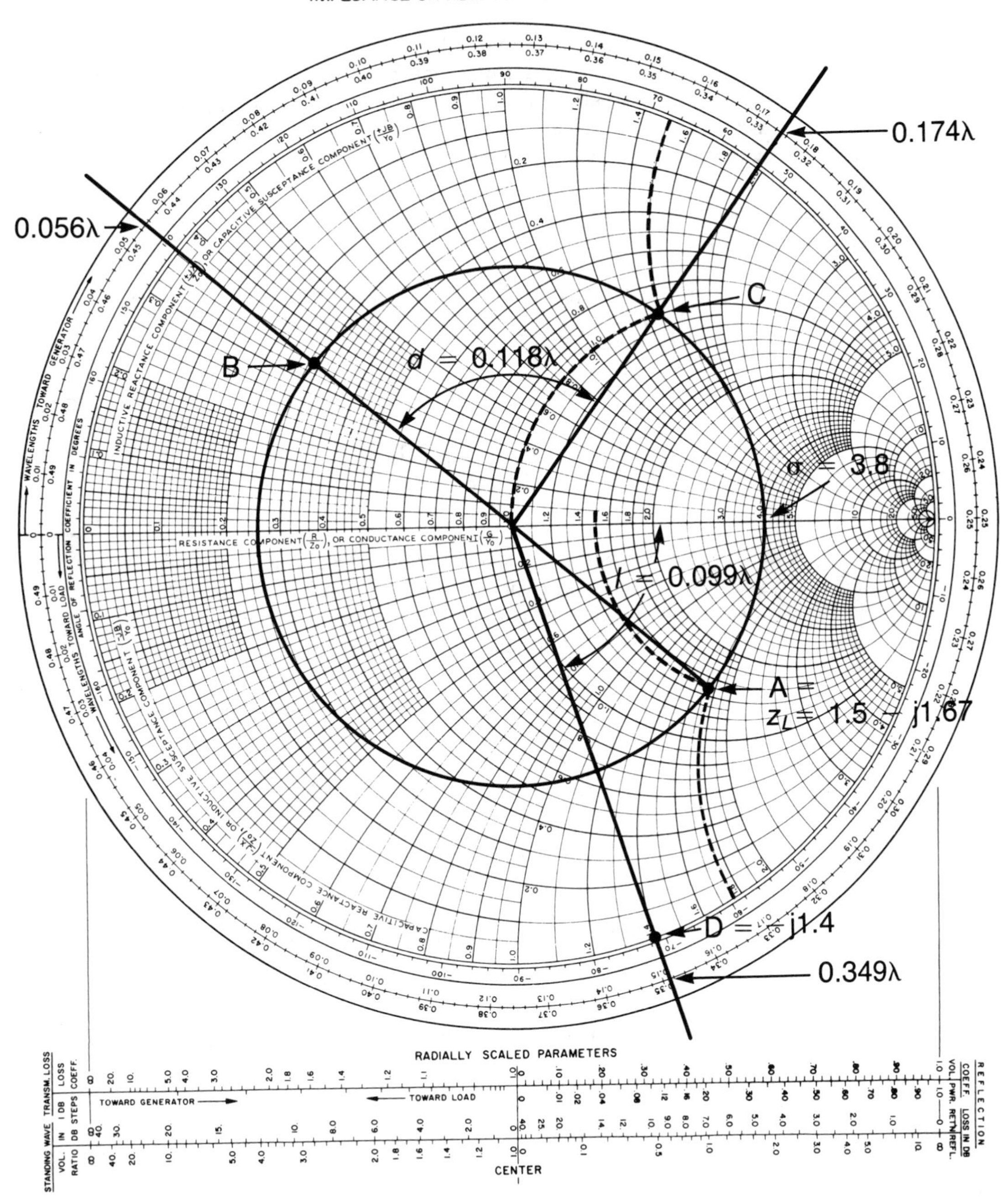

FIGURE 1–27
Smith chart solution: R = 45 ohms; C = 26.5 pF; f = 0.12 GHz; Z_0 = 30 ohms; l = 0.099λ; d = 0.118λ; $\Gamma_L = 0.58/-40$ degrees.

TIME-DOMAIN REFLECTOMETRY

Time-domain reflectometry (TDR) had its beginnings in the power distribution industry where, for years, technicians used it to establish the location of broken lines and other major discontinuities. The idea behind TDR (also called one-dimensional or closed-loop radar) is that of sending a pulse down the line and converting its return time into distance. A typical set-up might look like that shown in figure 1–28.

FIGURE 1–28
Typical TDR set-up.

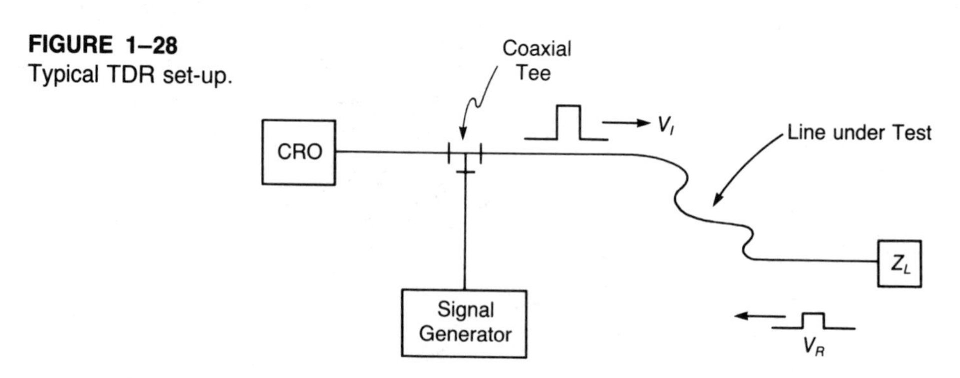

A return pulse (V_R) will occur at any location along the cable's length where either (1) a physical discontinuity exists or (2) the load is any impedance other than a pure resistance equal to Z_0 as discussed earlier. This pulse, whose characteristics will be altered in some important way depending on the nature of the discontinuity, may be viewed on an oscilloscope, and the distance determined from this information. Figure 1–29 represents a typical scope display obtained from an open line ($Z_L = \infty$).

FIGURE 1–29
Scope display.

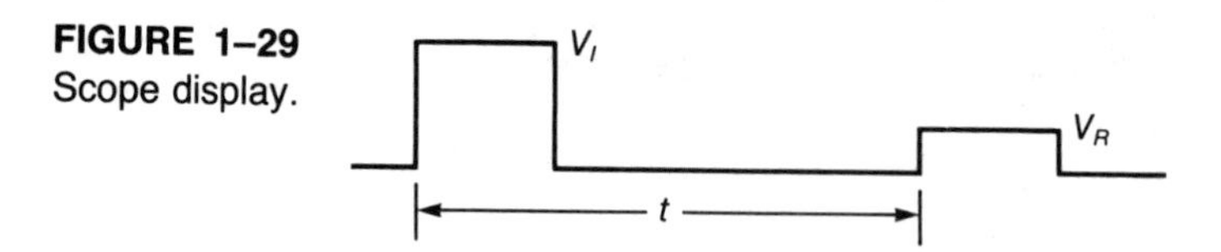

The distance (d) to Z_L may be determined as follows:

$$\text{distance } (d) = \text{velocity} \times \text{time}$$

In this case, the velocity is the phase velocity as defined by equation (1–7):

$$v_p = \frac{c}{\sqrt{\epsilon_r}}$$

where ϵ_r is the dielectric constant of the cable.

Note that the phase velocity may be determined experimentally by using a known length of cable as the same type under test and solving equation (1–58) for v_p.

The time is actually $t/2$ since t is the round-trip time of the pulse. Therefore, the distance is

$$d = v_p\left(\frac{t}{2}\right) \tag{1–58}$$

As stated, the return pulse will not appear the same as the incident pulse since the nature of the discontinuity will alter one or more of its characteristics. To see why this is so, recall equation (1–38) for the voltage reflection coefficient:

$$\Gamma_L = \frac{Z_L - Z_0}{Z_L + Z_0} = \frac{V_R}{V_I}$$

It may be seen, for example, that if the line was terminated in an open circuit, then $Z_L = \infty$. Rearranging equation (1–38) by dividing numerator and denominator by Z_L, we obtain,

$$\frac{V_R}{V_I} = \frac{1 - \dfrac{Z_0}{Z_L}}{1 + \dfrac{Z_0}{Z_L}} = +1$$

which is true only in the case of an open-circuit termination. Therefore, the scope display in this particular case is as shown in figure 1–30.

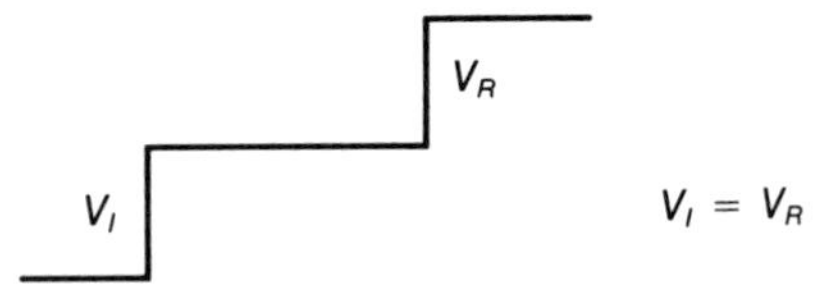

FIGURE 1–30
Scope display for open-circuit termination.

Note that the display shown in figure 1–30 is what would be encountered in practice and differs from the idealized display shown in figure 1–29.

In the case of a line terminated in a short circuit, equation (1–38) becomes

$$\frac{V_R}{V_I} = \frac{0 - Z_0}{0 + Z_0} = -1$$

The scope display for this case is shown in figure 1–31.

FIGURE 1–31
Scope display for short-circuited line.

V_I $-V_R$

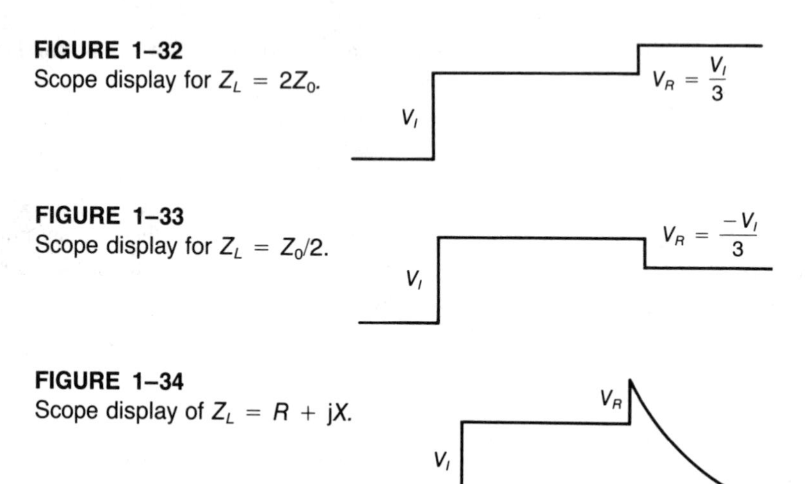

FIGURE 1–32
Scope display for $Z_L = 2Z_0$.

FIGURE 1–33
Scope display for $Z_L = Z_0/2$.

FIGURE 1–34
Scope display of $Z_L = R + jX$.

For pure resistance terminations, the magnitude of V_R is determined by the relative values of Z_L and Z_0. For example, if $Z_L = 2Z_0$, equation (1–38) becomes

$$\frac{V_R}{V_I} = \frac{2Z_0 - Z_0}{2Z_0 + Z_0} = \frac{Z_0}{3Z_0} = +\frac{1}{3}$$

The scope display is shown in figure 1–32. In the case of $Z_L = Z_0/2$, equation (1–38) becomes

$$\frac{V_R}{V_I} = \frac{\frac{Z_0}{2} - Z_0}{\frac{Z_0}{2} + Z_0} = -\frac{1}{3}$$

The scope display is shown in figure 1–33.

If the line termination is a complex load of the form $Z_L = R + jX$, the scope display will not be the simple rectangular form shown for the previous four cases. For example, if the load consisted of a series resistance and inductance, the display would be like that shown in figure 1–34. An analysis of such complex loads is beyond the scope of this book, but there are several excellent references.[14]

SCATTERING PARAMETERS

The complete characterization of a microwave device depends on more than just the specification of its scalar quantities. While gain and VSWR, for example, provide important magnitude-only data, phase information is missing. Consequently, the complex impedance of the device as well as its reflection coefficient, for example, remain unspecified, thus giving one an incomplete view of the overall package. At lower frequencies (up to about 100 MHz), active devices like transistors have their performance specified in terms of *h, y,* or *z* parameters[15] which require the device to be operated under various shorted and open conditions while measurements are taken. In the microwave region, parasitic effects (stray *L* or *C*) make it inconvenient if not impossible to maintain these conditions, making measurement of voltage and current impractical at best. At microwave frequencies, a short may act as an inductor and an open behaves as if it were capacitive, thus encouraging the device to break into spontaneous oscillation.

As a consequence of the problems inherent with measurement of conventional transistor parameters, a new set of parameters was devised for microwave work that makes use of the reflected ("scattered") waves moving to and from the device under test. These scattering or "S" parameters, as they are called, require the network to be terminated in a pure resistance instead of an open or short. This termination scheme lessens the tendency of the device to oscillate, and only the magnitude of incident and reflected traveling-wave voltages need be measured.

S Parameters

We may consider the passive resistor circuit shown in figure 1–35 as a two-port network. In a similar way, we may regard a transistor as a two-port device, as shown in figure 1–36. In general, any two-port black-box device may be represented as shown in figure 1–37.

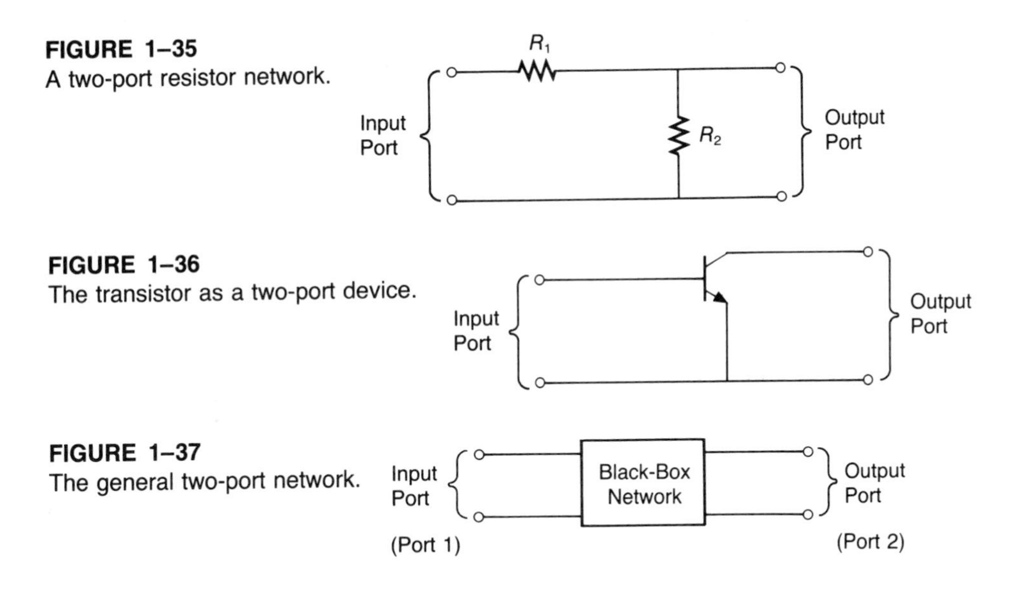

FIGURE 1–35
A two-port resistor network.

FIGURE 1–36
The transistor as a two-port device.

FIGURE 1–37
The general two-port network.

---

[14]Laverghetta, T. *Handbook of Microwave Testing*. Norwood, Massachusetts: Artech, 1981, pp. 426–32. See also Hewlett-Packard, Application Note 62, in *Time Domain Reflectometry*, 1964; and Hewlett-Packard, Application Note 67, in *Cable Testing with TDR*, October 1965.

[15]For a review of *h* and *y* parameters, see Bogart, T. *Electronic Devices and Circuits*. Columbus, OH: Merrill Publishing Co., 1986. Chapter 9.

Suppose that a voltage $V_1 = 1$ is applied across the input-port terminals of the black-box device shown in figure 1–37. Suppose further that a measured current $I_1 = 1$ is obtained at port 1. It should be obvious that there is some opposition presented by port 1 whose value is 1 ohm. If we were to obtain the same results at port 1 of the devices shown in figures 1–35 and 1–36, we may conclude that insofar as the parameter V_1/I_1 is concerned, all input ports (port 1) of these three networks are equivalent despite the actual components that comprise the device. The importance of this statement in practice is that for any given set of parameters, a mathematical model of a device may be obtained that is completely independent of specific circuit elements. Such a model allows the circuit designer to predict the total performance of a device from a knowledge of its parameters alone and to evaluate the effect of the device on a larger system by using the equivalent circuit model to represent the network. Of course, the simpler the parameter set used to characterize the device ports, the easier it will be to measure these parameters and the more useful they will be to the designer. The parameters used depend on which network characteristics one wishes to specify. In microwave work, it is often useful to characterize the reflection and transmission coefficients of a device since these provide information on such things as input impedance and power gain. We will now define the basic two-port microwave network and its associated S parameters.

The S-parameter model is shown in figure 1–38. By convention, a represents an input and b represents an output. We may regard a_1 and a_2 as *incident* voltage waves on their respective ports, 1 and 2. Similarly, b_1 and b_2 may be considered as *reflected* voltage waves from their corresponding ports. The following mathematical definitions of a and b will now be presented, and the justification of these definitions will become clear shortly.

$$a_1 = \frac{1}{2}\left(\frac{V_1}{\sqrt{Z_0}} + \sqrt{Z_0}\, I_1\right) \tag{1–59}$$

$$b_1 = \frac{1}{2}\left(\frac{V_1}{\sqrt{Z_0}} - \sqrt{Z_0}\, I_1\right) \tag{1–60}$$

$$a_2 = \frac{1}{2}\left(\frac{V_2}{\sqrt{Z_0}} + \sqrt{Z_0}\, I_2\right) \tag{1–61}$$

$$b_2 = \frac{1}{2}\left(\frac{V_2}{\sqrt{Z_0}} - \sqrt{Z_0}\, I_2\right) \tag{1–62}$$

From figure 1–38, we observe that part of a_1 is transmitted to port 2 (S_{21}) and part is reflected from port 1 (S_{11}). By convention, the S-parameter numbering system is:

1st number = port where energy *emerges* from device
2nd number = port where energy *enters* the device

Therefore, the designation S_{11} means that part of the energy of a_1 which emerged from port 1 also came from port 1. Likewise, S_{21} means that part of the energy coming from port 2 came from the energy a_1 into port 1. Thus, S_{11} is a *reflection* coefficient, and S_{21} is a *transmission* coefficient.

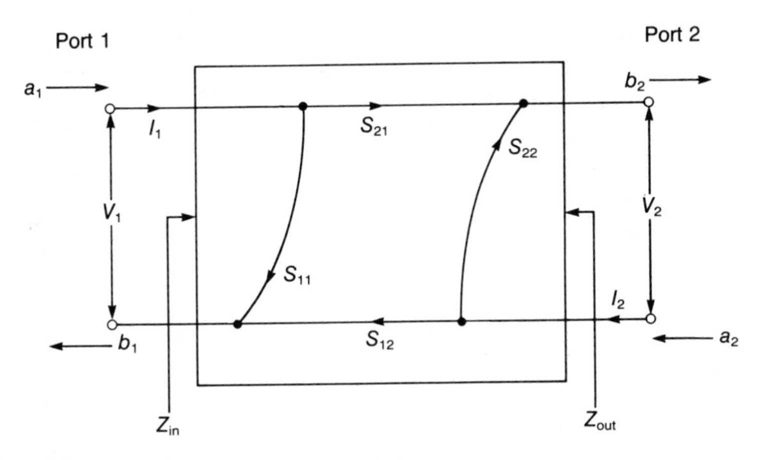

FIGURE 1–38
S-parameter model of 2-port linear microwave network.

Note that if b_1 is divided by a_1, the result is[16]

$$\frac{Z_{\text{in}} - Z_0}{Z_{\text{in}} + Z_0}$$

which we recognize from equation (1–38) as a reflection coefficient. Since this result involves the input of the device, we designate this as the input reflection coefficient S_{11}.

$$S_{11} = \frac{b_1}{a_1} = \frac{Z_{\text{in}} - Z_0}{Z_{\text{in}} + Z_0} \qquad \textbf{(1–63)}$$

A careful study of figure 1–38 reveals the following relationships:

$$b_1 = S_{11}a_1 + S_{12}a_2 \qquad \textbf{(1–64)}$$

$$b_2 = S_{21}a_1 + S_{22}a_2 \qquad \textbf{(1–65)}$$

For example, equation (1–64) states that the output of port 1 is the sum of the fraction of the energy ($S_{11}a_1$) reflected from port 1, plus the fraction ($S_{12}a_2$) transmitted from port 2. It should be apparent from equation (1–64), then, that the results given by

---

[16]The following details of the division may begin to show why the values of a and b were so chosen.

$$\frac{b_1}{a_1} = \frac{\dfrac{V_1 - Z_0I_1}{\sqrt{Z_0}}}{\dfrac{V_1 + Z_0I_1}{\sqrt{Z_0}}} = \frac{V_1 - Z_0I_1}{V_1 + Z_0I_1}$$

$$= \frac{\dfrac{V_1}{I_1} - Z_0}{\dfrac{V_1}{I_1} + Z_0} = \frac{Z_{\text{in}} - Z_0}{Z_{\text{in}} + Z_0} = \Gamma_{\text{in}}$$

equation (1–63) apply only if $a_2 = 0$, and that this only happens if there is no reflected energy coming from the load connected to port 2. Of course, the only way to guarantee this is to terminate port 2 in a resistance load equal in value to Z_0. This will be made clearer by an observation of figure 1–39.

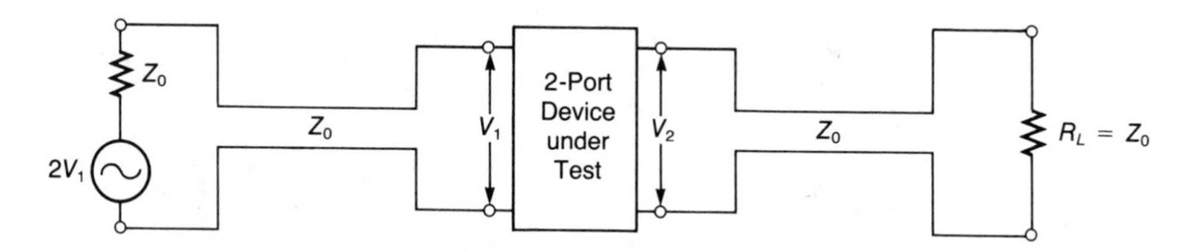

FIGURE 1–39
Using Z_0 transmission line to connect source and load to the device under test.

From figure 1–39, it will be noted that if the load resistance equals the characteristic impedance of the line, any wave traveling toward the load will be completely absorbed by the load, and none will be reflected. Implicit in the foregoing statement is that the output impedance (Z_{out}) of the device does *not* need to match Z_0 of the line.

An analysis similar to the above may be made for the calculation of S_{22} when $a_1 = 0$:

$$S_{22} = \frac{b_2}{a_2} = \frac{Z_{out} - Z_0}{Z_{out} + Z_0} \tag{1–66}$$

We see, therefore, that the reflection coefficients of both the input and output of the device are completely determined by the S parameters S_{11} and S_{22}. We also note that the device must be driven at both its input as well as its output with the other port suitably terminated in Z_0 in order to obtain these measurements.

We now turn our attention to the *transmission coefficients* S_{21} and S_{12}. From equation (1–61), if $a_2 = 0$, then

$$S_{21} = \frac{b_2}{a_1} \tag{1–67}$$

Note that by the definition of a_1 given by equation (1–59), we may write

$$a_1 = \frac{V_1}{2\sqrt{Z_0}} + \frac{\sqrt{Z_0}I_1}{2} = \frac{V_1 + Z_0I_1}{2\sqrt{Z_0}} = \frac{2V_1}{2\sqrt{Z_0}} = \frac{V_1}{\sqrt{Z_0}} \tag{1–68}$$

Now, since a_2 is required to be zero, then by equation (1–61)

$$a_2 = 0 \text{ iff } \frac{V_2}{\sqrt{Z_0}} = -\sqrt{Z_0}I_2 \tag{1–69}$$

Moreover, by equation (1–62),

$$b_2 = \frac{1}{2}\left(\frac{V_2}{\sqrt{Z_0}} - \sqrt{Z_0}I_2\right)$$

Substituting $V_2/\sqrt{Z_0}$ from equation (1–69) for $-\sqrt{Z_0}I_2$ in equation (1–62) we obtain

$$b_2 = \frac{1}{2}\left(\frac{V_2}{\sqrt{Z_0}} + \frac{V_2}{\sqrt{Z_0}}\right) = \frac{V_2}{\sqrt{Z_0}} \tag{1–70}$$

Finally,

$$S_{21} = \frac{b_2}{a_1} = \frac{\dfrac{V_2}{\sqrt{Z_0}}}{\dfrac{V_1}{\sqrt{Z_0}}} = \frac{V_2}{V_1} \tag{1–71}$$

It may also be shown in a similar manner that

$$S_{12} = \frac{b_1}{a_2} = \frac{V_1}{V_2} \text{ when } a_1 = 0 \tag{1–72}$$

We may now summarize the various coefficients as follows:

$$S_{11} = \frac{b_1}{a_1} \text{ (when } a_2 = 0) = \Gamma_{\text{in}} = \frac{Z_{\text{in}} - Z_0}{Z_{\text{in}} + Z_0}$$

$$S_{22} = \frac{b_2}{a_2} \text{ (when } a_1 = 0) = \Gamma_{\text{out}} = \frac{Z_{\text{out}} - Z_0}{Z_{\text{out}} + Z_0}$$

$$S_{21} = \frac{b_2}{a_1} \text{ (when } a_2 = 0) = \frac{V_2}{V_1} \quad \text{forward transmission coefficient}$$

$$S_{12} = \frac{b_1}{a_2} \text{ (when } a_1 = 0) = \frac{V_1}{V_2} \quad \text{reverse transmission coefficient}$$

From the manner in which the a and b voltage waves were defined by equations (1–59) through (1–62), their squares may be interpreted as power waves. For example,

$$|b_2|^2 = \frac{1}{4}\frac{V_2^2}{Z_0} - \frac{1}{2}V_2I_2 + \frac{1}{4}Z_0I_2^2$$

$$= \text{power from output of network delivered to } Z_0 \text{ load}$$

Similarly,

$|b_1|^2$ = power available from Z_0 source less power delivered to input of network (i.e., power reflected from input port)

$|a_2|^2$ = power incident on output of network

$|a_1|^2$ = power incident on input of network

From the above descriptions, we may form the following ratio

$$|S_{21}|^2 = \left|\frac{b_2}{a_1}\right|^2 = \frac{\text{power delivered to load}}{\text{power incident on input port}}$$
$$= \text{power gain (with } Z_0 \text{ source and load)}$$

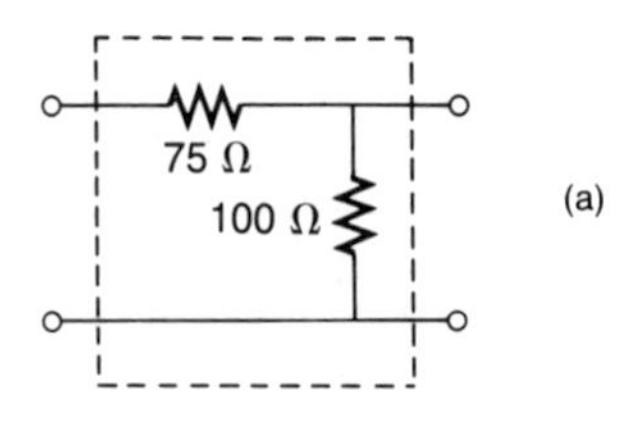

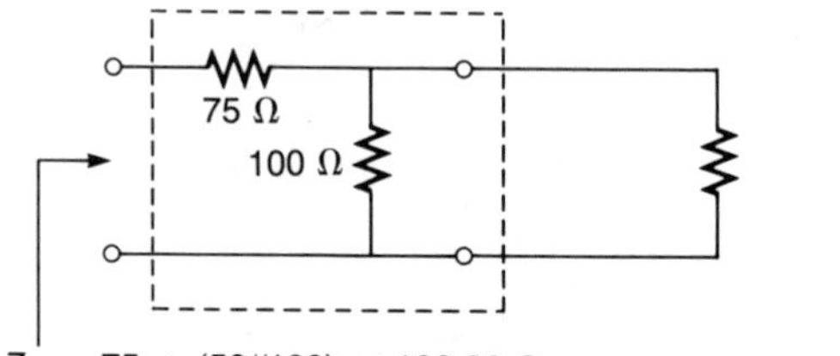

$Z_{in} = 75 + (50//100) = 108.33\ \Omega$

$$S_{11} = \frac{Z_{in} - Z_0}{Z_{in} + Z_0} = \frac{108.33 - 50}{108.33 + 50} = 0.368$$

In polar form, $S_{11} = 0.368\angle 0°$

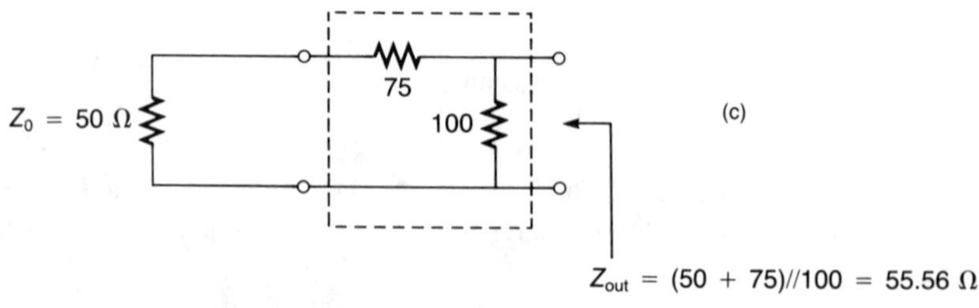

$Z_{out} = (50 + 75)//100 = 55.56\ \Omega$

$$S_{22} = \frac{Z_{out} - Z_0}{Z_{out} + Z_0} = \frac{55.56 - 50}{55.56 + 50} = \frac{5.56}{105.56} = 0.053$$

In polar form, $S_{22} = 0.053\angle 0°$

FIGURE 1–40
Analysis of a passive resistor network: (a) 2-port network; (b) calculation of $S_{11} = \Gamma_{in}$; (c) calculation of $S_{22} = \Gamma_{out}$

A simple example using a passive resistor network may make clear some of the more pertinent ideas. Consider the circuit shown in figure 1–40(a) for $Z_0 = 50$ ohms.

If port 2 is terminated in Z_0, the input reflection coefficient (S_{11}) is found as shown in figure 1–40(b). To calculate S_{22}, port 1 is terminated in Z_0, and the reverse reflection coefficient is calculated as shown in figure 1–40(c). If port 1 is driven with a source voltage of $2V_1$ with impedance of $Z_0 = 50$, the forward transmission coefficient (S_{21}) is calculated as shown in figure 1–40(d). If port 2 is now the driven port and port 1 is terminated in Z_0, S_{12} may be calculated as shown in figure 1–40(e).

If one attempts to compute S_{21}, for example, as the ratio V_2/V_1 using Ohm's law, one discovers with some disappointment that the ratio is 0.308 instead of 0.421 as shown in figure 1–40(d). Why should this contradiction exist? The answer is that unless $Z_{in} = Z_0$, some reflection will occur, and V_1 will not be the same as predicted by the

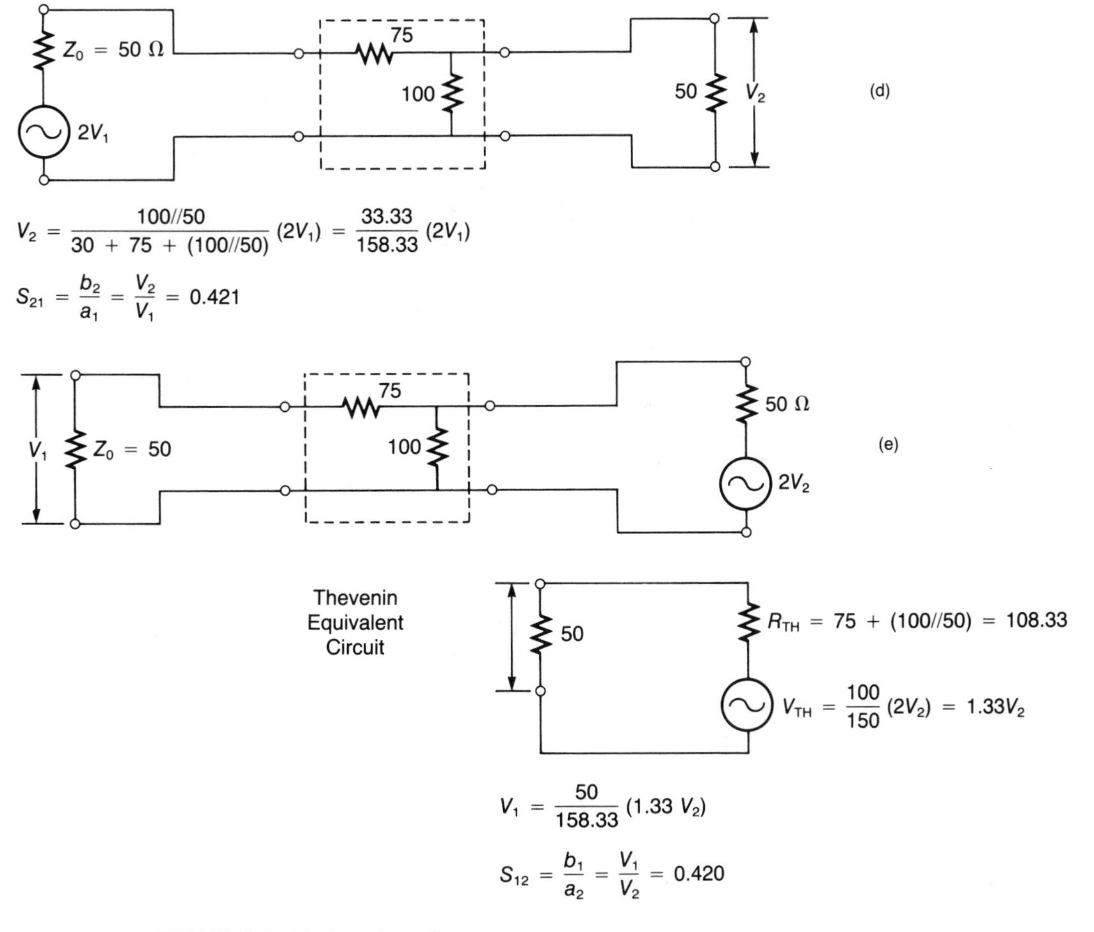

FIGURE 1–40 *(continued)*
(d) calculation of S_{21}; (e) calculation of S_{12}.

simple application of Ohm's law. Remember, we are dealing with *traveling* and *reflected* voltage waves. If, however, $Z_{in} = Z_0$, then no power will be reflected from port 1 (input port) and V_1 will be exactly as predicted by DC theory—that is, one-half the source voltage. This is why the generator voltage is specified as $2V_1$.

We may compute the power gain of the network shown in figure 1–40(a) as

$$\text{power gain} = |S_{21}|^2 = (0.421)^2 = 0.177$$

In decibels, the power gain is given by

$$G = 10 \log |S_{21}|^2 = 10 \log 0.177 = -7.5 \text{ dB}$$

The voltage ratios representing the S parameters of a device may be determined with the use of an instrument known as the network analyzer such as the one shown in figure 1–41.

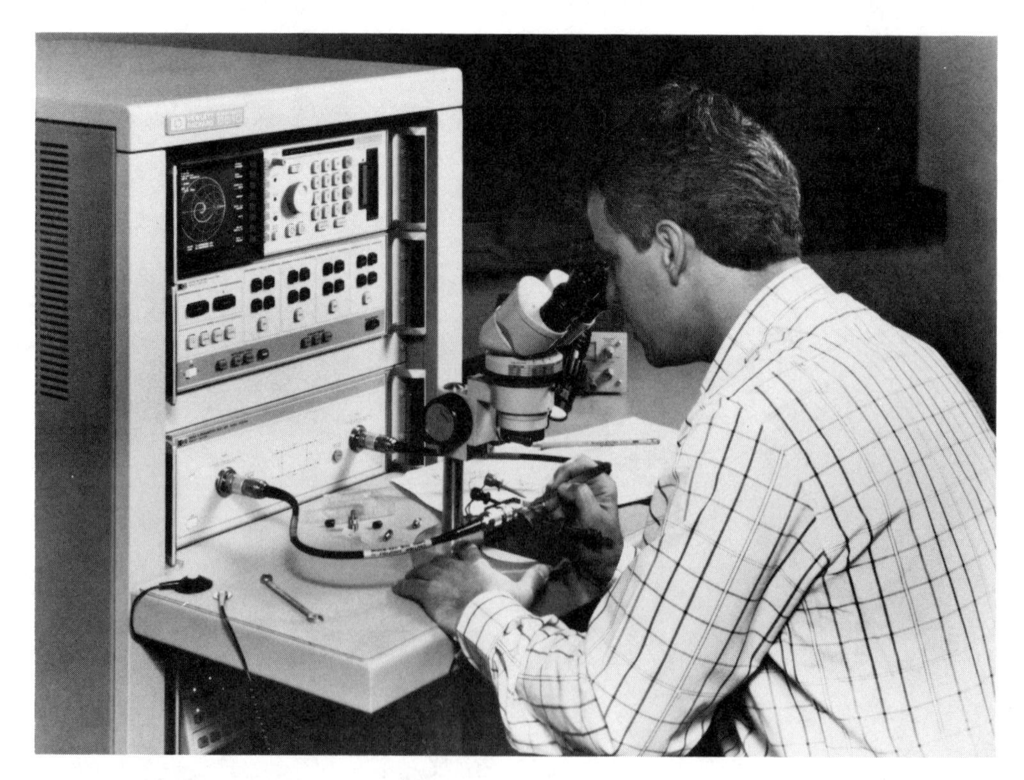

FIGURE 1–41
The Hewlett-Packard 8510E network analyzer system (45 MHz–20 GHz).

For further reading on S parameters, consult,

1. Hewlett-Packard Application Notes: #95, "S-Parameters—Circuit Analysis & Design," Sept. 1968.
2. Hewlett-Packard Application Notes: #117–1, "Microwave Network Analyzer Applications," June 1970.
3. Hewlett-Packard Application Notes: #154, "S-Parameter Design," April 1972.
4. Hewlett-Packard Bulletin #5954–8355, "Vector Measurements of High-Frequency Networks," March 1, 1987.
5. Laverghetta, T. *Handbook of Microwave Testing*. Dedham, Mass.: Artech House, 1981.

SUMMARY

The opening of the chapter introduced the *transmission line* as an arrangement of conductors capable of propagating power in the form of TEM waves from source to load. By virtue of the cable's cross-sectional geometry, certain distributed parameters were seen to exist that imparted a characteristic impedance to the cable quite independent of either physical length or frequency. It was seen that the value of the characteristic impedance (Z_0) could be determined from a knowledge of either the distributed parameters themselves (equation 1–16), the dielectric properties of the cable (equation 1–17), or its cross-sectional geometry (equations 1–21 and 1–24).

Standing waves were shown to exist on a mismatched line and represented power reflected from the load. The fraction of the incident voltage reflected was determined by the reflection coefficient (equation 1–38), and the magnitude of the mismatch was calculated as the standing-wave ratio (equation 1–40).

The transmission line was shown to possess reactive properties, and even exhibited characteristics of resonant circuits. The quarter-wave transformer (equation 1–47) was introduced as an impedance-matching device, and use of the Smith chart was developed as a line-load impedance-matching calculator (equation 1–53).

The chapter concluded with a brief discussion of time-domain reflectometry as a technique of identifying and isolating line discontinuities. Various scope displays of typical mismatched line conditions were illustrated. Scattering parameters were introduced and defined mathematically.

PROBLEMS

1. A pulse requires 553 nanoseconds to travel from one end of a 116 meter coaxial cable to the other. What is the dielectric constant of the cable? [2.04]

2. The dielectric constant of a certain medium propagating a TEM wave is 1.55. The wavelength of the signal within that medium is 2.4 cm. What is the frequency of the signal? [10 GHz]

3. A transmission line has a distributed inductance of 11.3 μH/m. Its characteristic impedance is 250 ohms. What is the value of the line's distributed capacitance? [181 pF/m]
4. What would be the characteristic resistance (R_0) of the DC transmission line shown in figure 1–9 if $R_s = 115$ ohms and $R_p = 927$ ohms? [389 ohms]
5. If a parallel-wire transmission line has a dielectric constant of 2.28, what is the absolute smallest value of Z_0 it can have? [55 ohms]
6. What is the smallest possible value of Z_0 for an air-dielectric parallel-wire transmission line? [83 ohms]
7. What is the diameter of the inner conductor of a coaxial cable whose dielectric constant is 1.12 and whose outer conductor diameter is 3/8 inch? [approx. 1/10 inch]
8. The distance between adjacent nodes of a standing wave in an air dielectric is determined to be 2.406 cm. What is the frequency of the generator? [6.23 GHz]
9. What is the value of the reflection coefficient Γ_L of a complex load $Z_L = 28 - j60$ ohms, terminating $Z_0 = 50$ ohms? [$0.65 \angle -72.5°$]
10. What is the VSWR for the line-load mismatch in problem 9? [4.71]
11. What is the impedance of the quarter-wave transformer required to match a 150-ohm line to a resistive load of 216 ohms? [180 ohms]
12. In figure 1–19, a complex load $Z_L = 70 - j55$ ohms was connected to $Z_0 = 50$ ohms. Using the Smith chart, plot the point of the load's normalized impedance (z_L) and find the value of the SWR. Use the Smith chart to determine the value of the reflection coefficient and the return loss (L_r). [VSWR = 2.6; $\Gamma = 0.44 \angle -45.4°$; $L_r = 7.1$]
13. Using the Smith chart, find the nearest point at which a quarter-wave transformer may be connected to a complex load $Z_L = 255 + j255$ ohms in order to get a VSWR = 1.0. Z_0 = 300. What is the impedance (Z'_0) of the transformer? [d = 0.105 wavelength; $Z'_0 = 474$ ohms]
14. Using the Smith chart, find the stub length (l) and distance (d) required to obtain a matched line-load condition between $Z_L = 41.25 - j22.5$ ohms and $Z_0 = 75$ ohms. [$d = 0.159$; $l = 0.351$]
15. Using the Smith chart, find the stub length (l) and distance (d) required to obtain a matched line-load condition between $Z_L = 31.25 + j10$ ohms and $Z_0 = 50$ ohms. [$d = 0.056$; $l = 0.326$]
16. Find the four S parameters and the power gain of the passive network shown if $Z_0 = 50$ ohms. [$S_{11} = .07 \angle 180°$; $S_{22} = 0.0074 \angle 180°$; $S_{21} = 0.219$; $S_{12} = 0.219$; power gain = -13.2 dB]

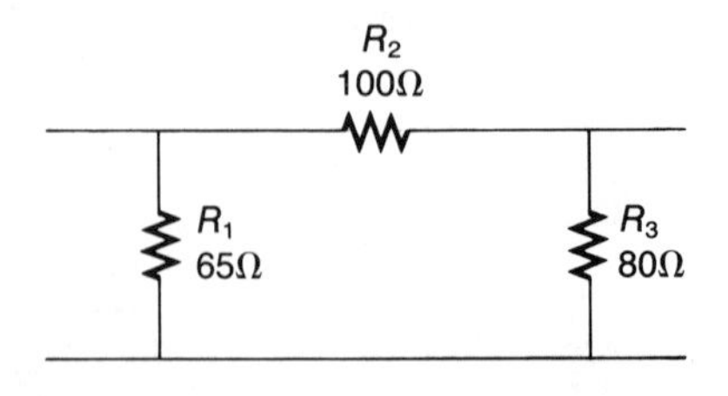

QUESTIONS

It is expected that students will answer these questions in their own words and not rely on the explanations already presented in the text.

1. What is a transmission line? How can you justify the assertion that a transmission line is not merely a wire?
2. What is a balanced line? An unbalanced line? Does one have an advantage over the other? Explain.
3. Define the term *characteristic impedance*. What is characteristic about it?
4. What role does the dielectric constant play in transmission lines?
5. How are standing waves produced? Why are they important?
6. Explain what the voltage reflection coefficient means. How is it calculated? What two components make up Γ_L, and what does each mean?
7. What does the term *phase velocity* mean? Why is it important?
8. Define the term *wavelength*.
9. What is meant by a complex impedance?
10. What is the mathematical relationship between impedance, admittance, resistance, conductance, reactance, and susceptance? Why are these terms important?
11. Exactly what is VSWR a measure of? What is the relationship between VSWR and Γ_L?
12. What is a quarter-wave transformer? Why is it important?
13. What is the principal difficulty encountered with the use of the quarter-wave transformer?
14. What is meant by normalized impedance? For what is it used?
15. What is a Smith chart?
16. What is a stub? When is it used?
17. What is meant by the reactive properties of a transmission line? What role do these play in connection with the Smith chart?
18. What is a VSWR circle?
19. Name at least three important transmission line factors that may be obtained from the radially scaled parameters of a Smith chart. Define each of these terms.
20. In making stub calculations using the Smith chart, it is more convenient to work with admittance rather than impedance. Why is this the case?
21. What does movement around the SWR circle represent in reality?
22. What do the complete circles on the Smith chart represent? The arcs?
23. What does each of the four scales drawn on the rim of the Smith chart represent?
24. Why is the intersection of the VSWR circle with the $r = 1$ circle such an important point? What does it represent?
25. If Z_0 is 50 ohms, what value would you expect to find at the exact center of the universal Smith chart? Why?
26. Of what importance are the S parameters in microwave work? Give an example.
27. What is the physical meaning attached to the S_{11} and S_{22} parameters?
28. What is the physical meaning attached to the S_{12} and S_{21} parameters?

2

WAVE PROPAGATION MODEL

Mathematical and physical models are used in engineering in order to *predict* the ideal behavior of complex events as they would occur in an ideal world. Obviously, since the utopian world does not exist, our models only approximate real-world behavior. As such, models are somewhat limited in their usefulness, but they do provide a starting point from which further refinements can be made in predicting actual event performance.

For example, in dealing with the predicted range of a projectile as a two-dimensional event, one might begin with the mathematical model expressed by the equation[1]

$$R = \frac{V^2 \sin 2\theta}{g}$$

where R = range; V = initial velocity; θ = angle to the horizontal; and g = acceleration due to gravity.

While this equation gives idealized results, it obviously does not take into account real-world variables such as wind velocity and direction, air density, ambient temperature, actual aerodynamic contours of the projectile, and many other variables that affect the trajectory. To include such variables in the initial model, however, would make the model so complicated that it would lose its usefulness. The key to successful engineering model design is *simplicity*.

In a similar way, the wave propagation model of electromagnetic wave phenomena allows us to make predictions as to how a wave will behave under idealized conditions. This prediction then gives us a starting point from which actual wave behavior may be deduced and refined.

[1]Ouseph, P. *Technical Physics*. New York: D. Van Nostrand, 1980.

ELEMENTS OF THE MODEL

Our model assumes an idealized wave environment wherein no other outside forces are at work. This environment is called *free space* and is the epitome of nothingness. In free space, there are no other TEM waves, no gravity, no obstructions, no atmosphere, no celestial events, no terrestrial events, no electrical noise, and no observers. In short, the wave environment is free of everything except the wave itself. Obviously, then, free space does not exist. However, it is important to visualize such an environment in order to speculate initially about ideal wave behavior.

In free space, a TEM wave propagates with the speed of light *(c),* 300 million meters per second (3×10^{10} cm/S),[2] or about 186,000 miles per second. In any other real-world environment, the velocity of propagation is always slower.

FIGURE 2–1
Vector diagram of a TEM wave.

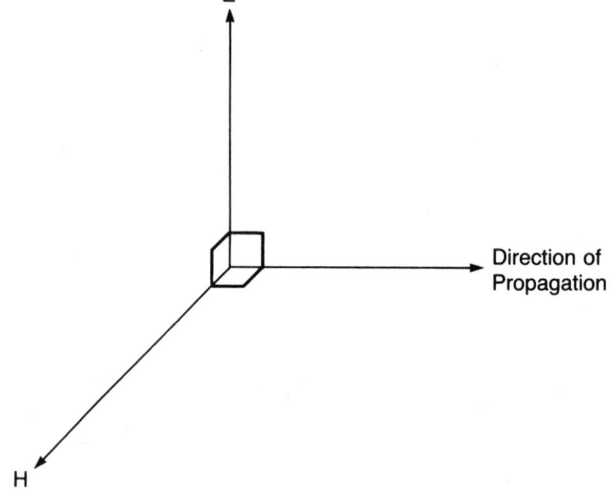

The wave is called TEM because it is a *t*ransverse *e*lectro*m*agnetic phenomenon wherein the electric (E) and magnetic (M) vectors are transverse (T) (cross one another) and are also mutually perpendicular to the direction of propagation. A vector drawing of the TEM wave is shown in figure 2–1, and an attempt to show such a wave as a sinusoidal event (for simplicity) is shown in figure 2–2.

A radiated TEM wave in free space is often spoken of as being in *time phase* and *space quadrature*. This means that the E (electric) and H (magnetic) vectors rise and fall together in time, but are 90 degrees apart in space. These conditions may be appreciated by again referring to figure 2–2.

Within the milieu of free space, our wave is thought of as emanating from a dimensionless source. Mathematically, such a zero-dimensional source is obviously a point source. Moreover, the waves are regarded as radiating *uniformly* in all directions from this point. Consequently, we call such a radiation point an *isotropic source*.

---

[2]The value of 300×10^6 m/S is a rounded-off value of the more precise value of 299.7925×10^6 m/S. (Gray & Issacs, Eds. *A New Dictionary of Physics*. London: Longman, 1975.)

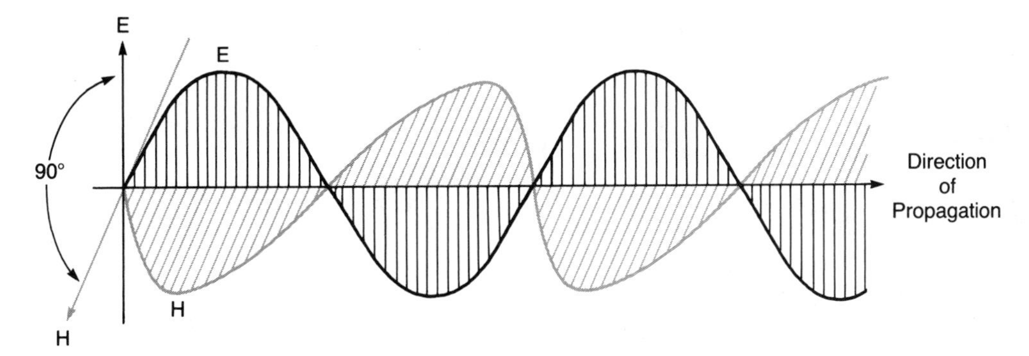

FIGURE 2–2
Spatial relationships of the electric (E) and magnetic (H) vectors of a sinusoidal TEM wave.

The choice of this name is important since it implies that if all points of a given intensity were connected together, the resultant three-dimensional shape would be a sphere whose surface area is given by $A = 4\pi r^2$, where r is any fixed distance from the source to where the intensity is measured. Since there are an infinite number of such fixed distances (r), we begin to see that TEM waves must leave the source as ever-expanding spheres like the layers of an onion. Moreover, each layer consists of *both* electric as well as magnetic vectors at right angles. This fact was pointed out in connection with figures 2–1 and 2–2. The isotropic model is shown in figure 2–3.

FIGURE 2–3
The isotropic model.

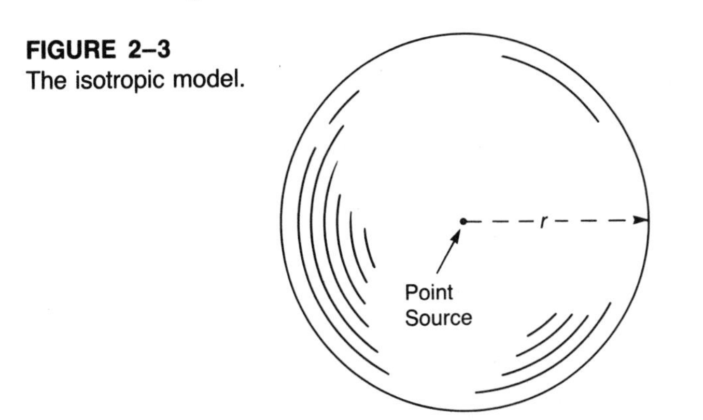

As one moves further and further away from the isotropic source, the spheres of radiated energy become larger and larger. Therefore, at some finite distance, any small, local region of the sphere will appear, for all practical purposes, to be flat. This is very similar to a small lake whose surface appears flat, but which, in fact, follows the curvature of the earth's surface. In other words, the lake is *locally* flat.

It is for this reason that we often refer to a wave *front* (small, flat, local region of a spherical TEM wave) as a *plane* wave, and treat this region as if it were, in fact, a two-dimensional shape. This treatment is useful in explaining the optical effects and behaviors of TEM waves under certain conditions.

The student should have very little difficulty understanding wave phenomena if the various pictures of the model as presented here are visualized and kept in mind.

POWER DENSITY AND ATTENUATION

In our previous discussion we mentioned that under the idealized conditions of our model, TEM waves emanate uniformly from an isotropic radiator. This assertion implied that the energy at any given distance (r) from the source would distribute itself uniformly in all directions over an area having the shape of a sphere $A = 4\pi r^2$. These conditions give rise naturally to one means of quantifying the intensity of the wave at a point in free space.

Note that if we divide the total power radiated (P_t) by the area over which it is distributed, we obtain the relationship

$$P_d = \frac{P_t}{4\pi r^2} \qquad \textbf{(2–1)}$$

where P_d is called the *power density;* P_t is in watts; and the denominator is given in units of area. For example, if the total radiated power (P_t) is 800 watts and the distance from the source (r) is 16,000 km, then the power density (P_d) at any point on the chosen sphere is

$$P_d = \frac{P_t}{4\pi r^2} = \frac{800}{3.2 \times 10^{15}} = 2.5 \times 10^{-13} \text{ W/m}^2$$

Note that P_d has units of watts per unit area of wavefront. It is *incorrect* to express power density simply in watts.

A closer look at equation (2–1) reveals an interesting fact. Since the reciprocal of 4π is a constant (k), we may rewrite the power density equation as

$$P_d = k\frac{P_t}{r^2}$$

This relationship shows more clearly that the power density is directly proportional to the transmitted power (P_t) and inversely proportional to the distance (r) squared. That is, for a fixed value of P_t, if we double the distance, the power density will become one-fourth as great. Conversely, halving the distance causes the power density to become four times greater. Students of physics will recognize this relationship as the familiar inverse square law which governs the behavior of so much of natural phenomena.

EXAMPLE 2–1 A certain moon-to-earth relay satellite has a receiving antenna whose effective area is 7.3 m^2. The satellite is located 36,000 km above the earth's surface, and receives a power of 2×10^{-12} watts from a lunar transmitter when the two bodies are closest. Assuming the moon-based transmitter is an isotropic source, what power must it radiate? (*Note:* Free-space conditions are approximated in satellite communication systems.)

Solution

We make use of equation (2–1). The power density (P_d) at the receiver is

$$P_d = \frac{2 \times 10^{-12}\ \text{W}}{7.3\ \text{m}^2} = 2.74 \times 10^{-13}\ \text{W/m}^2$$

The distance (r) between the satellite and the moon may be calculated as follows:

$$\begin{aligned} \text{average earth-to-moon distance}^3 &= 384{,}000\ \text{km} \\ \text{earth-to-satellite distance} &= 36{,}000\ \text{km} \\ \text{closest satellite-to-moon distance is } 384{,}000 - 36{,}000 = r &= 348{,}000\ \text{km} \end{aligned}$$

We now solve equation (2–1) for P_t and substitute the values of P_d and r obtained above

$$\begin{aligned} P_t &= P_d\, 4\pi r^2 \\ &= (2.74 \times 10^{-13}) \times 4\pi \times (348 \times 10^6)^2 \\ &= 417\ \text{kW} \end{aligned}$$

EXAMPLE 2–2 While maintaining the same transmitted power (P_t) from an isotropic source, it is desired to increase the power density (P_d) by 15%. By what percent must the distance (r) be reduced?

Solution

Solving equation (2–1) for r, we obtain

$$r = 0.282\sqrt{\frac{P_t}{P_d}}$$

Under the new requirement, equation (2–1) becomes $P_d + 0.15P_d = P_t/(4\pi r^2)$, which may be solved for the new distance (r') to obtain

$$r' = 0.263\sqrt{\frac{P_t}{P_d}}$$

Therefore, $r - x\%r = r'$ or $0.282 - 0.282x = 0.263$, from which, $x = 6.7\%$ reduction in distance.

While the isotropic source is useful in conveying the conceptual idea of power density, any such practical radiator is not usually useful. For example, consider the situation of the satellite communication network mentioned in Example 2–1. Here, it was seen that 417 kW had to be distributed in *every possible direction* in order to achieve the desired

[3] Average earth-to-moon distance is 238,857 miles. (Sterling & Monroe. *The Radio Manual,* 4th ed. New York: Van Nostrand, 1950.)

power reception at only one particular point. Such an inefficient system represents a monumental waste of energy. Indeed, if the transmitted power could have been concentrated somehow, much less would have been needed to obtain the required receiver power.

As we shall discover in chapter 3, most practical antennas are *not* isotropic, but rather radiate better in one particular direction than any other. These antennas are said to have *directivity* or *directive gain*[4] compared to an isotropic source. For example, the power density of a test antenna $P_{d(\text{test})}$ as measured along the axis of maximum radiation will be greater than that of an isotropic antenna $P_{d(\text{iso})}$ when both are fed the same input power and measured at the same distance. We then define the directive gain (G_d) of the test antenna as

$$G_d = \frac{P_{d(\text{test})}}{P_{d(\text{iso})}} \tag{2–2}$$

The ratio expressing antenna gain is more often given in decibels as:

$$G_d(\text{dB}) = 10 \log \frac{P_{d(\text{test})}}{P_{d(\text{iso})}} \tag{2–3}$$

If the transmitting antenna has a directive gain ratio G_{tx}, then the power density along the axis of maximum radiation will be

$$P_{d(\text{tx})} = P_d G_{\text{tx}} \tag{2–4}$$

The power received (P_r) by an antenna oriented for maximum reception depends upon its aperture (i.e., effective area, A_p), and is given by

$$P_r = P_d G_{\text{tx}} A_p \tag{2–5}$$

Since it may be shown that the ratio of aperture to directive gain (A_p/G_d) is the same as the ratio ($\lambda^2/4\pi$), we may relate equations (2–1) and (2–5) to obtain a useful expression for signal attenuation in free-space transmission. We proceed as follows: Given

$$\frac{A_p}{G_d} = \frac{\lambda^2}{4\pi} \tag{2–6}$$

from equation (2–5)

$$A_p = \frac{P_r}{P_d G_{\text{tx}}} \tag{2–7}$$

---

[4]Directive gain does not take into account antenna losses, as will be discussed in chapter 3. In order to consider these losses, we vary the input power while maintaining the same power density for each antenna. This provides us with a practical measure of gain called *power gain* (G_p).

and from equation (2–1)

$$P_d = \frac{P_t}{4\pi r^2} \tag{2–8}$$

Substituting for P_d in (2–7)

$$A_p = \frac{P_r 4\pi r^2}{G_{tx} P_t} \tag{2–9}$$

Substituting for A_p in (2–6) and using G_r for the directive gain ratio of the receiving antenna

$$\frac{P_r 4\pi r^2}{G_{tx} P_t G_r} = \frac{\lambda^2}{4\pi} \tag{2–10}$$

Rearranging,

$$\frac{P_r}{P_t} = \frac{\lambda^2 G_{tx} G_r}{4\pi(4\pi r^2)} \tag{2–11}$$

Substituting $c/f = \lambda$ from equation (1–8), we obtain

$$\frac{P_r}{P_t} = \frac{c^2 G_{tx} G_r}{f^2 4\pi(4\pi r^2)} = G_{tx} G_r \left(\frac{c}{fr4\pi r}\right)^2 \tag{2–12}$$

If f is given in MHz and r in kilometers, then (2–12) may be written as

$$\frac{P_r}{P_t} = G_{tx} G_r \frac{0.57 \times 10^{-3}}{(fr)^2} \tag{2–13}$$

where,

$$0.57 \times 10^{-3} = \frac{300 \times 10^6 \text{ m/S}}{4\pi(1 \times 10^6 \text{ Hz})(1{,}000 \text{ m})}$$

Upon expressing the power ratios of equation (2–13) in decibels, we obtain equation (2–14)

$$\left(\frac{P_r}{P_t}\right)\text{dB} = G_{tx}(\text{dB}) + G_r(\text{dB}) - (32.5 + 20 \log r + 20 \log f) \tag{2–14}$$

The last parenthetical term of equation (2–14) is the isotropic spreading loss of the signal as the wave spreads out from the source. It is this spreading out that accounts for most of the signal loss in free space. If f is in GHz and r in miles, the last term of

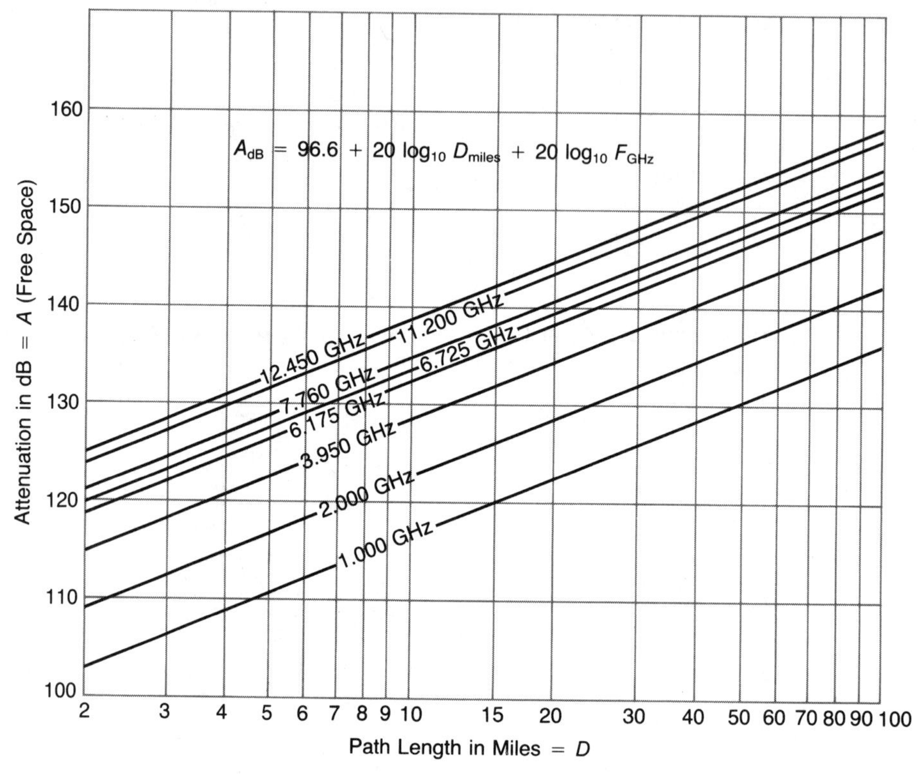

FIGURE 2–4
Free-space attenuation between isotropic antennas. (Reprinted from *Engineering Considerations for Microwave Communications Systems* with permission of Siemens Transmission Systems, Inc.)

equation (2–14) gives us a useful expression for determining the free-space attenuation between isotropic antennas

$$\text{dB loss} = 96.6 + 20 \log r + 20 \log f \qquad \textbf{(2–15)}$$

This loss term is conveniently found in engineering charts such as the one in figure 2–4.

Note in figure 2–4 that the *difference* in loss between any two points is the same regardless of the frequency. However, the *absolute* value of the loss is greater for higher frequencies.

EXAMPLE 2–3 Three geostationary observation satellites are separated by 120 degrees and "parked" 36,000 km above the earth's surface. One satellite has a transmitting antenna with a gain of 21 dB. The next orbitally-adjacent satellite has a receiving antenna whose gain is 65 dB. If the frequency is 6.2 GHz, find:

(a) the free-space attenuation. (b) the received power if the transmitted power is 150 watts.

Solution

From simple trigonometry, the line-of-sight distance between orbitally-adjacent satellites is 62,354 km.

(a) From equation (2–14), the loss term is:

$$\text{loss (dB)} = 32.5 + 20 \log 62354 + 20 \log 6200$$
$$= 204 \text{ dB}$$

(b) From equation (2–14), (P_r/P_t)dB = 21 + 65 − 204 = −118 dB, which represents a power ratio of 1.58×10^{-12}. Therefore, $P_r/150\ \text{W} = 1.58 \times 10^{-12}$, from which $P_r = 2.38 \times 10^{-10}$ W.

EXAMPLE 2–4 At 42 miles from an 11.2 GHz signal source in free space, the signal attenuation is 150 dB. At a previous point, the signal attenuation was less by 22.5 dB. By how many miles are the two points separated?

Solution

150 dB − 22.5 dB = 127.5 dB. Drawing a horizontal line from this point over to the 11.2 GHz line on the chart in figure 2–4 shows a distance of 3 miles. Therefore, 42 mi − 3 mi = 39 miles between the two points.

FIELD STRENGTH

In chapter 1, we derived a value of 377 ohms in connection with an intrinsic property of free space called the wave impedance (Z_w) [see equations (1–12) and (1–13)]. We will now apply this value in an expression that will give us yet another means of quantifying signal intensity under free-space conditions at a distance (r, in meters) from an isotropic radiator. This new expression will allow us to calculate the electric field strength (V_i).

Just as the formula $P = V^2/R$ gives us a measure of the power in electric circuits, so, too, may we derive an expression for the power density (P_d) in terms of the field strength (V_i volts/meter) and wave impedance (Z_w = 377 ohms). We write

$$P_d = \frac{V_i^2}{Z_w} \qquad \textbf{(2–16)}$$

We may solve equation (2–16) for V_i to obtain:

$$V_i = \sqrt{P_d Z_w} \qquad \textbf{(2–17)}$$

If we now substitute equation (2–1) for P_d and 377 ohms for Z_w, we obtain a more useful expression for field strength in terms of transmitted power (P_t, in watts) and distance (r, in meters) from the source under free-space conditions

$$V_i = \frac{\sqrt{30P_t}}{r}\ \text{V/m} \qquad \textbf{(2–18)}$$

Note that the units of equation (2–18), volts/meter, are logically consistent since power is proportional to voltage squared and the square root of V^2 is simply V. Moreover, the denominator is given in rectilinear units.

Any measuring instrument that purports to measure *absolute* values of field strength must do so in units of volts/meter to be consistent with equation (2–18). Such measurements are accomplished by noting the voltage induced in an antenna one meter in length at a fixed distance from the source. Alternately, any antenna calibrated against the one-meter standard may be used and conversion factors applied to the instrument readings.

Field strength meters without such internal calibration standards measure *relative* signal intensity and are useful only for making comparisons. Such meters are employed, for example, in orienting an antenna in the direction of maximum signal reception.

ATMOSPHERIC SIGNAL ATTENUATION

In free space where there is no signal absorption due to the presence of an atmosphere, our wave propagation model allows us to predict signal attenuation. In equation (2–1), we established that the power density (P_d) varied inversely with the distance squared. That is, P_d is proportional to $1/r^2$. Also, in equation (2–18), it was shown that the field strength (V_i) varied inversely as the distance. In other words, V_i was proportional to $1/r$.

If we express the ratio of two power densities P_{d_1} and P_{d_2} at distances r_1 and r_2 of a given signal in decibels, we obtain

$$G_d = 10 \log \left(\frac{r_2}{r_1}\right)^2$$

which may be rewritten as

$$G_d = 20 \log \frac{r_2}{r_1} \qquad \textbf{(2–19)}$$

Similarly, if we express the ratio of two field intensities V_{i_1} and V_{i_2} at distances r_1 and r_2 of a given signal in decibels, we obtain

$$A_{v_i} = 20 \log \frac{r_2}{r_1} \qquad \textbf{(2–20)}$$

A glance at equations (2–19) and (2–20) will show that the decibel value of both expressions of attenuation are identical. So, for example, if the distance between the two points along a signal path were doubled, both G_d and A_{v_i} would be 6 dB down from their previous values.

EXAMPLE 2–5 A communications satellite transmits a signal isotropically with an initial power of 30 watts. At a distance of 27 km from the satellite, the signal is 5.4 dB less than some previous point. What is the power density at that previous point?

Solution

From equation (2–19), we write

$$5.4 \text{ dB} = 20 \log \frac{27{,}000}{r_1}$$

Solving for r_1

$$r_1 = \frac{27{,}000}{\log^{-1}\left(\frac{5.4}{20}\right)} = 14.5 \text{ km}$$

Substituting r_1 into equation (2–1) gives

$$P_{d_1} = 1.14 \times 10^{-8} \text{ W/m}^2$$

Note that the results obtained from equations (2–19) or (2–20) are consistent with those obtained from figure 2–4 or equation (2–15), which takes frequency and receiving antenna aperture into account. For example, if $r_1 = 10$ mi (16 km) and $r_2 = 50$ mi (80.5 km), then from equation (2–19)

$$G_d = 20 \log \frac{r_2}{r_1} = 20 \log \frac{50}{10} = 14 \text{ dB}$$

From figure 2–4 ($f = 1$ GHz for example), dB(10 miles) = 117 dB, and dB(50 miles) = 131 dB. Therefore, 131 dB − 117 dB = 14 dB.

In situations where the effects of an atmosphere on attenuation must be taken into account, empirical data must be heavily relied upon since theoretical calculations lose their usefulness rapidly. In figure 2–5, for example, the chart shows attenuation for various degrees of precipitation ranging from a light drizzle to heavy rain and fog. As an example of how the chart is used, consider the attenuation of a 30 GHz signal. For a drizzle (chart line A), the attenuation is about 0.04 dB/km. However, for a heavy rain (chart line D), the attenuation is about 3 dB/km. Notice that a heavy fog (chart line H) causes higher attenuation than a moderate rain (chart line C) at all frequencies shown.

It must be understood that the formulas developed previously for power density and field intensity assumed free-space conditions. For terrestrial environments, signal parameters depend on many variables including ground wave, reflected and refracted

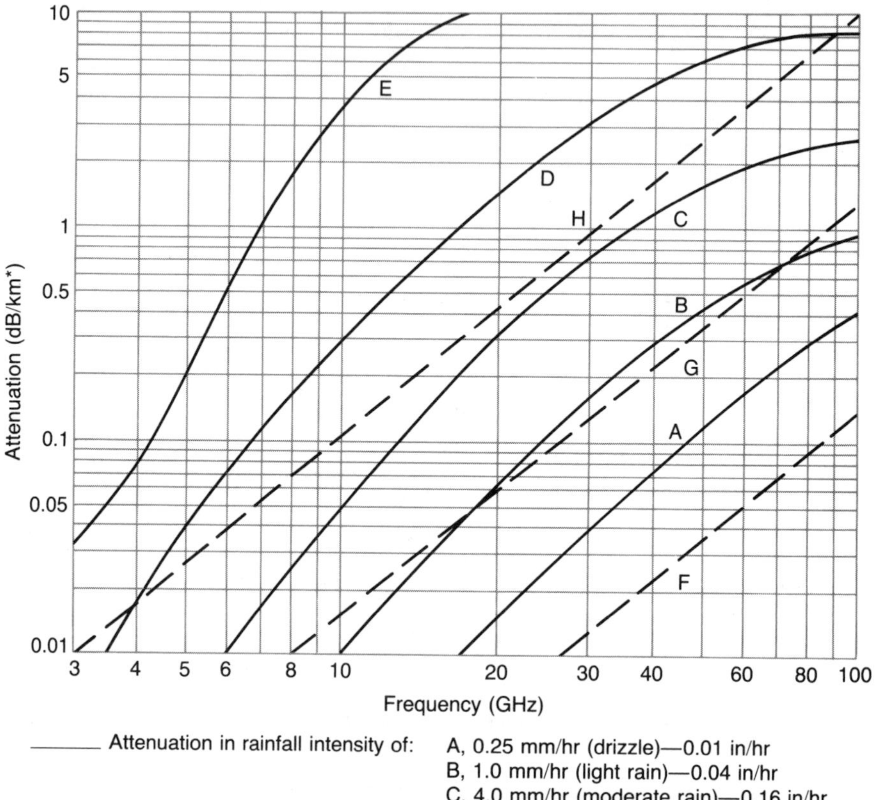

——— Attenuation in rainfall intensity of:
A, 0.25 mm/hr (drizzle)—0.01 in/hr
B, 1.0 mm/hr (light rain)—0.04 in/hr
C, 4.0 mm/hr (moderate rain)—0.16 in/hr
D, 16 mm/hr (heavy rain)—0.64 in/hr
E, 100 mm/hr (very heavy rain)—4.0 in/hr

— — — Attenuation in fog or cloud:
F, 0.032 gm/m^3 (visibility greater than 600 meters)
G, 0.32 gm/m^3 (visibility about 120 meters)
H, 2.3 gm/m^3 (visibility about 30 meters)

*Attn.—dB/mile = 1.61 × (Attn. in dB/km)

FIGURE 2–5
Attenuation due to precipitation. (Reprinted from *Engineering Considerations for Microwave Communication Systems* with permission of Siemens Transmission Systems, Inc.)

waves, and time of day unless the signal path is extremely short. The time of day affects such factors as atmospheric density, water vapor content, and temperature, all of which contribute to the departure from free-space wave predictions. For these and other reasons, signal intensity and similar parameters follow empirical formulae for earth-based situations rather than mathematically derived expressions as is the case when free space is assumed.

BEHAVIOR OF WAVES

As mentioned at the beginning of this chapter, electromagnetic waves leave an isotropic antenna uniformly in all directions. However, it is customary to show the progress of a TEM wave through its environment as a straight line. The line actually represents the axis through a progression of wave fronts in the direction of interest. It must be clearly understood, however, that any actual radiated wave has both electric and magnetic components in space quadrature which array themselves spherically about an isotropic source. One might well imagine the complicated picture that would result if such a drawing were to be attempted each time it became necessary to talk about wave propagation behaviors. A ''wedge'' section (solid angle) of such a spherical wave front is shown in figure 2–6.

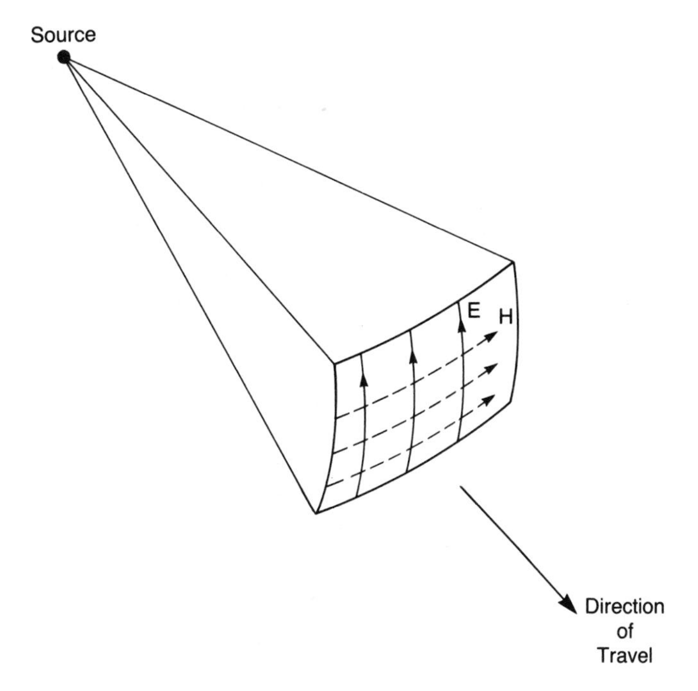

FIGURE 2–6
Spherical wave front of propagating TEM wave.

Even the depiction of such a small portion of the signal as shown in figure 2–6 is often too complicated for most everyday engineering purposes. Therefore, the situation is most often simplified in the extreme, and the wave progression is usually shown merely as a straight line. For example, in figure 2–7, a straight line depicts a wave front traveling from a transmitting antenna (Tx) to a receiving antenna (Rx).

FIGURE 2–7
Straight line showing progression of wave front.

Tx —————➤————— Rx

Again, it must be emphasized that the straight line shown in figure 2–7 simply represents a wave front progression in the direction of interest, that is, from Tx to Rx. However, a more accurate, though more complicated, depiction might be shown as a ball with Tx in the middle and Rx as a dot somewhere on the surface. As long as the student keeps the three-dimensionality of the idea in mind, nothing is lost by simplifying the situation as we have done.

In connection with wave propagation, we speak of three types of waves: *ground* (or surface) wave; *sky* wave; and *space* (or direct) wave. Basically, a ground wave travels over, and quite near, the surface of the earth; a sky wave bounces back and forth between the earth and ionosphere; and a space wave travels in a straight (line-of-sight) path. Figure 2–8 shows the basic propagation paths of the three types of waves.

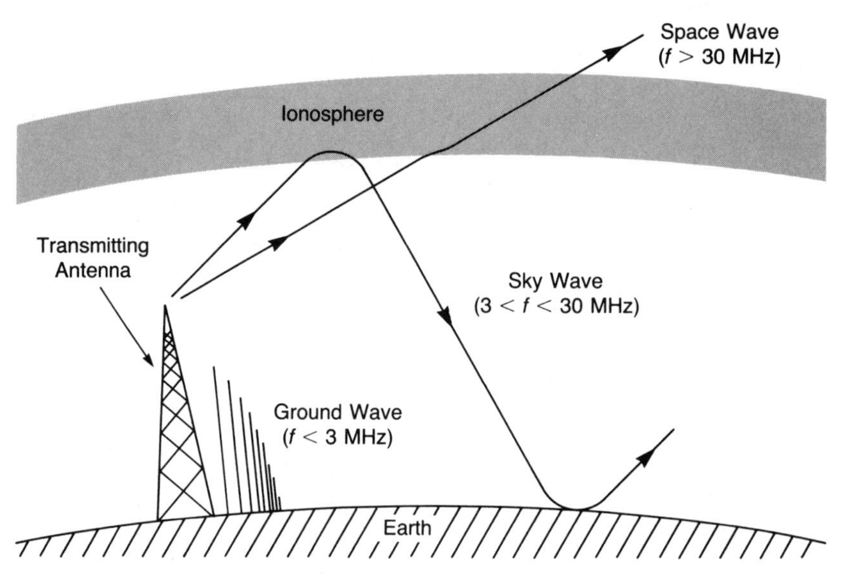

FIGURE 2–8
Propagation paths of TEM waves.

In connection with figure 2–8, it must again be emphasized that waves leaving the antenna do so as spherical ''shells''[5] arrayed in a three-dimensional fashion. Which of the three waves dominate, however, depends largely on the frequency of the signal being transmitted. The magic number appears to be about 30 MHz. For frequencies in the HF range (3–30 MHz), propagation is largely by means of sky waves since signals are reflected between the ionosphere and the earth. Below 3 MHz, propagation is usually by means of ground waves which follow the earth's curvature and, if the initial transmitter power is sufficient, may even circumnavigate the globe. Above 30 MHz, signals are mostly unaffected by the ionosphere and pass right through it.

[5]In the case of a Marconi antenna (discussed in chapter 3), the wave pattern of radiation resembles ''half-shells'' or a ''half-donut.''

In space or on bodies lacking an atmosphere, ground waves and sky waves are not possible. In space, all signals (regardless of the frequency) travel in straight lines as direct waves. In the microwave region,[6] frequencies are above 1 GHz and propagate by direct, line-of-sight waves which are unaffected by the ionosphere.

Although we have stated that waves whose frequencies exceed 30 MHz travel in straight lines, we must qualify this assertion somewhat. For, as with light waves, radio waves are subject to the laws of reflection and refraction,[7] both of which alter the path of the signal. One implication of such path alteration by reflection is shown in connection with figure 2–9 where two signals interfere with each other. Signal path A is the direct path and path B is by a reflected route. It is possible, then, for the two signals to arrive completely out of phase with one another, resulting in a virtual cancellation at the receiving antenna. Since the receiving antenna is located in a "dead zone," no signal will be detected. One is, therefore, left with the alternatives of either changing the antenna site or altering the terrain to eliminate the reflection. Both are expensive operations. In reality, however, good engineering practice would have anticipated the problem.

FIGURE 2–9
Interference of direct and reflected signal paths.

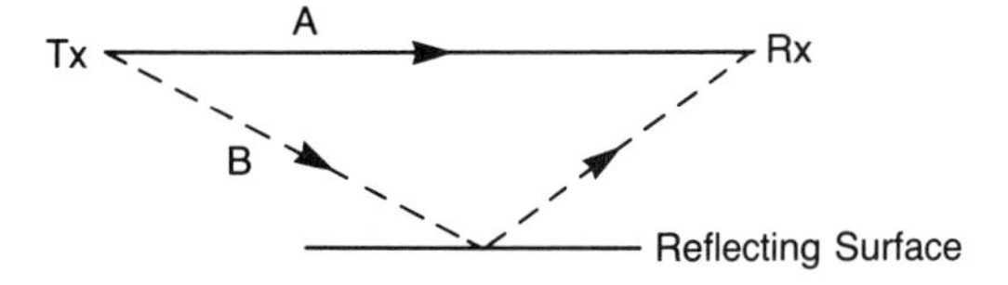

In a similar manner, if an oblique signal approaches a boundary between two mediums of different densities, the signal path will undergo a bending or refraction. Such a situation may occur when a signal traverses a path through the atmosphere having different densities due to thermal effects or precipitation. An example of signal refraction is shown in figure 2–10.

FIGURE 2–10
Refraction of signal.

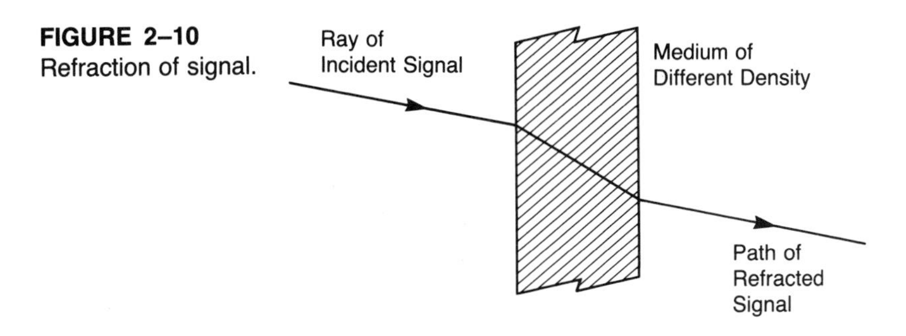

---

[6]There is no universal agreement as to what range of frequencies constitute the "microwave band." According to one author (Cooke & Markus. *Electronics Dictionary,* McGraw-Hill, 1945), a microwave is any wave shorter than 1 meter (300 MHz). Most recent authors, however, have accepted the microwave region to be 1–100 GHz (30 cm–3 mm). The actual range is unimportant except from a taxonomy point of view.

[7]A discussion of these basic laws may be found in any standard college physics textbook, for example, Ouseph, P. *Technical Physics*. New York: Van Nostrand, 1980. Chapters 15 through 18.

THE IONOSPHERE

The ionosphere is a region of ionized atmosphere extending up to about 250 miles. The ions are formed by the sun's incident ultraviolet radiation as well as cosmic rays and other extraterrestrial radiation. Obviously, then, the greatest degree of ionization is in the uppermost regions of the atmosphere and is greatest on the "day-side" of the earth. Consequently, during any given 24-hour period, the ionosphere has two distinct distributions of ionized gaseous layers, as shown in figure 2–11.

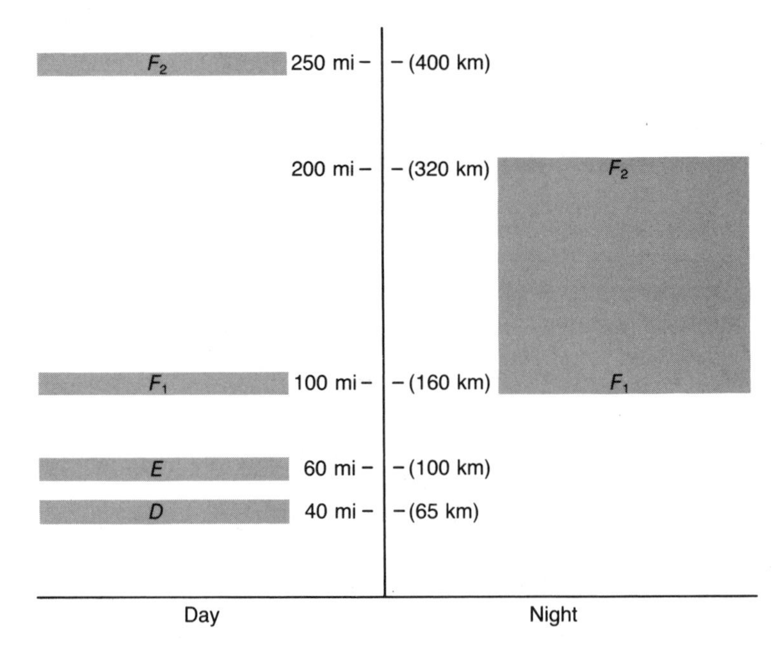

FIGURE 2–11
Day and night distribution of ionospheric layers.

As seen in figure 2–11, four distinct layers (D, E, F_1, and F_2) exist during the day. Consequently, any signal managing to penetrate the first three layers before being returned to earth by the F_2 layer are attenuated a second time on the return trip. As a result, daytime signal reception may be weak at remote locations. During the night, however, the D and E layers dissipate, and the F_1 and F_2 layers combine to form a single strong reflecting layer. Therefore, no signal attenuation occurs by absorption in lower layers, and remote night reception is quite good.

In the early days of communication where frequencies below 30 MHz dominated, virtually everyone expressed a major concern over the ionosphere and sun-spot activity that affected the ionized layers and, therefore, the ability to communicate. Today, however, with satellite communication at microwave frequencies, not much concern exists within the commercial communication industries. We have mentioned the ionosphere briefly here, however, since it does form a part of the communication environment.

RADIO HORIZON

Because of signal-path bending due to atmospheric refraction, a direct wave may actually bend well beyond the horizon. Thus, the "radio horizon" is well in excess of the "true" (geometrical) horizon. The situation is shown in figure 2–12.

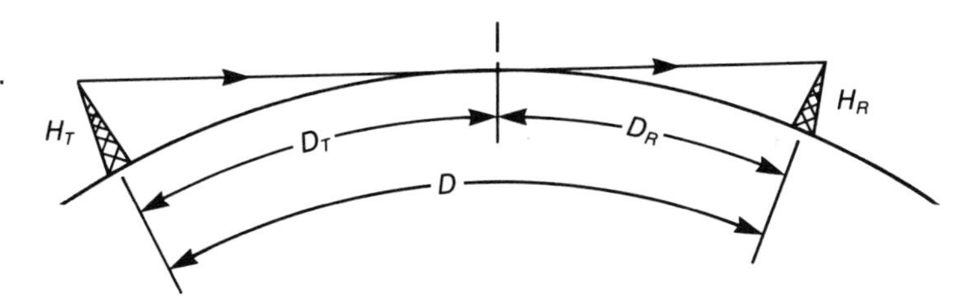

FIGURE 2–12
Direct-wave radio horizon.

The curvilinear distance (D) between transmitting and receiving antennas is derived empirically and given by

$$\begin{aligned} D &= D_t + D_r = 4\sqrt{H_t} + 4\sqrt{H_r} \\ H_t &= \text{height of Tx antenna} \\ H_r &= \text{height of Rx antenna} \end{aligned} \qquad \textbf{(2–21)}$$

where D, D_t, and D_r are in kilometers and H_t and H_r are both in meters. It must be emphasized that equation (2–21) gives a good approximation, but other factors might cause considerable variation in the results obtained.

Note that the radio horizon is about 1⅓ times further than the optical horizon.

EXAMPLE 2–6 The curvilinear distance (D) between the Tx and Rx antennas is 67.2 km. The distance (D_r) from the horizon to the Rx antenna is 36.2 km. What is the height of the Tx antenna?

Solution

From equation (2–21), we know that

$$D = 4\sqrt{H_t} + D_r$$

which may be solved for H_t

$$H_t = \frac{(D - D_r)^2}{16} = 60 \text{ m}$$

SUMMARY

In this chapter we saw how a mathematical model could be used to make predictions concerning electromagnetic wave behavior. The basic element of the model was the isotropic source whose energy was assumed to be spread uniformly over the surface area of a sphere according to equation (2–1). A principal tenet of the model was the postulated existence of free space wherein wave propagation is unimpeded.

The major components of a radiated TEM wave were shown to be electric and magnetic vectors mutually perpendicular and orthogonal to the direction of propagation. Such waves were regarded as plane waves, which travel in straight lines away from the source.

The concept of antenna directionality was developed, and we saw how antenna gain could be envisioned and calculated using equation (2–2). A fundamental free-space signal-transmission equation was developed as equation (2–13), and the calculation of attenuation due to isotropic wave spreading was discussed.

An expression for field strength was developed in equation (2–17), and atmospheric attenuation due to precipitation was discussed in connection with practical engineering charts.

The characteristics of ground waves, sky waves, and space waves were briefly discussed, and the ionosphere was introduced. The possibility of wave interference was discussed in connection with reflected waves. The radio horizon was mentioned, and equation (2–20) was presented as an empirical device for calculating the distance between over-the-horizon antennas.

PROBLEMS

1. The power density measured at a certain distance from a 1 kW isotropic source in free space is 3.18×10^{-14} W/m. How far away is the source? [50,000 km]
2. The distance between an isotropic source and a receiving antenna in free space is increased by 20%. If the power density is to remain the same, by what percent must the transmitted power be changed? [increased 44%]
3. A power density of 1.68×10^{-4} W/m is measured 30 meters from a test antenna whose directive gain is 2.8 dB. How much power was fed into the test antenna? [approximately 1 watt]
4. Four equally-spaced satellites orbit a common center of gravity. The line-of-sight distance between adjacent satellites is 50,000 km. Each satellite is equipped with a 4 GHz, 130-watt transmitter feeding a 30 dB antenna. What gain must the receiving antennas have to insure a minimum power reception of 250 picowatts? [54.5 dB]
5. Derive the expression for isotropic wave spreading loss as given by equation (2–15) and shown in figure 2–4.
6. The field strength of a free-space signal is measured as 2 μV/m at a distance of 85,000 km. What is the transmitter power? [963 watts]

7. An isotropic transmitter in free space sends a signal whose power is 50 watts. At a point B located 18 km from the transmitter, the signal is 9.3 dB less than at some previous point A. What is the field strength at point A? [6.28 mV/m]
8. According to the chart in figure 2–5, does a medium-visibility fog or a light rain affect X-band radar less?
9. Derive an expression for over-the-horizon distance between two UHF antennas of equal height.
10. A 6-foot tall sailor stands erect on the deck of a boat 4 feet above the water line. Approximately how many miles away is his horizon on open seas? [about 3-1/4 miles]

QUESTIONS

1. What is a mathematical model? Give an example of such a model not in this text. Define the model mathematically.
2. Why is the idea of free space important in the wave-propagation model? What real-life situation approximates the conditions of free space?
3. Define a TEM wave in your own words. Using figure 2–2 as a guide, draw a wave in which the E field is lagging 90 degrees in time phase, but in phase quadrature.
4. Why must an isotropic source be considered as a point?
5. Explain the statement that spherical waves are locally flat.
6. Of what is power density a measure?
7. Describe in words the relationship between power density and transmitted power and between transmitted power and distance.
8. An antenna is a passive device in the sense that it does not add energy to the signal. How, then, can an antenna be said to have gain?
9. Directive gain does not take antenna losses into account. Define a new gain that would take such losses into account. (*Hint:* For G_d, the input power to both antennas is the same and the power densities are not.)
10. Explain what is meant by isotropic spreading signal loss.
11. In figure 2–4, why are the diagonal chart lines straight?
12. Can you explain why the distance 22,364 miles is significant in microwave communication work?
13. Define field intensity in your own words.
14. What type of signal attenuation is absent in free space?
15. The graphs in figure 2–5 are based on empirical data. What is meant by empirical data?
16. Explain the difference between ground, sky, and space wave in terms of wave behavior and frequency.
17. Why do TEM waves of any frequency travel only in straight lines in free space?
18. Explain the mechanism of wave interference and signal cancellation in your own words.
19. What is the difference between reflection and refraction of TEM waves?
20. How do you explain the disappearance of the *D* and *E* ionospheric layers during the evening?

3

ANTENNAS

In chapter 1, in connection with our discussion of TDR, it was mentioned that the "signature" of an open-circuited transmission line was based on equation (1–38) for the voltage reflection coefficient

$$\Gamma = \frac{Z_L - Z_0}{Z_L + Z_0} = \frac{V_R}{V_I}$$

This equation was rearranged to show that a *true* open-circuited line ($Z_L = \infty$) would have unity coefficient

$$\Gamma = \frac{V_R}{V_I} = \frac{1 - \dfrac{Z_0}{Z_L}}{1 + \dfrac{Z_0}{Z_L}} = 1$$

In practice, however, even when dissipative line losses are taken into account, the coefficient is substantially less than unity. How do we explain the additional loss? The answer is that there is some loss of energy due to radiation from the open end of the line. Some small amount of the signal energy "leaks" from the line and radiates into space, as shown in figure 3–1.

Such loss also explains why open-circuited transmission line stubs are not generally used in line-load matching situations. Recall that only shorted stubs were considered in connection with Smith chart solutions in chapter 1.

If one were to somehow alter the open end of a transmission line to optimize this loss of radiation, one would have an efficient radiator. In fact, the bulk of this entire chapter focuses on the various means of optimizing end-of-line loss by radiation. Such line-termination devices designed for the express purpose of increasing loss due to radiation are called *antennas*.

FIGURE 3–1
Radiation loss from open-circuited line.

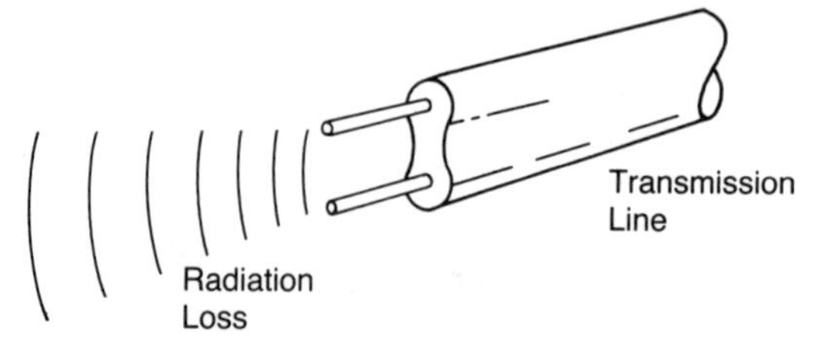

From the standpoint of communication, the primary purpose of an antenna is to provide a means of coupling transmitter and receiver to the intervening space link. This chapter will emphasize that application.

As will be seen in this chapter, antennas have a variety of salient characteristics which are almost invariably specified in decibels. We have already encountered one such important characteristic, *directive gain,* in chapter 2 (see equation 2–2). A complete discussion of decibels is given in appendix D, "How to Speak dB-ese." It is *strongly* recommended that the student review appendix D at this time.

THE MECHANICS OF ANTENNA RADIATION

If we were to spread the ends of a parallel-wire transmission line 180 degrees apart as shown in figure 3–2, we would have a simple antenna. For the moment, we will not make any constraints as to the ideal length of the separated wires forming this crude radiator. We will merely note that an electric field exists between the two wire elements (as shown in figure 3–2) due to the alternating supply voltage V_s.

FIGURE 3–2
A simple antenna.

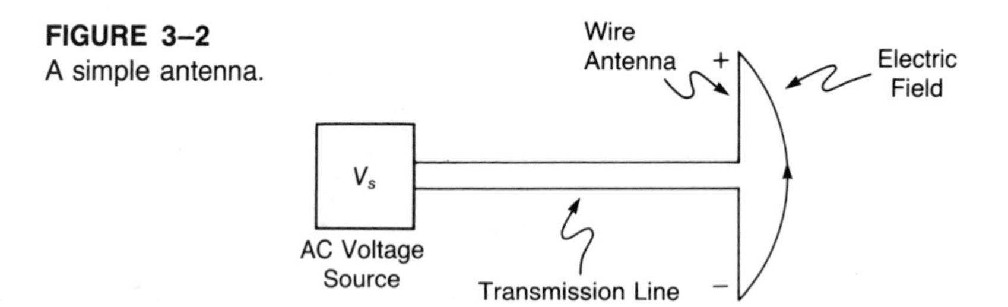

Note that only one electric line of force is shown in figure 3–2 in the interest of simplicity. In truth, however, an infinite number of such lines would surround the antenna, and magnetic lines of force would encircle these electric lines at right angles. The polarity signs (+ and −) represent an arbitrary instantaneous polarization of the voltage on the antenna. Throughout our discussion, we will be assuming sinusoidal voltages for convenience and simplicity.

In figure 3–2, it was assumed that the value of V_s was at a maximum. As V_s decreases, the electric line(s) of force will begin to collapse, as shown in figure 3–3(a), (b), and (c). Finally, as shown in part (d) of figure 3–3, V_s drops to zero, and the electric line of force collapses completely, forming a loop or curl. This loop tends to

FIGURE 3–3
Formation of the radiation electric field.

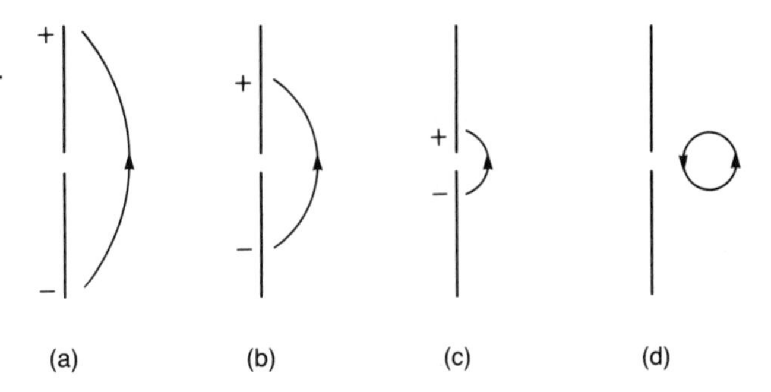

FIGURE 3–4
Radiation field leaving the antenna.

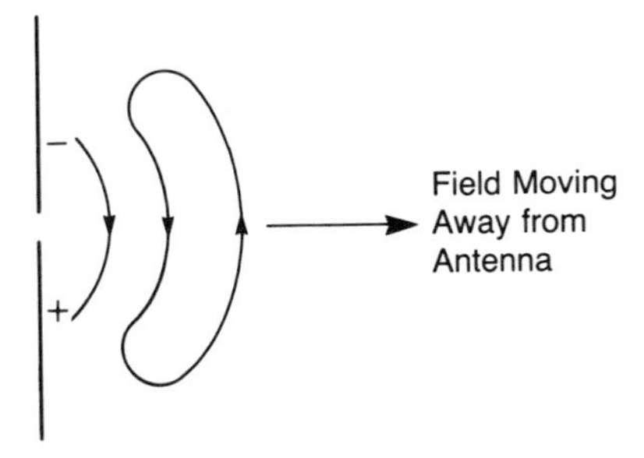

FIGURE 3–5
Complete electromagnetic radiation field leaving the vicinity of an antenna.

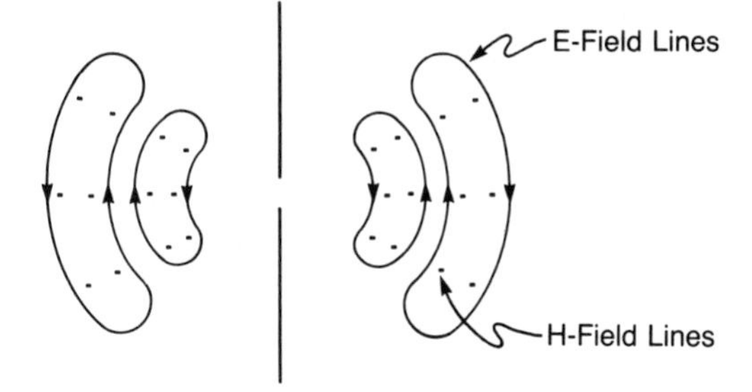

fall back onto the antenna, but the polarity of V_s reverses too quickly, forming a new electric field opposite to those shown in figure 3–3. The direction of this newly formed electric field repels the closed loop away from the antenna and distorts its shape into the elongated form, as shown in figure 3–4.

The distorted closed loop now leaves the vicinity of the antenna at the speed of light, 3×10^8 m/S. Again, it must be emphasized that this event completely encircles the antenna as a 3-dimensional phenomenon, and that a magnetic field exists simultaneously with the electric vector, as shown in figure 3–5.

It may have occurred to the reader to ask by what mechanism a TEM wave sustains itself after it leaves the antenna. The answer is contained in the Maxwell equations,[1]

[1]See Collin, R. *Foundations of Microwave Engineering*. New York: McGraw-Hill, 1966.

and was expressed earlier in connection with equations (1–3) and (1–4). Specifically, the answer is that either field, once set in motion, will always induce the other in time phase and space quadrature.

Implicit in the above discussion is the existence of two completely separate electromagnetic fields. The first, called the *induction* field, remains attached to the antenna. Its effect is quite local, being a matter of a few wavelengths, because its strength varies inversely with the square of the distance from the antenna.

The other field, which is actually created from the induction field, is called the *radiation* field. It is this field that leaves the antenna and propagates to astronomical distances. Its strength decreases directly as the distance increases, as shown in equation (2–17).

In figure 2–2, it was shown that a radiated TEM wave has both electric and magnetic components in time phase and space quadrature. The total E-M field, however, leads the antenna current by 90 degrees since the maximum-strength field does not leave the antenna until the current falls to zero.

On the other hand, the E-M components of the induction field are *both* in time *and* space quadrature. However, the total E-M field is in phase with the antenna current since the field rises and falls in step with the current.

THE HERTZIAN ANTENNA

At the beginning of the previous section, we imposed no constraints on the length of the wire elements comprising our simple antenna. This was done to avoid confusion. Now, however, it has become important to point out that the length of the elements is, in fact, of paramount importance. If the antenna elements are each cut to one-quarter wavelength (at the generator frequency[2]), as shown in figure 3–6, the resultant antenna is called a half-wave *dipole*. More commonly, the dipole is referred to as a *Hertz antenna* after the German physicist, Heinrich Hertz (1857–1894).

When such an antenna is excited by an AC source, the following conditions will exist. Figure 3–7 shows that current flows in the antenna in the direction indicated due

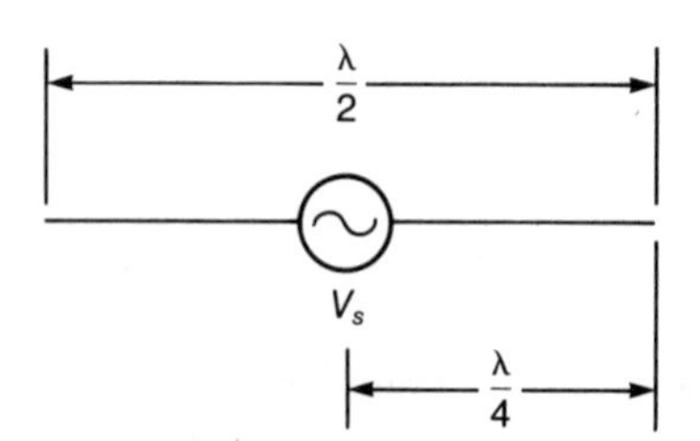

FIGURE 3–6
A half-wave (Hertzian) dipole.

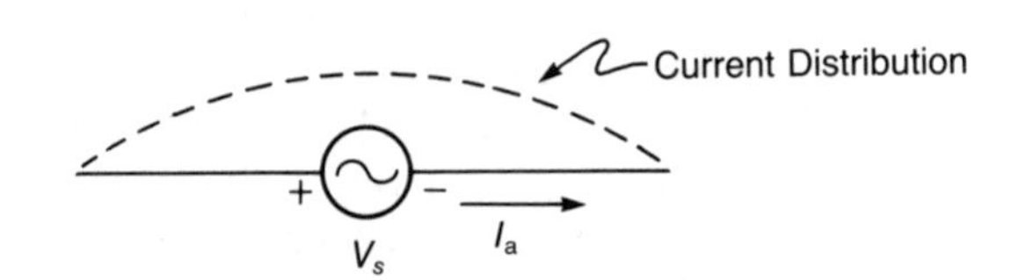

FIGURE 3–7
Current distribution on a half-wave dipole.

[2]Wavelength may be calculated from equation (1–8): $\lambda = v_p/f$.

to the polarity of V_s. As a result, negative charges build up on the right half of the antenna, and positive charges accumulate on the left. While these charges are building up, the current is distributed as shown. The greatest current flow is at the feed point (center) of the antenna, while it becomes zero at the ends because there is no place for the electrons to go. The magnitude of the distribution of these charges (hence voltage) is clearly shown in figure 3–8.

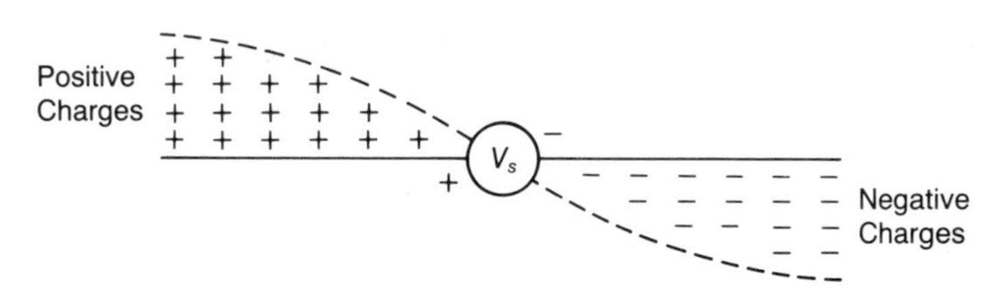

FIGURE 3–8
Voltage distribution along a half-wave dipole.

At the instant depicted in figure 3–8, the antenna current (I_a) is zero and the supply voltage (V_s) as well as the voltage across the antenna is at a maximum. That is, the supply voltage lags the antenna current by 90 degrees.

Note that if the antenna elements were not cut to the proper length, the electrons would still be moving as the supply voltage reversed its polarity. The electron flow must just be slowing to a stop as the voltage reaches its peak. Otherwise, the conditions described above would not exist, and the antenna would not be as effective as a radiator.

The conditions discussed above should have a slightly familiar ring. Chapter 1 pointed out that an open-circuited transmission line one-quarter wavelength long behaved very much like a series resonant circuit. That is, from equation (1–47), $Z = Z_0^2/Z_L$, an open circuit load ($Z_L = 0$) causes the input impedance to the line (Z) to be zero. This is exactly the case we have with the Hertzian dipole, which is merely a quarter-wavelength stub opened up 180 degrees. That is, from figures 3–7 and 3–8, we observe that at the feed point

$$R_{\text{ant}} = \frac{V}{I} = \frac{LOW_V}{HIGH_I} = LOW_R$$

The theoretical value of R_{ant} for a Hertz antenna in free space is 73 ohms. This value varies slightly in practice. Because of this impedance property and the fact that electrons oscillate back and forth on the line 90 degrees out of phase with the voltage, the half-wave Hertzian dipole is often called a *resonant* antenna.

Note that the antenna wire elements have standing waves along their length. However, if the characteristic impedance (Z_0) of the transmission line feeding the antenna is the same as the antenna resistance calculated above, R_{ant}, there will be no standing waves on the transmission line, and hence no reflected power from the antenna (load). In other words, standing waves on this particular antenna are very desirable, while standing waves on the line are always to be avoided. Later in this chapter we will see exactly why standing waves on the antenna itself are desired.

In connection with our half-wave dipole, we mentioned that the electrical length could be calculated from equation (1–8), $\lambda = v_p/f$. In free space, $v_p = c$, and the electrical length is the same as the antenna's physical length. In practice, however, the antenna is almost never completely isolated from outside influences; hence, the velocity of the wave is always *less* than that in free space. Consequently, a correction factor must be applied to the equation for wavelength. For frequencies greater than 30 MHz, the correction factor has been found empirically to be about $0.95c$. That is, the velocity of propagation is about 95% the speed of light. For example, the overall length of a half-wave dipole operating at 250 MHz is

$$\frac{\lambda}{2} = \frac{0.95c}{2(250 \times 10^6)} = 0.57 \text{ meters (1.9 ft)}$$

Note that, predictably, the physical length is about 5% less than the free-space dimension of 0.6 meters (2 ft).

In practice, the dipole elements would be cut to the length given by $c/2f$ and then trimmed down to resonant length to suit specific environmental conditions.

THE MARCONI ANTENNA

The Hertz antenna discussed in the previous section is complete unto itself. On the other hand, the *Marconi* antenna[3] requires the earth or similar large conductor to make it a complete radiator. That being the case, one might wonder what the advantage of this antenna is over the Hertzian dipole if, indeed, any advantage exists at all. In order to answer that question, consider the overall length required for a Hertz antenna at the bottom end of the commercial AM broadcast band, 550 kHz. Using the equation length $= 0.95c/2f$, we find the length to be about 259 meters (850 feet), or almost 2/10 mile! While this length is certainly possible technologically, the cost per foot of such a radiator is high. A Marconi antenna, on the other hand, is only one-quarter ($\lambda/4$) wavelength long (high) and utilizes the earth's reflection of itself to provide the missing half. This situation is shown in figure 3–9.

The height of a Marconi antenna at 550 kHz is 425 feet, only half that of the dipole. Moreover, the antenna resistance at the feed point (between the bottom of the antenna and the earth) is one-half that of the Hertz, or 36.5 ohms.

From the geometry of the situation shown in figure 3–10, one may appreciate how an apparent image of the missing $\lambda/4$ section may be visualized. If the surface is a perfect reflector, an observer located at some distant point will perceive two rays: the direct wave from the actual antenna and a reflected wave from the same point on the apparent image located directly beneath the antenna.

In those locations where the earth does not provide a good conductive surface, such as on dry sand or granite, *radials* are employed to form an artificial ground plane called an *earth mat*. Such radials consist of heavy copper or copper-clad conductors ideally about $\lambda/2$ in length and laid out around the base of the antenna like spokes in a

[3]Named in honor of the Italian physicist, Guglielmo Marconi, 1874–1937.

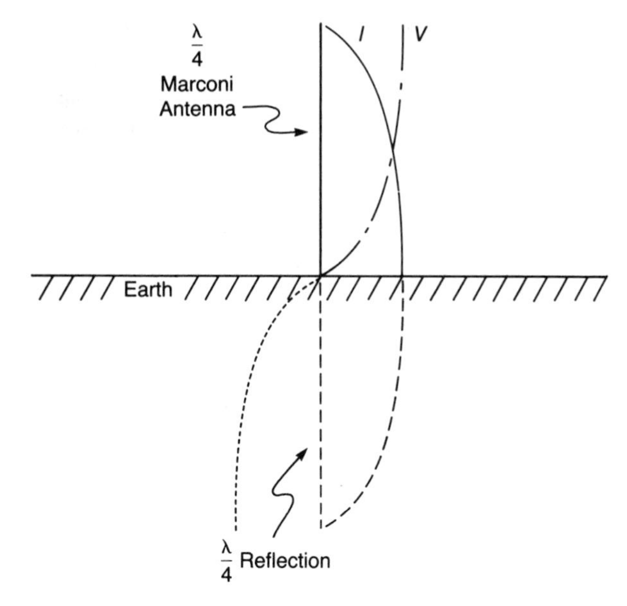

FIGURE 3–9
Marconi radiator showing voltage and current distributions.

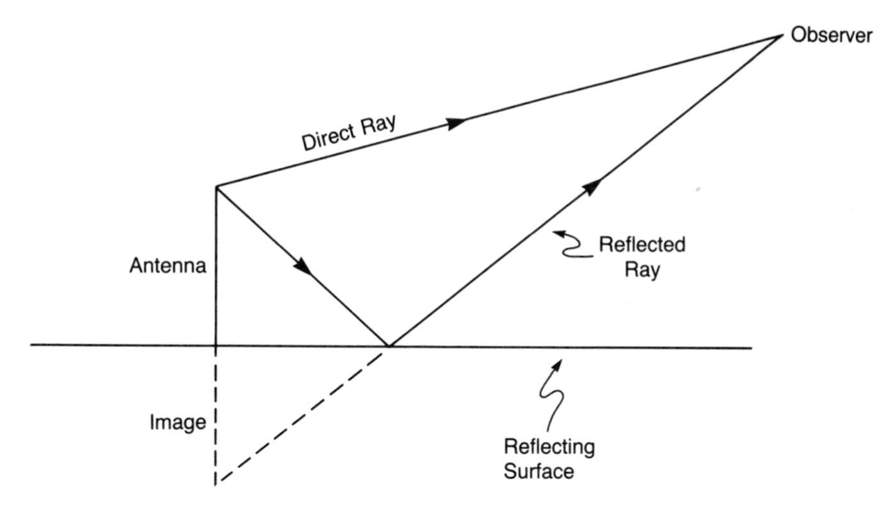

FIGURE 3–10
Reflected image of a vertical antenna.

wheel. The ends of the radials are connected together and buried a few feet below the surface. The antenna, which may weigh several tons, is supported by a high-strength insulator which electrically insulates the base of the antenna from the ground. The transmission line from the transmitter is connected between the base of the antenna and the earth mat radials.

Where a Marconi antenna is to be located atop a building, well beyond the influence of the earth, a *counterpoise* is used to establish the necessary conductive surface. The counterpoise (also called a ground plane) consists of a few radial elements similar to an earth mat, but is above the ground and insulated from it. If a Marconi ''whip'' is used on a truck or van for VHF mobile service or CB operation (just below 30 MHz), the metal skin of the vehicle itself is usually sufficient to provide the necessary counterpoise or ground plane.

Large Marconi antennas are either free standing (figure 3–11) or supported by guy wires (figure 3–12) which are insulated from each other and cut to lengths that are not an integral multiple of the fundamental frequency. Such a requirement is necessary to keep the guy cables from acting as parasitic antenna elements which would alter the antenna's radiation pattern.

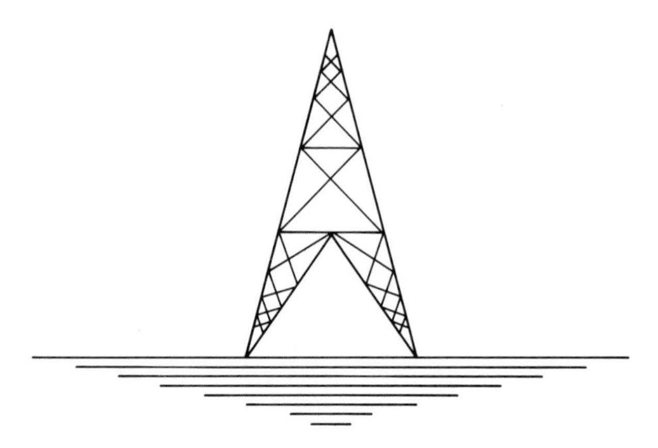

FIGURE 3–11
Free-standing Marconi antenna.

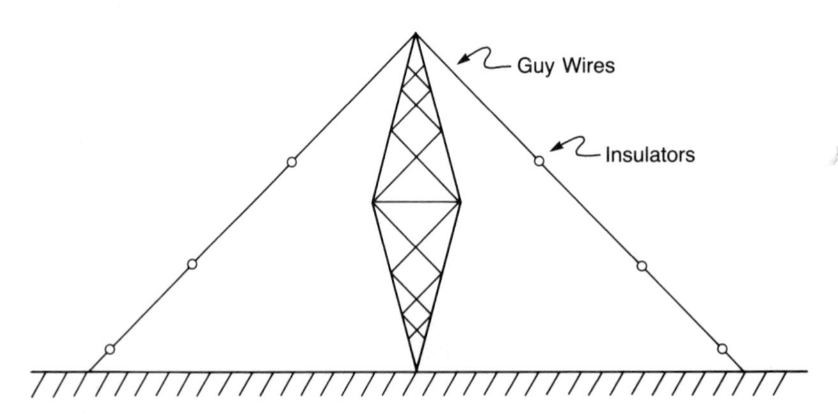

FIGURE 3–12
Guyed Marconi antenna.

When Marconi antennas are used at lower frequencies, it is often impractical to construct the antenna to its full quarter-wave height, thus causing the antenna feed-point impedance to assume a large capacitive reactance component.[4] The antenna will not radiate effectively under this condition. In such cases, *top loading* may be used to extend the antenna's electrical length. In these situations, a short horizontal conductor (wire or "top hat" wheel) is used to effectively add inductance to the antenna, which offsets the capacitive component. The horizontal section is not long enough to affect the radiation pattern substantially, but will provide the necessary low-current node required at the antenna base.

From the distribution of voltage and current patterns shown in figure 3–9, it may be apparent why the so-called "grounded" or quarter-wave Marconi is a preferred antenna. It is evident that at the feed point, the voltage is virtually zero, but may exceed several hundred kilovolts near the top of the antenna structure itself. Therefore, connection to the transmitter is a much less complicated affair.

WAVE POLARIZATION

A TEM wave leaving an antenna has both an electric vector as well as a magnetic component at right angles. It may seem a difficult task, then, to specify the wave's polarization. However, since the electric vector of the wave is *always* parallel to the antenna structure, we speak of its polarization using this reference. For example, the signal of a vertical Marconi is said to be vertically polarized. On the other hand, a horizontal Hertz antenna is said to have a horizontally polarized signal. Of course, the terms *vertical* and *horizontal* have meaning only in reference to a larger body (the earth, for example).

In the United States, commercial AM broadcast signals are vertically polarized (Marconi antennas are used). Most television signals (both the aural and video portions) are horizontally polarized using Hertzian antennas. Commercial FM signals are also horizontally polarized. In recent years, however, many TV and FM stations have switched to circularly polarized signals which allow for better reception even when the actual plane of signal polarization has changed. This unintentional repolarization of the transmitted signal is common as waves undergo reflection and refraction along their path. In contrast, television signals (and antennas) in the United Kingdom are nearly always vertically polarized.

It must be mentioned that receiving antennas must have the same polarization (orientation) as the transmitting antenna. Otherwise, practically no voltage will be induced in the receiving antenna and little, if any, signal will be detected.

In some situations, the electric vector may rotate as it leaves the antenna. Such rotation, called the *Faraday effect,* is due to an interaction of the wave with the F_2 layer of the ionosphere and the earth's magnetic field as the signal leaves or enters the vicinity of the planet. One consequence of this effect is the complete lack of reception if, say, a signal was vertically polarized as it left the transmitter, but became horizontally po-

[4] From figure 1–16 and equation (1–45), we see that an open-circuited line less than $\lambda/4$ acts like a capacitance.

larized in passing through the ionosphere. Since frequencies above 1 GHz are less affected by Faraday rotation, their use in satellite communication represents one solution for the problem. Another solution is contained in the use of a *helical* antenna shown in figure 3–13, which circularly polarizes the wave, thus making reception or transmission possible despite rotation of the wave vectors.

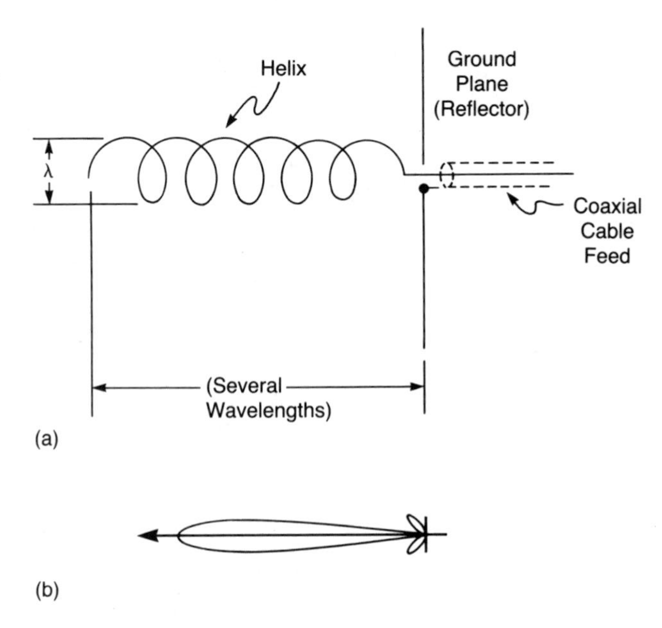

FIGURE 3–13
An end-fire helical antenna.

In closing this section, we should mention that from a theoretical viewpoint, transmitting and receiving antenna functions are interchangeable. This is known as the principle of reciprocity, and states that any antenna that functions as a transmitting device may function equally well as a receiving antenna, and vice versa. Of course, the principle does not address the consequences of sending large amounts of transmitter power into a receiving antenna, which may not have been designed to support the high voltages encountered without arcing.

RADIATION PATTERNS

A knowledge of the current and voltage distributions on an antenna does not provide sufficient information to completely characterize its performance. What is needed is a 3-dimensional model of the antenna's radiated energy showing the direction and the magnitude of the radiation. Such a model is called a *radiation pattern* and is obtained as follows.

Suppose the antenna whose radiation pattern we want to evaluate is located at the center of an imaginary sphere. Now suppose we took a field-strength meter out to the surface of this sphere and made an infinite number of readings all over the surface. If these various intensity readings at each corresponding location were then plotted and displayed on, say, a computer screen, the resultant 3-D image obtained would be the radiation pattern.

For example, the radiation pattern of an ideal Hertzian antenna in free space looks like a donut or torus with the antenna centered along the axis of the hole, as shown in figure 3–14.

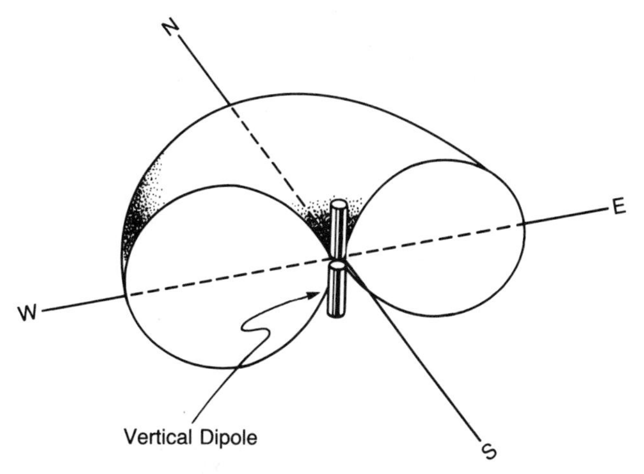

FIGURE 3–14
A 3-dimensional representation of the radiation pattern for a Hertzian antenna.

A truly excellent desk-top model of such a dipole radiation pattern was made by soldering together the ends of a "Slinky."[5] The proportions of the resultant torus quite accurately reveal the Hertz dipole radiation pattern.

For a Marconi antenna, the radiation pattern looks like half a donut, as shown in figure 3–15.

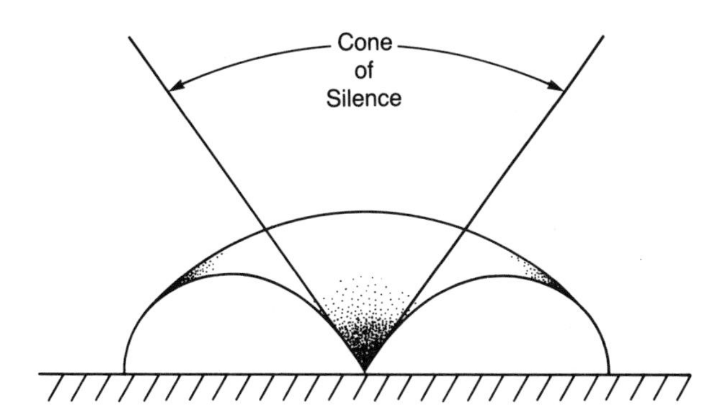

FIGURE 3–15
A 3-dimensional representation of the radiation pattern for a Marconi antenna.

---

[5]"Slinky" is the registered trademark of James Industries, Inc., Hollidaysburg, PA 16648.

Note that there is no energy radiated in a direction directly over the Marconi antenna for several degrees of solid angle. This dead zone is often referred to as the "cone of silence" since signals from the antenna cannot be detected by aircraft flying directly over the antenna site.

While the 3-D model of a radiation pattern gives some insight into the shape, extent, and magnitude of the radiated energy, its usefulness is limited since quantitative information is missing. In practice, therefore, radiation patterns are usually displayed on polar coordinate paper which gives the data in vector form showing magnitude as well as direction. For example, a vertically-oriented Hertz antenna (located well above ground) has a figure-8 shaped radiation pattern in the vertical plane, and its polar diagram is shown in figure 3–16.

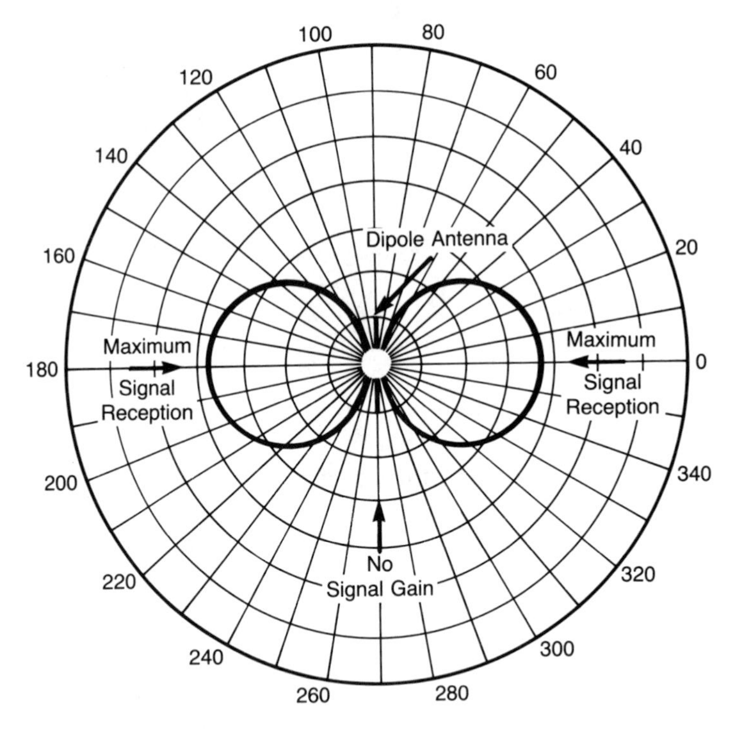

FIGURE 3–16
Polar diagram of a vertical Hertz antenna in a vertical plane.

Note that the figure-8 polar diagram shown in figure 3–16 could have been obtained by an appropriate section through the 3-D pattern shown in figure 3–14. Each portion of the "8" is called a *major lobe* and extends along the axis of maximum radiation called the *major axis*. As we shall see later, it is possible to have one or more minor lobes exist in the same plane of a given radiation pattern along some minor axis.

In interpreting such radiation patterns, a few ideas must be kept in mind. First, the polar diagram is a 2-dimensional representation of a 3-dimensional phenomenon taken at a *fixed* distance. This distance, whatever the actual value, remains the same for any given pattern. Secondly, different patterns will be obtained in different planes. For example, the horizontal polar diagram of the vertical Hertz antenna is a circle, not a figure 8 as shown in figure 3–16. Finally, the magnitude of the intensity is usually shown along the abscissa in decibels, while the direction is shown in degrees left and right (up and down) from some arbitrarily specified reference point. The polar diagram of a Marconi antenna is shown in figure 3–17.

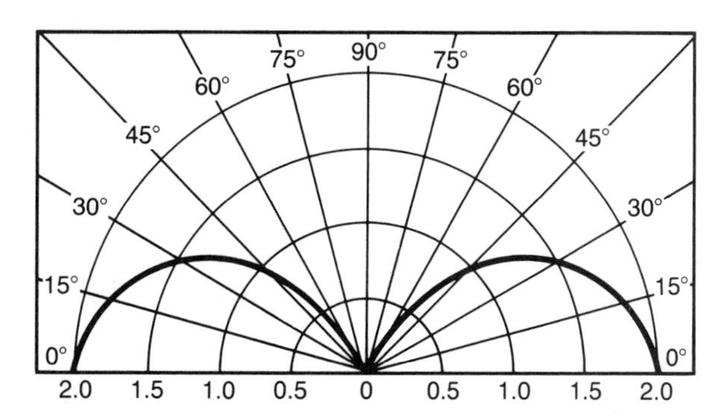

FIGURE 3–17
Polar diagram of a vertical Marconi antenna in the vertical plane.

Note that the radiation pattern of a vertical Marconi antenna in the horizontal plane is a circle. A Marconi antenna is often referred to as an omnidirectional antenna since its signal radiates outward 360 degrees in the horizontal plane. Marconi antennas used by commercial AM broadcast stations are the obvious choice because of their non-directional radiation characteristics.

The two antennas discussed thus far, the Hertz and Marconi, have directional patterns that are of an all-directional nature. However, it is not always desirable to have a radiation pattern that is omnidirectional in a given plane. There are various ways of achieving this result and they will be discussed in a later section.

BEAMWIDTH AND BANDWIDTH

It is often desirable to specify how narrow the major axis lobe of radiation is, especially in the case of those antennas specifically designed to have highly directional radiation patterns. In order to achieve such a specification on a comparative basis, a standard for measuring this lobe width has been adopted throughout the industry. The width of the major lobe is called *beamwidth* and is obtained by measuring the angle (in degrees) between lines drawn through the 3 dB points on the antenna radiation pattern, as shown

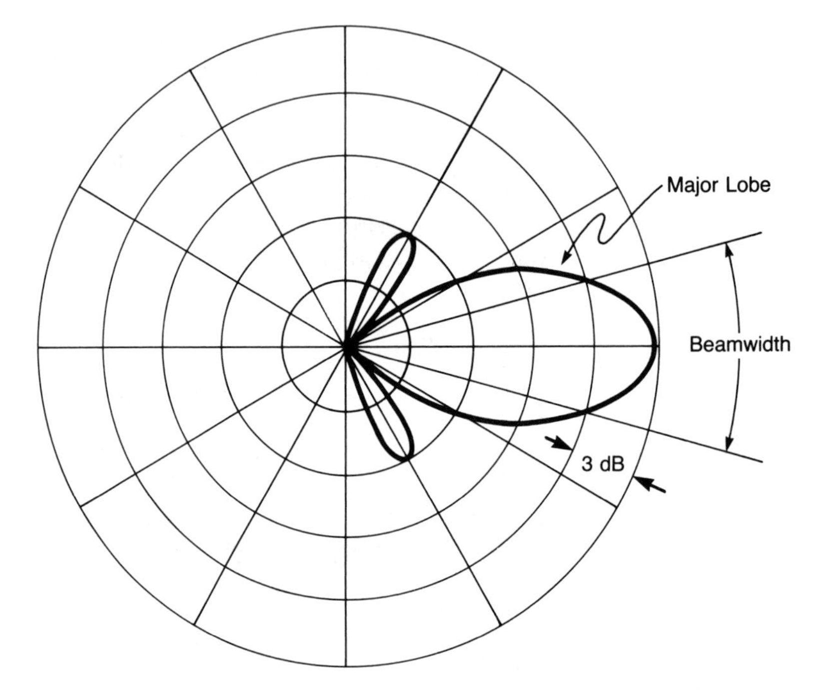

FIGURE 3–18
Measurement of antenna beamwidth.

in figure 3–18. For example, the Winegard model CA-8013 television antenna has its beamwidth specified as 36 degrees.

Another useful antenna specification is the *bandwidth* figure, which is a measure of the difference between the upper and lower useable frequencies as measured between the 3 dB points of a frequency-versus-attenuation graph. One such graph is shown in figure 3–19.

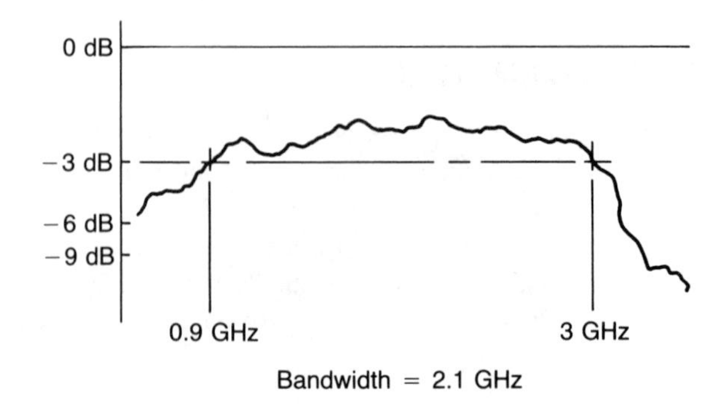

FIGURE 3–19
Frequency-versus-attenuation characteristic showing bandwidth measurement.

ANTENNA IMPEDANCE

Antenna impedance (Z_a) is a complex quantity given by

$$Z_a = R_a + jX_a \tag{3–1}$$

where R_a is an AC resistance called the *antenna resistance* and X_a is the *antenna reactance*. That an antenna may have a reactance should come as no surprise since it was shown in equation (1–45) that open-circuited transmission lines could acquire reactive properties if their length was not exactly $\lambda/4$ at the frequency of operation. Moreover, it should be intuitively obvious that such reactive properties must exist due to the electric and magnetic fields surrounding the antenna which could alternately store energy and then return it to the line.

In the case of resonant antennas like the Hertz or Marconi, the reactive component is tuned out by an equal, but opposite, type of reactance; hence, the name resonant antenna. Such being the case, equation (3–1) reduces to

$$Z_a = R_a = R_d + R_r \tag{3–2}$$

where R_d is called the *loss resistance* and is due to actual resistance losses in the antenna wires themselves. Such losses appear in the form of heat and represent a loss of power. R_d is usually quite small, usually less than a few percent of the total antenna resistance.

The major portion of the antenna resistance is called the *radiation resistance* (R_r), and is a fictitious resistance which can only be calculated, not measured. We may derive the value of R_r from the formula

$$P = I^2R$$

where P is the power fed into the antenna and I is the antenna current measured at the feed point. On solving for R, we get

$$R_a = \frac{P}{I^2} \tag{3–3}$$

The value of the resistance given by equation (3–3) is the free-space resistance and, for a simple dipole, equals 73 ohms. If the dipole is located close to the earth, however, a portion of the radiated wave will be reflected back to the antenna. Such a reflected wave will induce a voltage in the antenna that will either aid or oppose that supplied by the transmitter. This voltage is small by comparison, but is significant. Obviously, then, the value of the current shown in the denominator of equation (3–3) will be affected, and the magnitude of R_a will be different from that predicted by theory. Figure 3–20 shows the effect of antenna height above a perfect reflecting surface on the value of R_a for a half-wave dipole. Note that the value of the radiation resistance varies cyclically above and below the nominal value of 73 ohms.

From a knowledge of R_a and R_d, we may calculate the efficiency (η) of the antenna. Since the power *fed* to the antenna is I^2R_a and the power *radiated* is I^2R_r, then

$$\eta = \frac{P_{\text{out}}}{P_{\text{in}}} = \frac{I^2R_r}{I^2R_a} = \frac{R_r}{R_d + R_r} \tag{3–4}$$

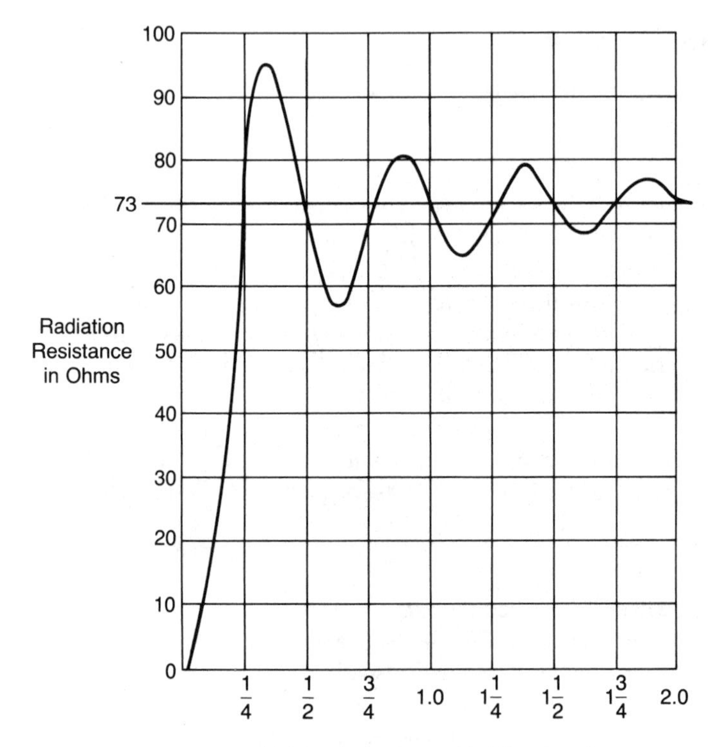

FIGURE 3–20
Effect of antenna height above ground on radiation resistance.

Note that if $R_d = 0$ in the denominator of equation (3–4), there are no losses in the antenna, so $\eta = 1.0$ (100%). For most practical antennas, $75\% < \eta < 95\%$.

We should emphasize again that whenever antenna impedance is mentioned without qualification, it is the impedance at the feed point that is implied. This is a necessary stipulation since, as shown earlier, the voltage and current vary along the length of the antenna wire. Therefore, the impedance will also vary with position along the antenna.

Note that if the frequency changes, there will be a net reactance that dominates the antenna. Therefore, the value of Z_a will change. It may be appreciated, then, that the simple dipole leaves a lot to be desired in terms of broadband performance. Later in the chapter we will see how this problem is lessened.

ANTENNA GAIN

By the strictest definition, antennas are passive devices. That is, antennas do not change the waveshape of the signals applied to them, nor do they add energy to the signal as an amplifier does. How, then, may we speak of the *gain* (a term reserved for *active* devices) of an antenna, and what new meaning is implied by the term?

In chapter 2, we mentioned that most practical antennas tend to concentrate their radiation more in one direction than another. This was in contrast to the isotropic source which, by definition, radiates its energy uniformly in all directions. Consequently, if we were to compare the radiated power from the practical antenna to that from an isotropic source as predicted mathematically, we would see that the practical antenna has some gain *over the isotropic source*. It is in this regard that we may speak of antenna gain. In particular, it was shown by equation (2–2) that a practical antenna had a *directive gain* (G_d) given by

$$G_d(\text{dB}) = 10 \log \frac{P_{d(\text{test})}}{P_{d(\text{iso})}}$$

where P_d is the power density (in watts/square meter). In calculating G_d, the distance from either antenna to the point where the energy is measured was kept the same. Moreover, the amount of power fed to both antennas was the same. These conditions were necessary in order to make logical comparisons. It must be understood that since an isotropic source does not exist in the real world, any values related to this antenna are those predicted mathematically from the model.

From figure 3–21, one may acquire an appreciation of this concept of directive gain. The dotted circle represents the radiation pattern of an isotropic source, while the solid line is the radiation pattern of a practical dipole antenna. Note that the radiation from the dipole is concentrated in two lobes along the major axis, whereas the isotropic source radiates equally in all directions. The shaded areas represent the additional gain of the practical dipole over an isotropic antenna in a *direction* along the major axis. In fact, the dipole has a voltage gain (A) of 1.28 (2.14 dBi) over the isotropic source. The equivalent power gain (G) is 1.64 (2.14 dBi). The i in dBi indicates that an isotropic source was the comparison antenna. Note that the gain in either case is 2.14 dB as one would expect from the relationship between voltage and power (power is proportional to voltage squared), as well as from the laws of exponents. In general, voltage decibel gain

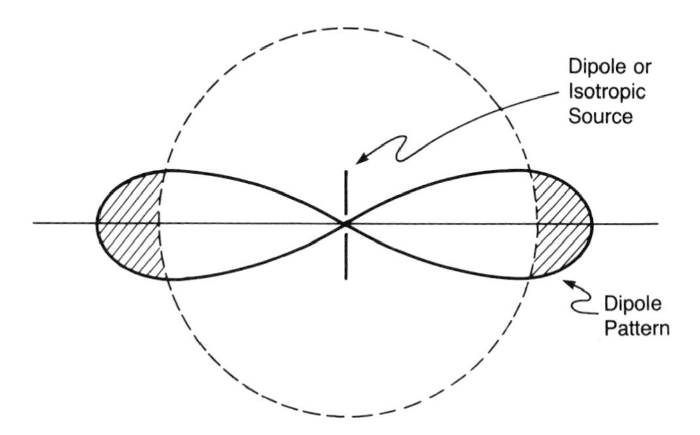

FIGURE 3–21
Comparison of radiation patterns of a simple dipole and isotropic source.

is given by $A = 20 \log (V_{out}/V_{in})$, and decibel power is given as $P = 10 \log (P_{out}/P_{in})$, where either the voltage ratio or power ratio is called the *gain factor*. Again, it should be emphasized that when speaking of antenna gain, a comparison device must always be specified. The comparison antenna may be the theoretical isotropic source or a practical dipole.

The problem with directive gain is that it does not take actual losses of the practical antenna into account. An inefficient antenna may require a great deal of input power to achieve even a modest output. For example, suppose the power density of an antenna under test was found to be $P_{d(\text{test})} = 8 \text{ mW/m}^2$ when $P_{\text{test}} = 2$ W and $r = 30$ m. Then, from equation (2–1), $P_{d(\text{iso})} = 1.77 \times 10^{-4} \text{ W/m}^2$. On substituting these values in equation (2–2), we obtain

$$G_d(\text{dB}) = 10 \log \frac{P_{d(\text{test})}}{P_{d(\text{iso})}} = \frac{8 \times 10^{-3}}{1.77 \times 10^{-4}} = 16.7 \text{ dB}$$

Now suppose that instead of comparing output characteristics (i.e., power density), we compared input parameters. For example, let us suppose that under the same conditions expressed above, it was found that 8 watts of test antenna input power was required to achieve the same power density as the isotropic source ($1.77 \times 10^{-4} \text{ W/m}^2$) because of losses in the test antenna. We now have

$$G = 10 \log \frac{P_{\text{test}}}{P_{\text{iso}}} = 10 \log \frac{8}{2} = 6 \text{ dB}$$

Obviously, this represents a sizeable drop in the gain of the test antenna. The new gain defined above is called the *power gain* (G_p) and is given by,

$$G_p(\text{dB}) = 10 \log \frac{P_{\text{test}}}{P_{\text{iso}}} \qquad \textbf{(3–5)}$$

Since power gain takes actual test antenna losses into consideration, it should be evident that

$$G_p(\text{dB}) = \eta G_d(\text{dB}) \qquad \textbf{(3–6)}$$

EFFECTIVE RADIATED POWER

Power gain (G_p) is a more practical expression of antenna performance than directive gain (G_d) because it takes into account actual losses introduced by the antenna itself. Any real antenna does not operate independently of its transmission line, however, and, quite obviously, there will always be some attenuation along the line. If this loss, as well as any other intervening *system* losses, are taken into consideration, a new system-gain figure results, which is an even more accurate picture of the situation.

In our previous example, it was found that 8 watts of antenna input power resulted in a power gain for the test antenna of 6 dB. Suppose, now, that we assume a line loss of 0.8 dB. What is the overall system gain? Such a situation is shown in figure 3–22.

From figure 3–22, it should be apparent that the effective gain of the antenna in the system will be 6 dB − 0.8 dB = 5.2 dB. Note that a power gain of 5.2 dB represents a power ratio or *gain factor* of 3.31:1. Therefore, the *effective radiated power* (ERP) of the system is 3.31 × 8 watts = 26.5 watts.

FIGURE 3–22
A practical antenna system.

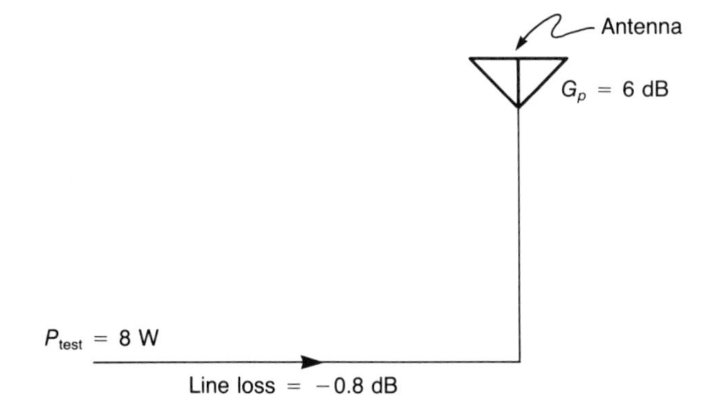

We may now define ERP as

$$\text{ERP} = \text{gain factor} \times \text{total power}$$

or

$$\text{ERP} = \log^{-1}(G_p'/10)P_t \tag{3–7}$$

where G_p' is the system gain (G_p − other losses) and P_t is the total input power to the system.

Note that as more and more actual conditions were taken into account, the gain of our antenna system declined

$$G_d = 16.7 \text{ dB}$$
$$G_p = 6 \text{ dB}$$
$$G_p' = 5.2 \text{ dB}$$

It should be mentioned in closing this section that although ERP shows a wattage value that exceeds the input power (P_t), there is no actual increase in the transmitted power. As mentioned earlier, an antenna is a passive device that does not change the amplitude of the input signal. What the ERP actually expresses is simply a more *effective* use of the input power in a particular *direction*.

NONRESONANT ANTENNAS

The antennas we have been discussing thus far comprise an important group of radiators called resonant antennas. Two primary characteristics usually associated with these antennas are (1) the existence of standing waves and (2) an omnidirectional radiation pattern in a plane perpendicular to the antenna's longitudinal axis. While these devices form a good theoretical basis for further study, they are *not* the only important group of radiators. Indeed, there is yet another entirely different group of antennas, called nonresonant antennas. Nonresonant antennas do not exhibit standing waves and have radiation patterns that are highly directional. We shall now investigate some of these antennas.

The Long-Wire Antenna

In connection with figure 3–2, it was shown that an open transmission line could form an efficient radiator. Such an extension of transmission line geometry, however, is not the only possible configuration. In figure 3–23, we note that a quasi-transmission line is formed by one conductor running parallel to the earth, which forms the other conductor. The "transmission line" thus contrived is terminated in a resistive load equal to its characteristic impedance (Z_0). Under these conditions of matched impedances, there are no reflected waves and, consequently, no standing waves. The only disturbance on the line is the traveling wave from source to load. The length of the conductor is on the order of several wavelengths and is not cut to any particular dimension. As a result, this antenna exhibits a unidirectional radiation pattern, as shown in figure 3–24 (a) and (b).

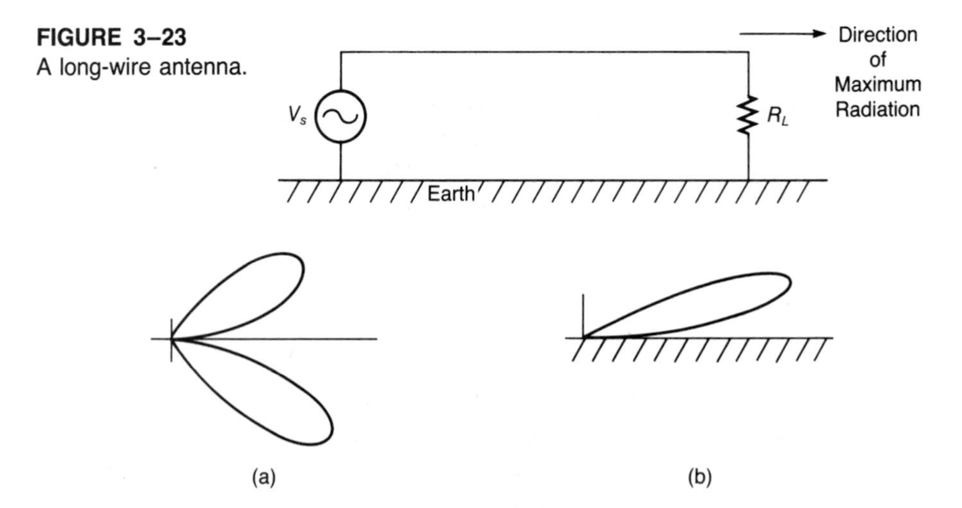

FIGURE 3–23
A long-wire antenna.

FIGURE 3–24
Radiation pattern of a long-wire antenna in (a) a horizontal plane and (b) a vertical plane.

Due to the proximity of the wire to the earth, there is a vertical tilt or upward component to the radiation pattern as shown in part (b) of figure 3–24, which makes this antenna ideal for sky-wave propagation of frequencies in the HF band (3–30 MHz). Moreover, since the antenna is nonresonant, it has a broadband frequency response provided its length exceeds about 2λ for the intended frequency of use in this range.

Rhombic Antenna

The rhombic antenna derives its name from its geometrical shape, the rhombus,[6] as shown in figure 3–25. Each leg of the radiator is essentially a long-wire antenna whose length and angle relative to other legs is chosen to give one major lobe to the radiation pattern which lies along the main axis of the antenna. The radiation pattern of such an antenna is highly directional and is shown in figure 3–26.

FIGURE 3–25
The rhombic antenna.

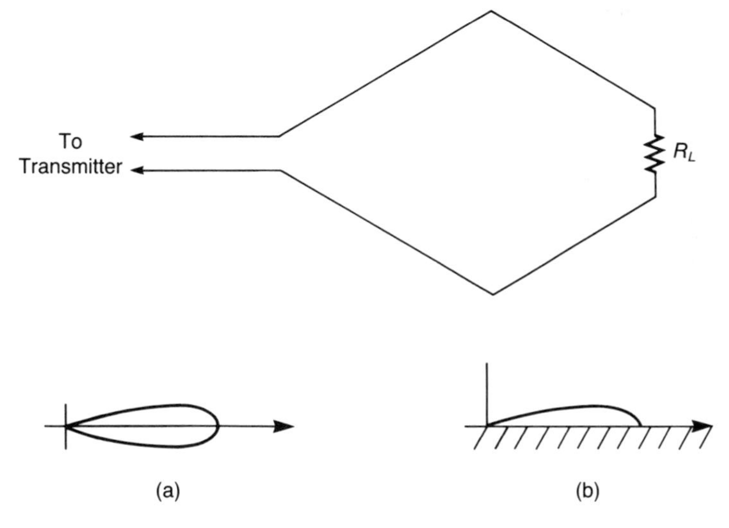

FIGURE 3–26
Radiation pattern of the rhombic antenna in (a) a horizontal plane and (b) a vertical plane.

As with the long-wire antenna, the rhombic antenna is located close to the earth, and the resultant lobe of the radiation pattern has an upward tilt which is ideally suited for sky-wave propagation of signals in the HF range.

Again it must be emphasized that since each leg of the rhombus is a long-wire antenna itself, each leg has a horizontal radiation pattern similar to that shown in figure 3–24(a). Consequently, the angle between legs must be chosen to cancel out unwanted side lobes, leaving only the desired main lobe along the major axis of the rhombus.

---

[6]The rhombus is a plane equilateral parallelogram having oblique angles.

The high input impedance of the rhombic antenna (over 600 ohms) allows for direct feed from a parallel-wire transmission line. The other end of the antenna is terminated in a resistance of comparable value which dissipates about 30% of the transmitted power. Thus, there are no reflected power and no standing waves.

As was the case with the long-wire antenna previously discussed, the rhombic antenna is a broadband, nonresonant device having a frequency range of about 10 to 1. The power gain of this antenna is in the range of 12 dB to 18 dB.

The structure of the rhombic is simple, consisting of four upright posts (one at each vertex), but actual dimensions may become unwieldy. For example, at the low end of the HF band, a typical rhombic may exceed 1,000 feet in length and about 800 feet in width. Obviously, then, such an antenna is not intended for rooftop operation. Rather, such antennas are usually constructed in large open fields.

ANTENNA ARRAYS

An antenna *array* is an assembly of two or more antenna elements (often half-wave dipoles) situated in close proximity to each other so that their induction fields interact. By an appropriate choice of spacing between elements, as well as proper phasing of the currents feeding individual sections, various specialized radiation patterns may be obtained.

The Broadside Array

One of the simplest arrays consists of a number of in-phase dipoles spaced $\lambda/2$ and situated in one plane, as shown in figure 3–27. Such an array is called a *broadside* (or sometimes *billboard*) array.

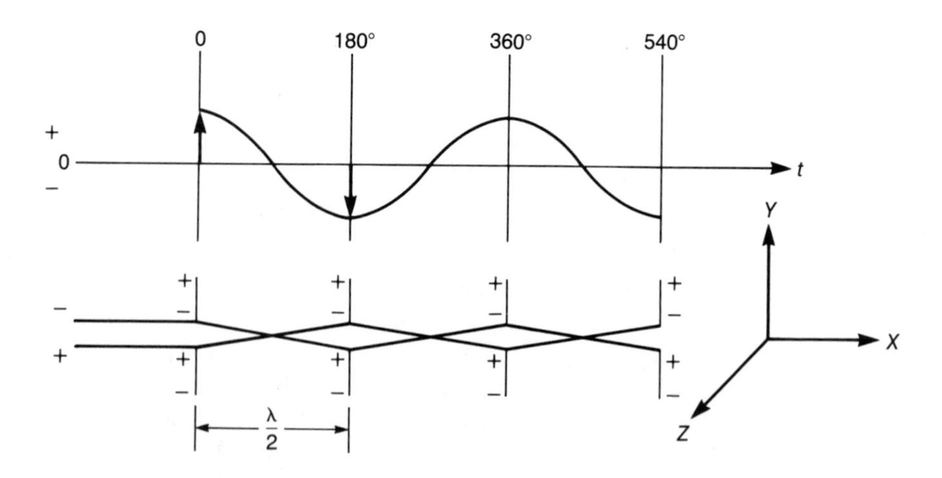

FIGURE 3–27
Broadside array.

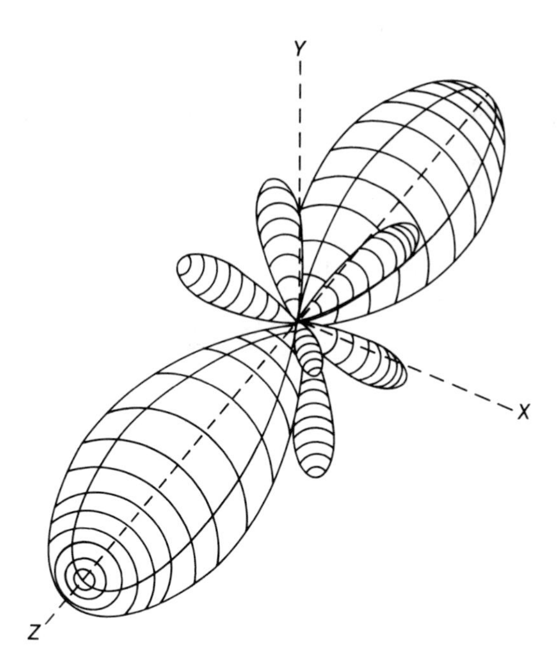

FIGURE 3–28
Solid radiation pattern of a broadside array.

Each individual antenna receives the same amount of energy from the transmitter, and each is excited in phase with the other elements. The proper phasing is achieved by spacing the elements $\lambda/2$ apart and crossing the feed points as shown in figure 3–27. Note that each of the four elements shown is in phase (+ along the top and − along the bottom). Therefore, each of the individual donut patterns combine, forming the resultant bidirectional radiation pattern which has maximum directivity along an axis perpendicular to the plane of the array. If, for example, the array is regarded as lying in the *X-Y* plane, then maximum radiation is in the *Z* plane, as shown in figure 3–28. The broadside array is a resonant device; hence, its usefulness as a broadband receiving antenna is limited.

The End-Fire Array

While the physical appearance of the *end-fire* array resembles that of the broadside array, a completely different radiation pattern results since the spacing between elements is only $\lambda/4$ and the elements are fed 90 degrees out of phase, as shown in figure 3–29.

The radiation pattern of an end-fire antenna is different from that of a broadside in that its beamwidth is somewhat broader for the same number of elements and the

FIGURE 3–29
An end-fire array.

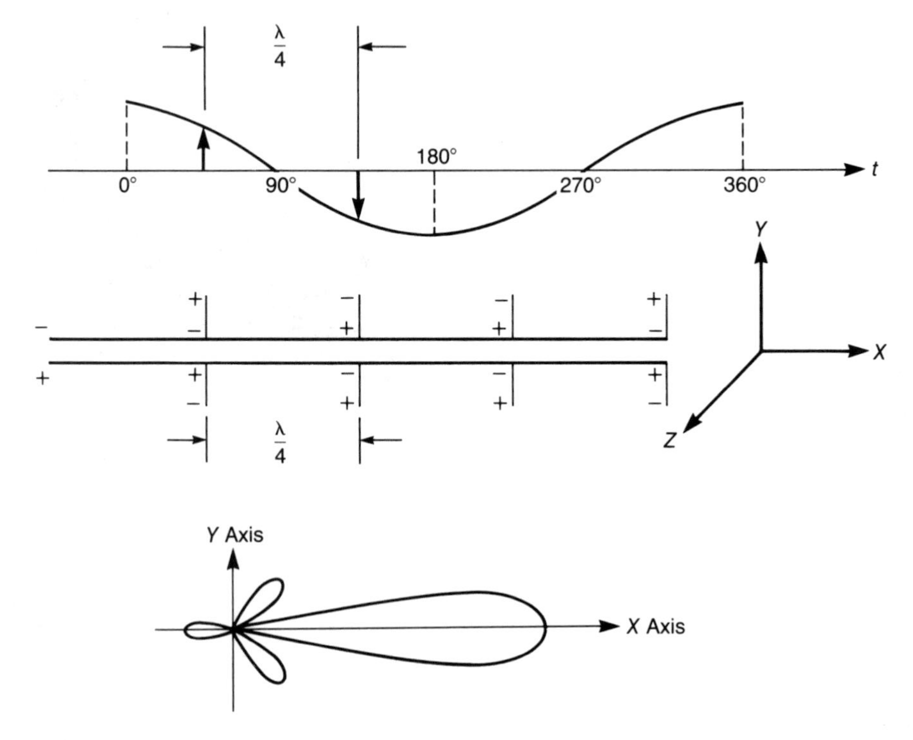

FIGURE 3–30
Radiation pattern of end-fire array.

direction of the main lobe is in the plane of the array, *not* at right angles. Moreover, the antenna is essentially unidirectional, with only a slight minor lobe, as shown in figure 3–30.

In order to explain the end-fire radiation pattern, we must account for its three characteristics (refer to figure 3–29):

1. Since the first and third (second and fourth) dipoles are out of phase, they cancel out each other's radiation at right angles to the array. This fact accounts for no radiation normal to the plane of the array.
2. A wave leaving the first dipole and moving to the *left* is $\lambda/4$ closer to some remote point than a wave leaving the second dipole and also moving to the left. Consequently, the direction of wave vectors arriving some time later at this point will be 180 degrees out of phase and will, therefore, cancel out each other. This accounts for virtually no radiation to the left.
3. A wave leaving the second dipole and moving toward the *right* will, in a similar way, arrive at a remote point in phase with the wave vector from the first dipole. Reinforcement will then result in this direction. This accounts for the unidirectional end-fire radiation pattern.

The end-fire array is, like the broadside, a resonant device with limited usefulness as a broadband receiving antenna.

The Turnstile

Another array which finds frequent application in commercial VHF communication is the *turnstile* antenna. The basic turnstile consists of two half-wave dipoles mounted at right angles to each other in the same horizontal plane. When the two antennas are excited by equal currents 90 degrees out of phase, their figure-8 radiation patterns merge to form an omnidirectional, almost circular, pattern. The basic turnstile antenna is shown in figure 3–31, and its radiation pattern in the horizontal plane is shown in figure 3–32.

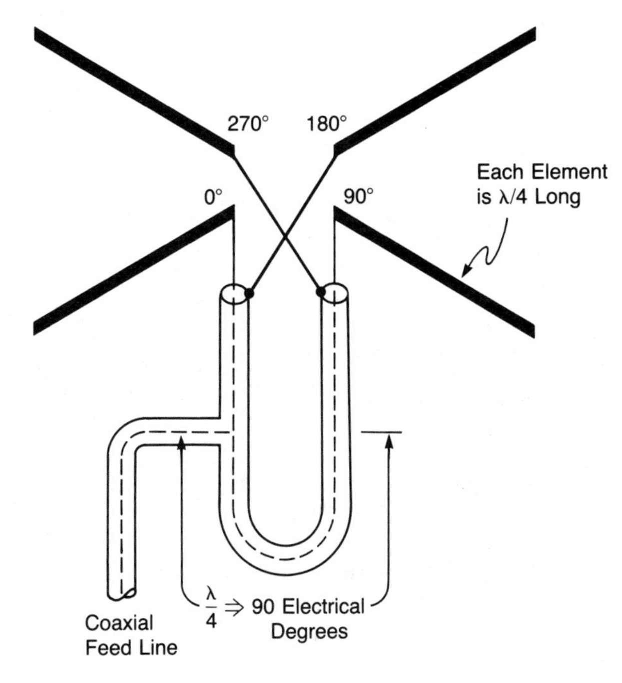

FIGURE 3–31
Basic turnstile antenna.

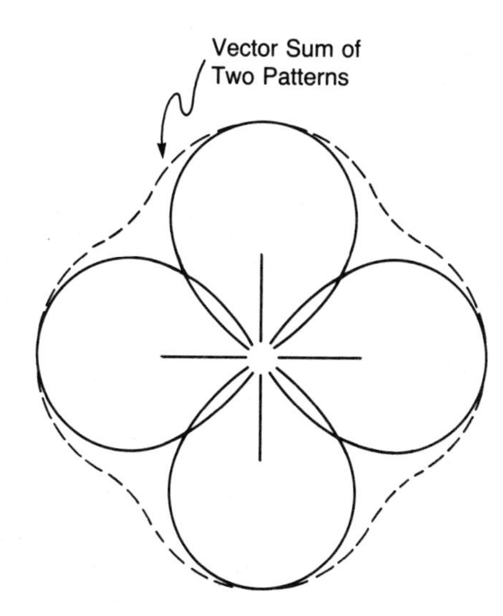

FIGURE 3–32
Vector sum of radiation patterns for turnstile antenna in a horizontal plane.

Note that the proper 90-degree phasing between adjacent sections is achieved by connecting a $\lambda/4$ transmission line section between the two feed points of the array. The characteristic impedance of the feed line must be about 35 ohms (roughly half that of a single dipole) since the two dipoles are fed essentially in parallel. This condition is necessary in order to eliminate standing waves on the main feed line from the transmitter.

Since VHF transmission is essentially a line-of-sight affair, any energy leaving the antenna at a steep vertical angle is wasted. Therefore, simple turnstile antennas are often stacked in "bays" to reduce vertical radiation. Such bays, stacked $\lambda/2$ apart and fed with the proper phase, will cause cancellation of the radiation vector in the vertical direction. Proper phasing between bays may be achieved by crossing the antenna feed line feeding individual bays, as shown in figure 3–33.

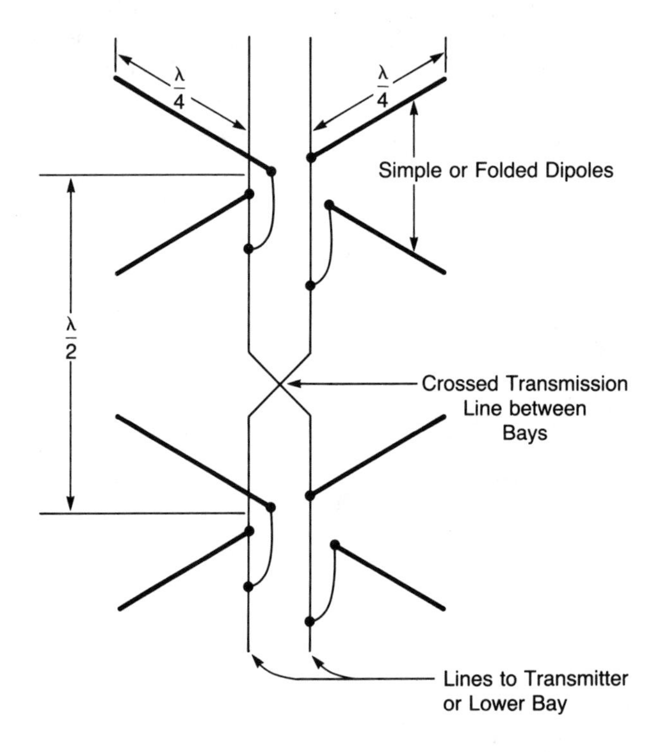

FIGURE 3–33
Stacked turnstile antenna.

The resultant vertical radiation pattern of several stacked bays is shown in figure 3–34. Note the highly narrowed arms of the figure-8 pattern.

In practice, several stacked sections of three or more bays each may be used to narrow the beamwidth in the horizontal plane. However, as more and more sections are added in parallel, the feed-point impedance decreases along with the bandwidth. Therefore, folded dipoles in the form of a "bat wing" are often used for each element to increase the input impedance without compromising bandwidth. The folded dipole has a radiation resistance of approximately 300 ohms, as may be seen from the following discussion.

Consider two half-wave antenna sections connected in parallel as shown in figure 3–35. Each element has half the total current flowing through it that would ordinarily flow through a simple dipole receiving the same amount of power. The power supplied to the folded dipole is then given by

$$P = \left(\frac{I}{2}\right)^2 R \tag{3–8}$$

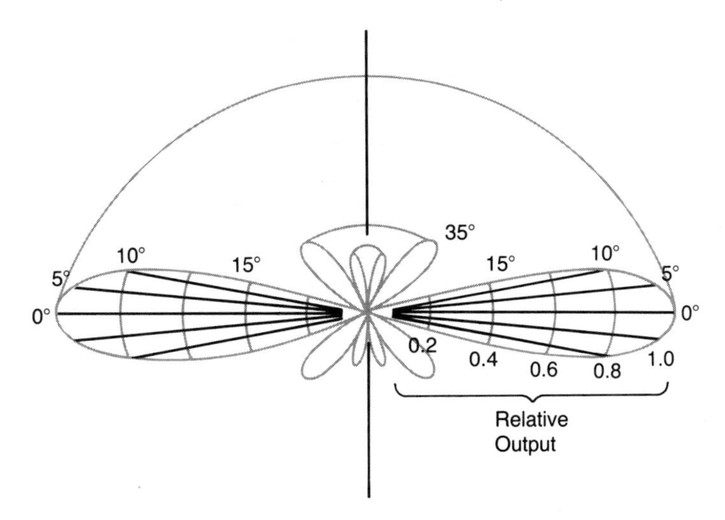

FIGURE 3–34
Radiation pattern of multi-bay stacked turnstile antenna.

FIGURE 3–35
The folded dipole.

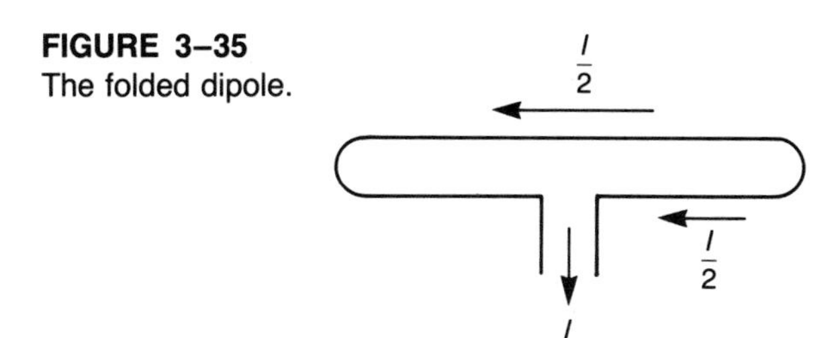

from which we get

$$R = 4\left(\frac{P}{I^2}\right) \qquad \textbf{(3–9)}$$

or four times the radiation resistance of a simple half-wave dipole: $4 \times 73 = 292$ ohms.

Since the resistance is now much higher than that of a simple dipole, the bandwidth is much broader.

The bandwidth of a resonant circuit is given by $BW = f_r/Q$, where f_r is the resonant frequency and $Q = X_L/R$. Whenever a resonant antenna is off frequency, its reactive components are no longer equal and opposite. Therefore, a net reactance will exist, and if R is low, Q will obviously be high, and the bandwidth (BW) will be narrow. Increasing the radiation resistance with a folded dipole will have the effect of making the bandwidth broader. (The development of the bandwidth formula is shown in Appendix F.)

In the United States, commercial television programs are frequently broadcast using stacked-bay turnstile antennas for the simultaneous transmission of both the video and FM audio portions of the signal. The two signals may not mix, however, and a network called a *diplexer*[7] is employed to keep the signals apart. Figure 3–36 shows such a diplexer network connected to one-half a bat-wing turnstile element. The other bat-wing section would be connected as shown in figure 3–31.

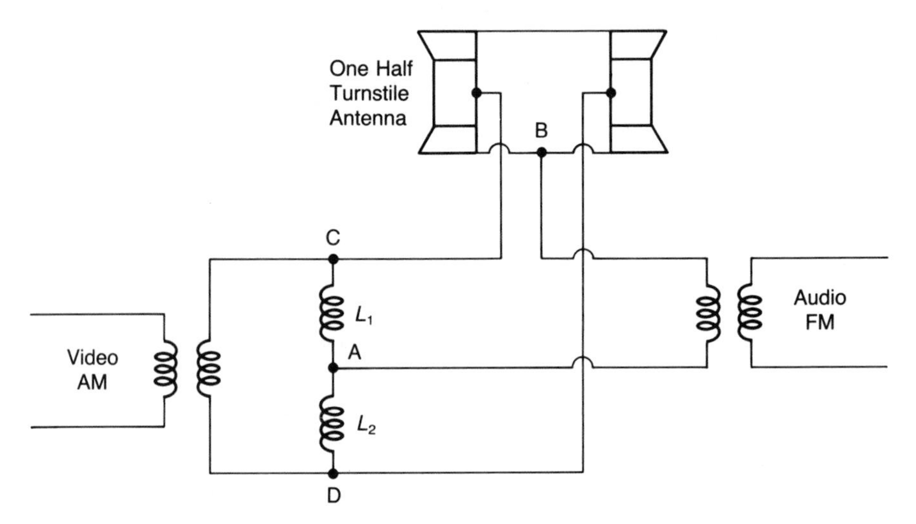

FIGURE 3–36
Diplexer network connected to a bat-wing section of turnstile TV broadcast antenna.

The circuit shown in figure 3–36 may be redrawn to show that it is actually a bridge network (see figure 3–37). If X_{L1} is equal to X_{L2}, the bridge is balanced and $V_{AB} = V_{CD} = 0V$. Consequently, the video signal appears across CD, and the audio appears across AB, and there is no intermixing between the two signals.

Turnstile antennas are mounted atop high towers in order to reduce ground reflections which cause phase interference as well as alter radiation resistance. Moreover, such antenna towers are usually placed atop high mountain peaks or tall buildings in order to increase the effective range of transmission which, as mentioned earlier, is a line-of-sight phenomenon (see equation 2–20). In Los Angeles, for example, the commercial television antennas (as well as others) are situated atop Mt. Wilson at an elevation of about 6,000 feet (1.8 km) above the city proper. One commercial television station (KCBS, channel 2) operates an antenna whose total height is close to 1,000 feet, giving a total elevation of nearly 7,000 feet or over 2,000 meters above the city. The radiation pattern for channel 2 is shown in figure 3–38. Note that the turnstile antenna used gives a radiation pattern that is approximately circular. The site of Mt. Wilson is roughly 19 miles (30 km) to the northeast of the downtown area.

[7]One must not confuse *di*plexer with *du*plexer. The *diplexer* is used for the simultaneous transmission of two different signals, whereas the *duplexer* allows for the simultaneous transmission and reception of only one signal. The duplexer circuit is widely used in radar, and will be discussed more fully in chapter 7.

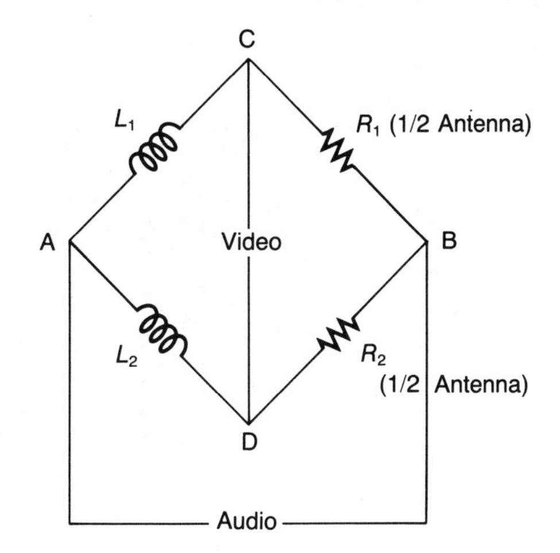

FIGURE 3–37
The diplexer bridge network.

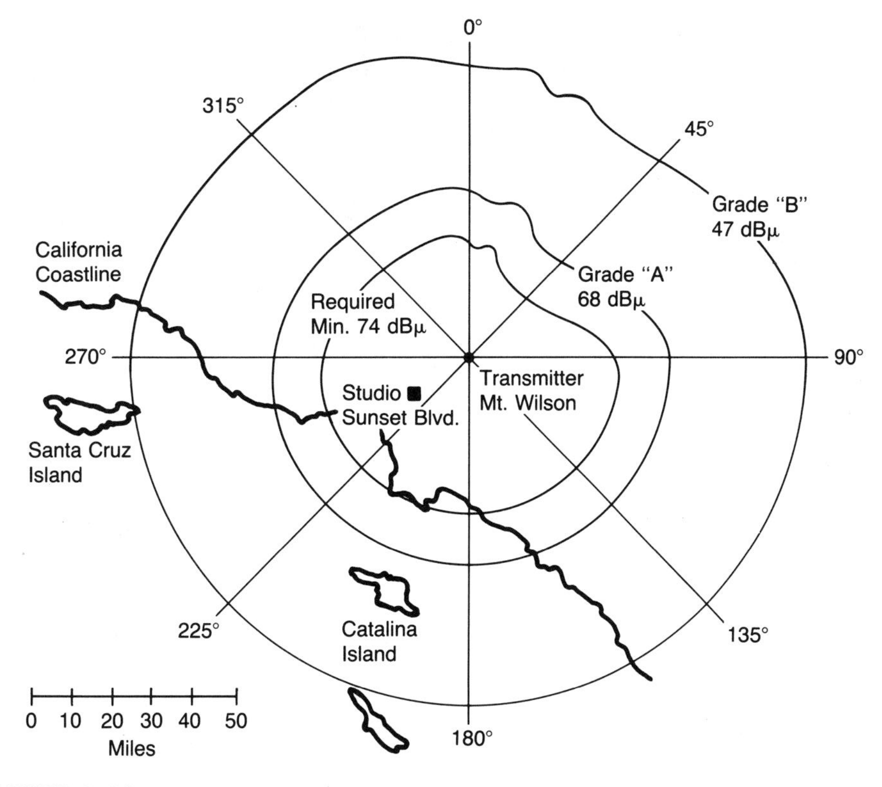

FIGURE 3–38
Radiation pattern of KCBS, channel 2, Los Angeles.

FIGURE 3–39
A 3-bay turnstile antenna.

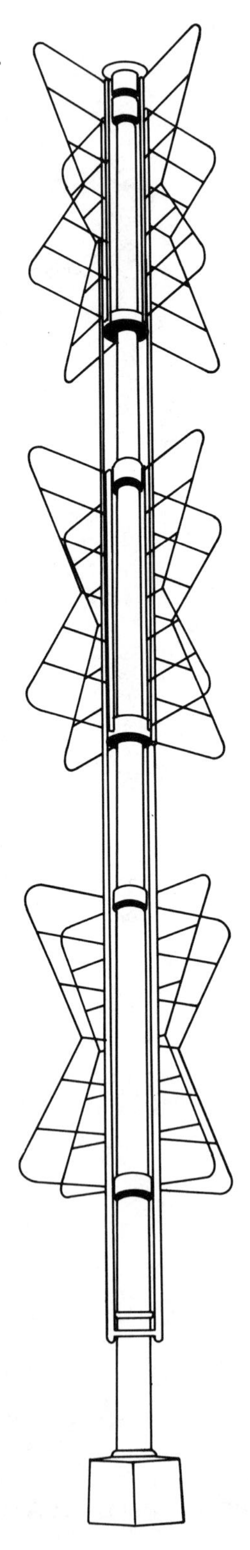

From a distance, VHF turnstiles resemble tall Marconi antennas used for commercial AM broadcasting. However, whereas the entire structure of the Marconi *is* the antenna, only the turnstile itself at the very top of the television tower is the active element. Figure 3–39 shows a 3-bay turnstile antenna.

The Parasitic Array

Recall that the radiation pattern of a horizontal Hertzian antenna in free space resembles a figure 8. Such an antenna is said to be bidirectional in the horizontal plane. It is possible, however, to truncate one of the major lobes and thus render the antenna highly directional. This is accomplished by means of various passive (i.e., nondriven) elements situated in the induction field of the dipole. Such elements have no direct connection to the driving source itself, but rather have currents induced in them by virtue of their proximity to the dipole element receiving the power. These currents, in turn, set up a complex interaction with the E-M vectors of the driven element which yields a directional radiation pattern. Such an antenna is called a Yagi-Uda[8] and is shown in its simplest form in figure 3–40.

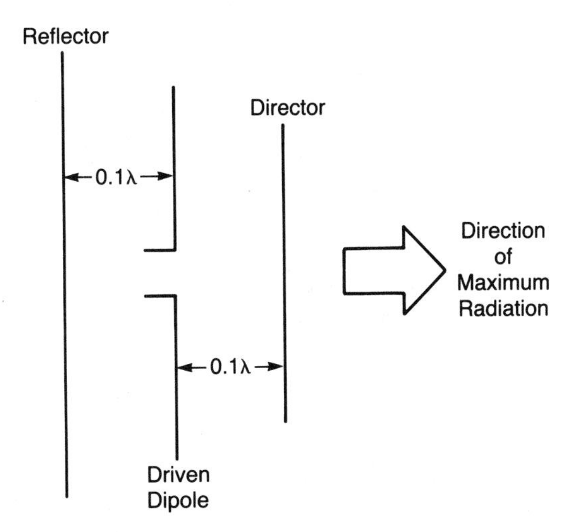

FIGURE 3–40
A simple Yagi-Uda antenna.

As shown in figure 3–40, two passive elements are located optimally at about 0.1λ in front and in back of the dipole. Actual spacing is determined empirically, but is in this range. The *director* is somewhat shorter than the half-wave dipole; hence, it acts inductively. Conversely, the *reflector* is longer than λ/2 and behaves capacitively. It is this difference in the phases of the currents of each passive element that is responsible for the directionality of the antenna. Both the reflector and director are called *parasitic* elements since they derive their energy from the only driven element in the array.

[8]Invented by the Japanese physicist Uda, but first translated into English by Hidetsugu Yagi in 1928. This antenna is often *incorrectly* called a "Yagi."

Most practical antennas of this type have several directors to improve directivity and beamwidth, but only one reflector, since additional reflectors do little to improve antenna performance. A single-channel *Winegard* Yagi-Uda for television reception is shown in figure 3–41.

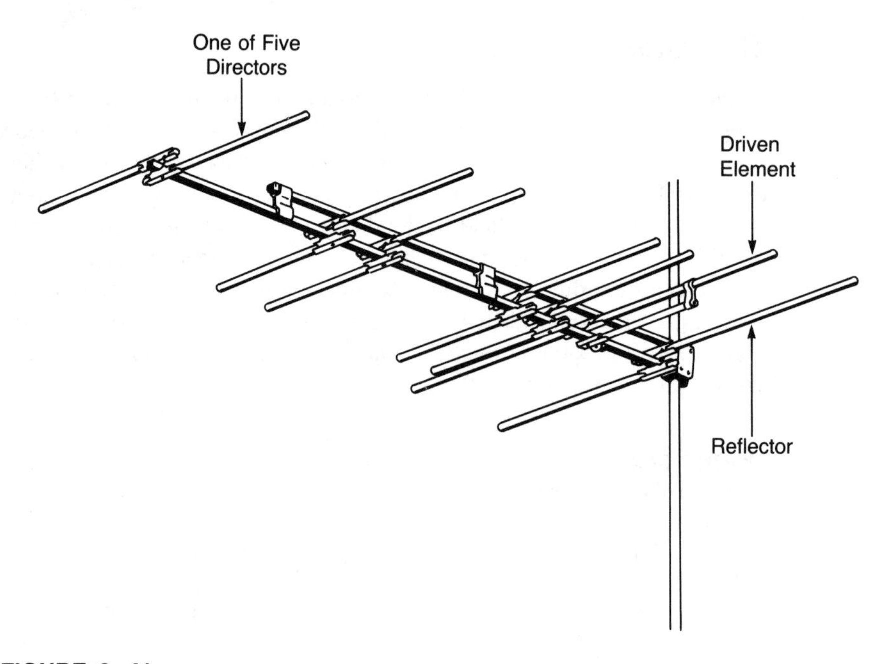

FIGURE 3–41
A single-channel television Yagi-Uda antenna. (From Winegard, 3000 Kirkwood St., Burlington, IA 52601)

Typical Yagi-Uda antennas have gains in the range of 7 to 15 dB over an isotropic source and front-to-back ratios near 30 dB. The front-to-back specification is a figure of merit for the antenna expressing the forward to reverse field strengths. The higher this figure, the more directive the antenna is. The Yagi-Uda finds extensive use as a television and FM receiving antenna. Such antennas may frequently employ a folded dipole as the driven element to increase bandwidth and are fed with 300-ohm twin-lead transmission line. Simple dipoles are fed with 75-ohm coaxial line for greater noise immunity. Beamwidths are in the range of 40 to 60 degrees.

MICROWAVE HORN ANTENNAS

As mentioned in chapter 1, signal attenuation due to dielectric and conduction losses seriously limit the usefulness of transmission lines in the microwave region. At these frequencies, it becomes necessary to convey energy by means of waveguides. Waveguide structures will be discussed fully in chapter 4. In the meantime, it must be mentioned that just as simple wire antennas were extensions of transmission line geometry, so, too, are microwave antennas extensions of waveguide configurations. For example, it was shown earlier that the dipole was merely the result of spreading apart the ends of a parallel-wire transmission line. In a similar way, microwave antennas are the result of flaring the open end of a waveguide structure to form a horn-like aperture. Such an opening provides a gradual transition for the wave into the milieu of free space, and thus prevents the reflection of energy back to the source.

As will be shown in chapter 4, the guide wavelength (λ_p) is always greater than that of the free-space wavelength (λ) by a factor of

$$\frac{1}{\sqrt{1 - \left(\frac{\lambda}{\lambda_0}\right)^2}} \tag{3–10}$$

where λ_0 is the cutoff wavelength and is determined by the cross-sectional dimensions of the guide. Moreover, the wave impedance of a rectangular waveguide is always different than the free-space impedance of 377 ohms by a similar factor. Quite obviously, then, as the aperture area of the guide increases as a result of flaring, which is to say the cutoff wavelength (λ_0) becomes very large, the radicand of the expression will approach unity and the free-space impedance of 377 ohms will be realized, thus providing a near-perfect match.

There are three basic types of *horn* or *aperture* antennas whose classification depends primarily on structural details of the flared end. Horn antennas are shown in figure 3–42.

As shown in figure 3–42, if the guide is flared in only one dimension, it is called a *sectoral* horn. If the flaring is in both directions, the horn is said to be *pyramidal*. Finally, the *conical* horn is used with round waveguides. Figure 3–43 shows the important dimensions related to horn antennas.

The beamwidth is narrow in the plane of the longer side and broader in the plane of the shorter side. Directivity is a function of both ϕ and L and increases as L increases.

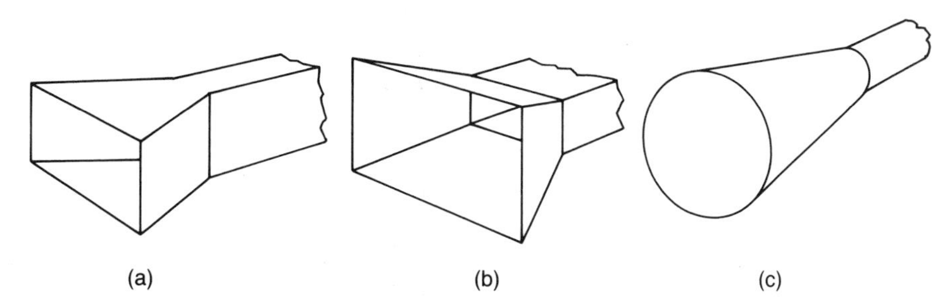

FIGURE 3–42
Horn antennas: (a) sectoral horn; (b) pyramidal horn; (c) conical horn.

FIGURE 3–43
Important horn dimensions.

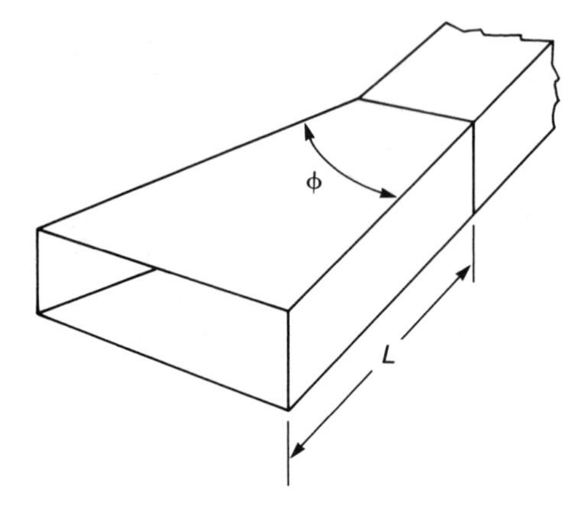

THE PARABOLOID ANTENNA

The horn antennas discussed above are often considered to be *primary* antennas in that they require no ancillary devices to radiate adequately and so are complete unto themselves. There is yet another class of microwave antennas called *secondary* radiators that do, however, make use of auxiliary structures in order to enhance their performance. By far the most frequently used of such secondary microwave antennas is the *paraboloid,* which employs a reflector in order to increase its power gain.

For an antenna of given dimensions, the maximum gain will be achieved whenever the waves leaving the radiator neither diverge nor converge, but rather travel away as plane waves whose radiation pattern is a right circular cylinder. Such a pattern is theoretically possible with a paraboloid reflector antenna. We will now prove the validity of this assertion.

As you may recall from earlier courses in analytic geometry, a *parabola* is the outline of the shape obtained by taking a section through a right circular cone in a plane parallel to its side, as shown in figure 3–44. Analytically, a *parabola* is defined as the locus (graph) of all points equidistant from a fixed point called the *focus* and a fixed line called the *directrix* (see figure 3–45).

From figure 3–45 and the definition of a parabola, we are given that PQ = PF. Expressed algebraically, this becomes

$$x + \frac{p}{2} = \sqrt{\left(x - \frac{p}{2}\right)^2 + y^2} \qquad \textbf{(3–11)}$$

Solving equation (3–11) for y^2 gives

$$y^2 = 2px \qquad \textbf{(3–12)}$$

which is the general expression for the parabola.

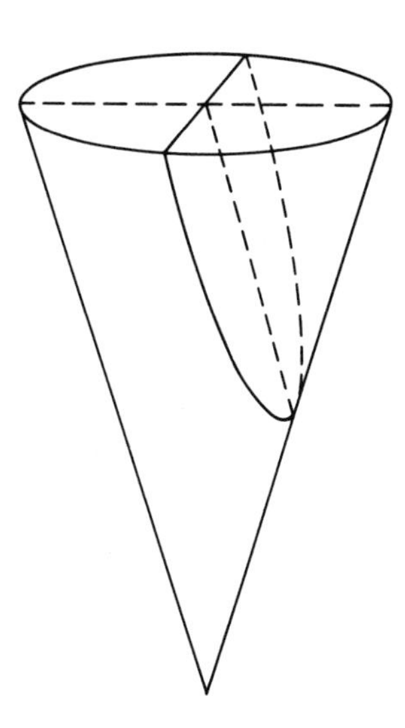

FIGURE 3–44
Parabola obtained from a conic section.

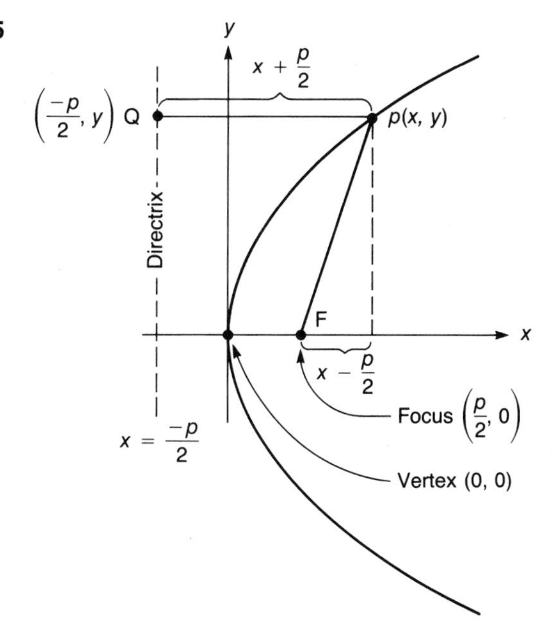

FIGURE 3–45
The parabola.

With this background definition and description, we now extend the parabola to a 3-dimensional figure constructed by rotating the locus about the X-axis. The resultant surface of revolution is called a paraboloid. This shape acts as the reflector for microwave energy exhibiting the characteristics of light rays in terms of reflection properties.

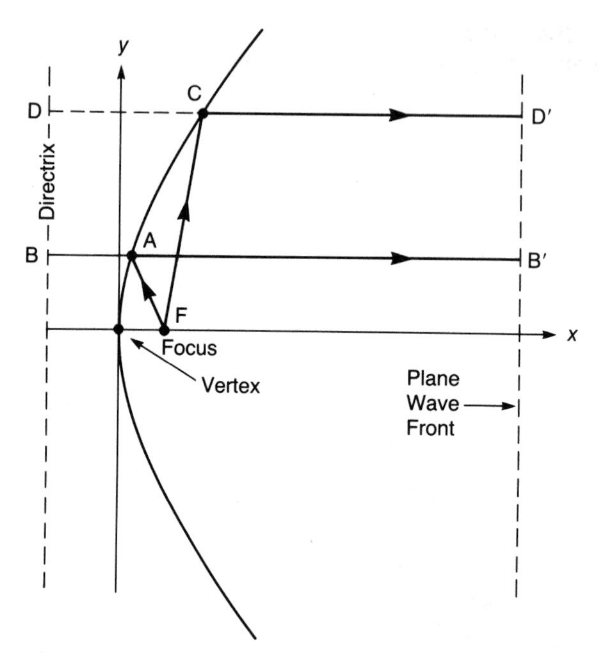

FIGURE 3–46
Plane waves leaving a parabolic surface.

It now remains only to show that a plane wave leaves such a surface when illuminated by an isotropic source at the focus. In figure 3–46, we will show that FA + AB′ = FC + CD′, and this will be sufficient to prove our assertion.

We will prove that

$$\mathrm{FA} + \mathrm{AB}' = \mathrm{FC} + \mathrm{CD}' \tag{3–13}$$

By definition

$$\mathrm{FA} = \mathrm{AB} \tag{3–14}$$

And

$$\mathrm{FC} = \mathrm{CD} \tag{3–15}$$

From figure 3–46

$$\mathrm{AB} + \mathrm{AB}' = \mathrm{CD} + \mathrm{CD}' \tag{3–16}$$

Upon substituting FA from (3–14) for AB in (3–16) and FC from (3–15) for CD in

(3–16), we obtain,

$$FA + AB' = FC + CD'$$

This proves our assertion, and it is now evident why the parabolic reflector is such a ubiquitous microwave device. Plane waves emanating from its surface travel in a narrow beam which not only increases gain, but also reduces susceptibility to noise.

In the above discussion, we assumed that the source of radiation was an isotropic radiator located at the focus. This was convenient from a mathematical viewpoint, but there are three problems associated with such an assumption. First, there is no such point source in reality capable of illuminating the paraboloid "dish." Secondly, even if it were possible to have such a source, rays escaping the reflector surface, called "spill-over," cause the resultant radiation pattern to diverge. Spillover is responsible for noise pickup when the paraboloid is used as a receiving antenna. Finally, the isotropic source is a disadvantage since its primary radiation in the direction along the major axis of the paraboloid (called "backlobe" radiation) interferes destructively with the main pattern.

In general, the actual size and shape of the primary antenna, called the "feed antenna," has a lot to do with the purity and homogeneity of the ideal radiation pattern of the paraboloid. Figure 3–47 shows the effect of various field characteristics of the primary feed antenna on the radiation pattern.

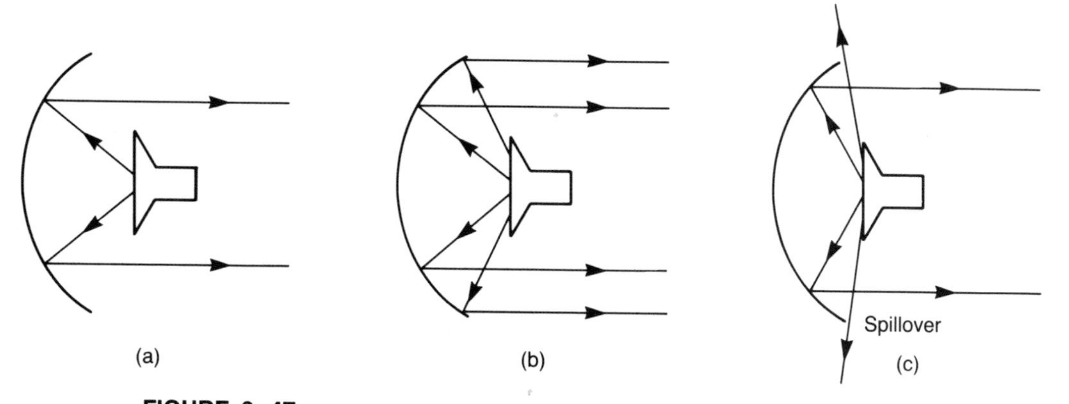

FIGURE 3–47
Various feed situations for a paraboloid: (a) insufficient illumination; (b) ideal illumination; (c) excessive illumination.

To reduce the effect of backlobe radiation, a spherical reflector is often used in conjunction with a simple dipole as the feed antenna. Errant backlobe radiation is then reflected back to the surface of the paraboloid and added in phase with the main radiation. These antennas require some minor tuning to insure proper phasing at the design frequency. Such an antenna, often referred to as a "center feed" antenna, is shown in figure 3–48. To reduce spillover effects, especially noise pickup in deep-space satellite tracking operations, the Cassegrain antenna is often used. Such an antenna is shown in figure 3–49.

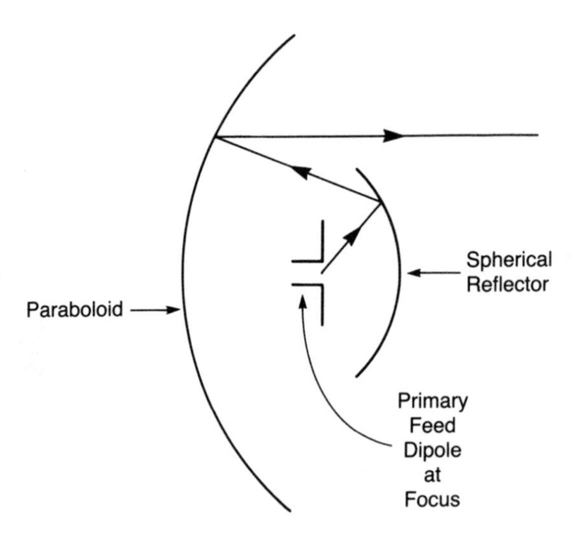

FIGURE 3–48
Spherical reflector used to reduce backlobe radiation.

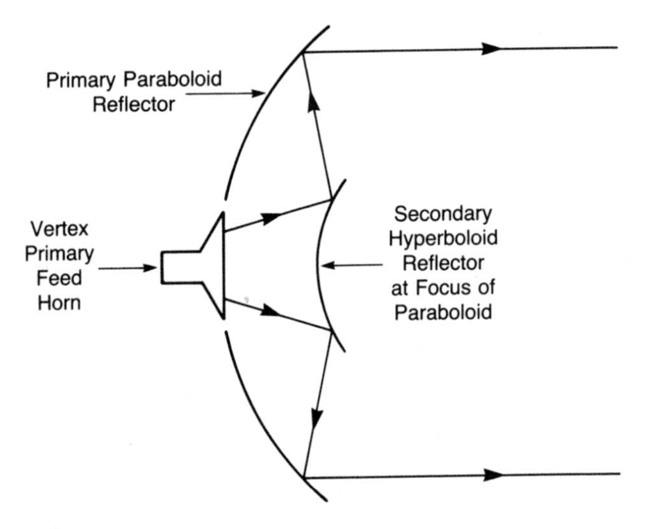

FIGURE 3–49
The Cassegrain antenna.

The hyperboloid reflector located at the virtual focus of the paraboloid illuminates the main dish almost uniformly and minimizes spillover effects.

It may be shown[9] that the gain of an ideal paraboloid with uniform illumination and no losses is given by

$$G = \left(\frac{\pi D}{\lambda}\right)^2 \tag{3–17}$$

where D is the maximum dish diameter and λ is the wavelength. If the aperture is not uniformly illuminated, an illumination efficiency term (η) must be introduced. Equation (3–17) then becomes

$$G = \eta\left(\frac{\pi D}{\lambda}\right)^2 \tag{3–18}$$

In practice, η may be on the order of 60%. On substituting in equation (3–18), we obtain the more generally accepted formula

$$G = 6\left(\frac{D}{\lambda}\right)^2 \tag{3–19}$$

where G is the power gain *ratio,* not the decibel power gain.

The beamwidth of an ideal paraboloid uniformly illuminated is given by

$$\text{BW} = \frac{70\lambda}{D} \tag{3–20}$$

where D is the maximum dish diameter and λ is the wavelength.

For aperture antennas of the type we have been discussing, there is a certain distance (R) from the antenna, called the *far-field* distance, at which one may be assured that the only radiation affecting the receiving antenna is the radiation field, not the induction field. At this distance, the angular distribution of radiated energy is entirely independent of the distance from the transmitting antenna. Moreover, a radiator at this distance may be regarded as a point or isotropic source. The far-field distance is given by

$$R = \frac{2D^2}{\lambda} \tag{3–21}$$

where D is the largest dimension of the aperture and λ is the wavelength.

---

[9] $G = (4\pi A/\lambda^2)$

Since A (aperture area) $= (\pi D^2)/4$, then $G = \left(\frac{\pi D}{\lambda}\right)^2$. Johnson and Jasik, eds. *Antenna Engineering Handbook,* 2nd ed. New York: McGraw-Hill, 1984.

DIELECTRIC LENS ANTENNA

Just as microwave energy exhibits the same reflective properties of light, so too does it conform to the properties of light refraction as governed by Snell's Law[10] shown in equation (3–22).

$$\frac{\sin i}{\sin r} = k \tag{3–22}$$

where i is the angle of the incident ray, r is the angle of the refracted ray, and k is the index of refraction. For any given material, regardless of the angle of incidence, the ratio of sines will always result in the value k. For air, k is approximately 1.0. Snell's Law is illustrated in Figure 3–50.

FIGURE 3–50
Illustration of Snell's Law.

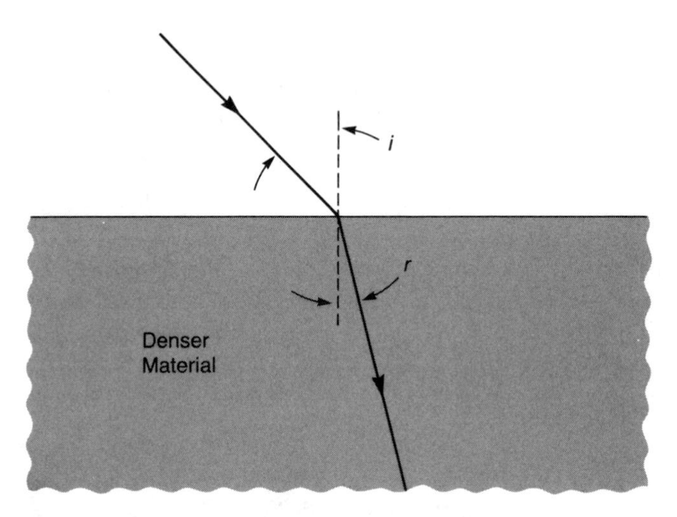

In microwave antenna applications, the lens functions as a wave collimator,[11] as shown in figure 3–51. Ray OA has traveled a further distance than ray OB in reaching the surface of the lens. This situation creates a slight time lag. However, within the lens, ray BB′ is slowed more than ray AA′ due to the change in phase velocity (see equation 1–7) at the air-lens boundary. As a result, the waves leave the lens as plane, parallel waves despite the curvature of the incident waves.

Dielectric lenses are made of polystyrene and other dense materials. Such materials are chosen because they produce the greatest amount of diffraction for the smallest size and weight. However, such materials also produce great attenuation of the signal as it passes through the lens. In order to offset this problem, dielectric lenses are typically stepped or zoned as shown in figure 3–52. It should be apparent that such stepped lenses have a great deal of their bulk removed with a consequent savings of both size and weight as well as a reduction in signal attenuation.

---

[10]Harris and Hemmerling. *Introductory Applied Physics*. New York: McGraw-Hill, 1955.

[11]*Collimate* means to make parallel.

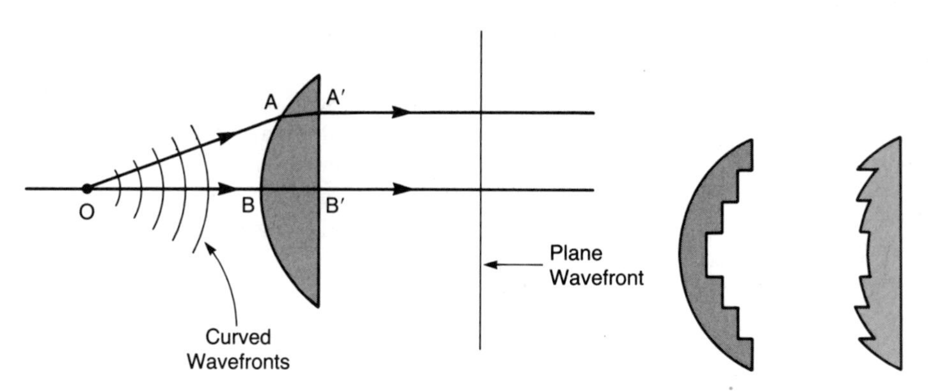

FIGURE 3–51
Principle of wave collimation using dielectric lens.

FIGURE 3–52
Cross-section of zoned lenses.

To work properly, zoned lenses must be uniformly illuminated. The horn antenna is ideally suited for this purpose and is often employed in such small, lightweight devices as portable police radar.

SUMMARY

This chapter began by extending the fundamental idea of a transmission line to the development of the dipole antenna. An antenna, it was pointed out, was designed expressly for the purpose of increasing the amount of radiation escaping into space and providing an interface with the space link between complementary antennas.

The distinction between the induction field and the radiation field of an antenna was discussed, and emphasis was placed on the mechanism of the radiation phenomenon.

Two important fundamental antennas, the Hertz and Marconi, were mentioned, and their differences discussed. It was mentioned that the Hertz antenna is complete unto itself, whereas the Marconi required the use of a reflecting plane in order to perform adequately as a radiator.

Wave polarization was defined and radiation patterns for the two basic antennas were discussed. Beamwidth was defined as the plane angle in degrees measured between the 3 dB down points along the axis of the major lobe. Bandwidth was defined as the useable frequency range between the half-power points.

The input impedance of an antenna was defined in terms of its radiation resistance, and the concept of antenna gain was discussed. It was shown that while directive gain was useful in explaining the concept of antenna gain, the most practical measures of antenna performance were power gain and ERP.

Two important nonresonant antennas, the long-wire and the rhombic, were shown to be highly directional in contrast to the Hertzian and Marconi antennas.

The concept of antenna arrays was introduced, and three useful types were the broadside, the end-fire, and the turnstile. The folded dipole and the diplexer network were introduced in connection with the turnstile as a broadband VHF transmitting device. The parasitic antenna was also introduced, and the Yagi-Uda was discussed as an important example of such an array.

Aperture antennas were defined and various horn designs were illustrated including the sectoral, the pyramidal, and the conical horn. In addition, the basic paraboloid antenna design was derived and discussed, and it was pointed out that there were two serious problems associated with illumination of the dish: backlobe radiation and spillover. To get around these problems, two other antennas, the center feed and the Cassegrain, were developed.

Finally, the idea of the dielectric lens antenna was explored as a means of producing plane, parallel waves from a spherical wave front.

PROBLEMS

1. A simple half-wave dipole receiving antenna is to operate at a frequency of 200 MHz. What is the length in feet and inches of each section of the antenna? [1 foot, 2 inches]
2. At what frequency will an ideal, 10 meter long Hertzian dipole have an input resistance of 73 ohms? [15 MHz]
3. A resonant half-wave dipole in free space has a feed-line current of 2 amperes. How much power appears at the input to the feed point? [292 watts]
4. The efficiency of an antenna is 82%. Its radiation resistance is 30 ohms. What is the value of its loss resistance? [6.6 ohms]
5. The power gain of an antenna is 15 dB. Theoretically, how much power must be fed into an isotropic comparison antenna if 2 watts were fed into the test antenna? [63 mW]
6. What is the ERP of a transmitting antenna whose gain is 64 dB and which is supplied by a transmitter delivering 1 kW through a transmission line known to have a total loss of 1.5 dB? [1,800 MW]
7. The chapter stated that Los Angeles television channel 2 (KCBS) had its turnstile antenna located at an elevation of 2,000 meters above the city. Given sufficient power, what is the expected radius of coverage in miles? [111 miles]
8. In figure 3–38, what is the approximate radius of actual coverage for grade ''B'' reception of channel 2 in Los Angeles? [slightly less than 100 miles]
9. A paraboloid is to have a power gain of 30 dB. If the frequency of operation is to be 5.5 GHz, what is the diameter of the parabolic reflector? [2.3 feet]
10. What is the beamwidth of the paraboloid antenna in problem 9? [5.4 degrees]
11. What is the far-field distance of the antenna in problem 9? [18 meters]
12. A paraboloid operating at 10 GHz is to have a beamwidth of 2.5 degrees. What is the dish diameter? [2.75 feet]

QUESTIONS

1. What is a TEM wave? How is it formed as a result of current flowing in an antenna?
2. What is the difference between the induction field and the radiation field of an antenna?
3. Why is a Hertz antenna considered complete unto itself, but a Marconi antenna is not?
4. What is the advantage of a Marconi antenna over a Hertz dipole?
5. Draw a diagram that illustrates why the antenna resistance of a half-wave dipole is higher near the ends than near the center.
6. What is the difference between a resonant antenna and a nonresonant antenna in terms of (a) standing waves; (b) directivity; (c) physical dimensions; and (d) bandwidth.
7. What is meant by the polarization of a wave?
8. Explain in detail why it is essential to keep each of the following three factors constant when deriving a radiation pattern of a transmitting antenna: (a) transmitter power; (b) distance from the transmitter; (c) plane of the measurements.
9. Define radiation pattern, and explain what information it conveys.
10. Draw a 3-dimensional sketch of the radiation patterns for a horizontal Hertzian antenna in free space and a Marconi antenna. Which antenna has an omnidirectional radiation pattern in the horizontal plane?
11. Define and illustrate beamwidth and bandwidth.
12. Distinguish among the terms antenna impedance, antenna resistance, radiation resistance, and loss resistance.
13. Describe how radiation resistance changes with elevation of an antenna above ground. What causes the change in this resistance?
14. Since an antenna is a passive device, how can it be said to have gain? Distinguish between directive and power gain. What is meant by ERP?
15. What makes an antenna an array? Describe the radiation patterns of the broadside, end-fire, and turnstile arrays.
16. What is a diplexer? How is it used in commercial television transmission?
17. Describe the important characteristics of a Yagi-Uda antenna. Why is this antenna referred to as a parasitic array?
18. What are three basic forms of horn antennas? Why is the horn flared?
19. What is a paraboloid antenna? Why is it so important in microwave work?
20. What problem does the center-feed paraboloid antenna solve? What is the principal advantage of the Cassegrain antenna?
21. What is a lens antenna? How does it provide wave collimation?
22. What advantages does a zoned lens offer over a solid dielectric lens?

4

WAVEGUIDES

TRANSMISSION LINE LOSSES

At frequencies within the microwave region (> 1 GHz) three common losses occur that render the conventional transmission line unacceptable as a means of conveying energy from one point to another. These three losses are conduction (or copper) loss; radiation loss; and dielectric loss. These will now be discussed.

Conduction Loss

Whenever a current (I) flows through a conductor of resistance (R), a certain amount of power will be lost as given by $P = I^2R$. As the cross-sectional area of the conductor decreases, it is well known that its resistance will increase, and the amount of power lost for a given current will increase. At microwave frequencies, the *effective* cross-sectional area varies inversely with frequency as a consequence of what is called ''skin effect.'' This effect is due to the tendency of electrons to travel closer to the conductor's surface (skin) as a result of small voltages induced in the wire itself. This voltage forces the electrons toward the surface of the conductor where they are confined to a thin-sheath region of the material despite the actual diameter of the conductor. This effect is illustrated in figure 4–1.

FIGURE 4–1
Skin effect at microwave frequencies.

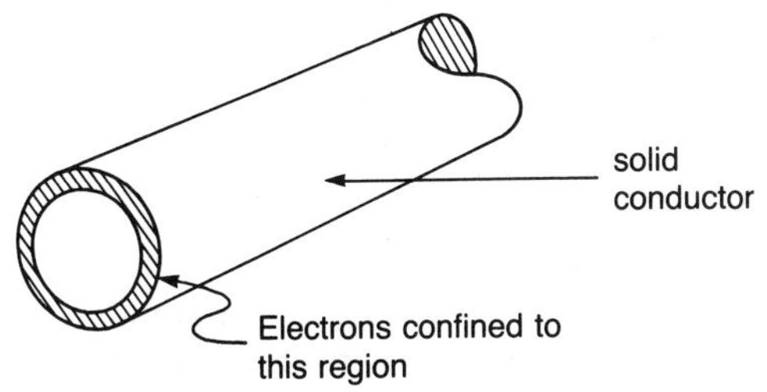

In practice, this effect may be slightly offset by increasing the surface area of the conductor (by increasing its diameter) and by making the conductor hollow rather than solid. This is often the case where high-power, semi-rigid coaxial cable is used at frequencies in both the VHF and UHF ranges.

Radiation Loss

Since the power of the microwave signal is contained in the TEM wave that carries it, any loss of the wave itself between source and load represents a loss of signal energy. Typically, loss of the TEM wave occurs as unintentional radiation from the transmission line. Loss by radiation is most noticeable in parallel-wire lines at frequencies whose wavelength is but a fraction of the spacing between the conductors.

In coaxial cable, however, the TEM wave is entirely confined between the inner and outer conductors, and little if any wave energy escapes by radiation. A comparison between the E-M fields about a parallel-wire and a coaxial line is shown in figure 4–2.

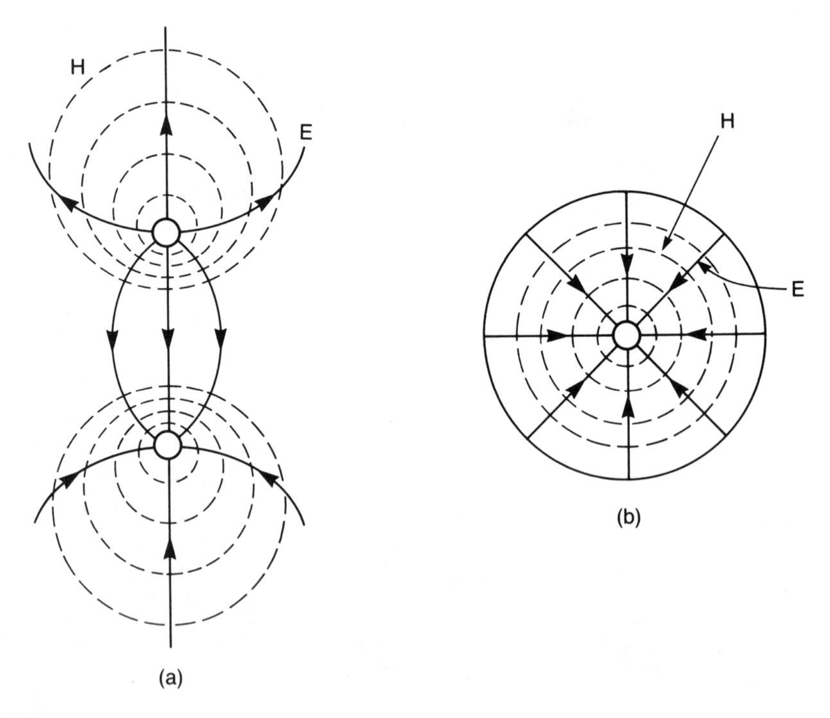

FIGURE 4–2
E-M fields about (a) parallel-wire and (b) coaxial cable transmission lines.

Dielectric Loss

The atomic profile of the individual atoms comprising the insulating material between any two conductors is constantly being stressed as the applied voltage reverses from

(+) to (−) and back again. Such distortion requires energy that is derived from the signal and manifests itself in the form of heat within the dielectric. The higher the frequency, the greater the heating. At microwave frequencies, the lost energy in the form of heat will quickly require us to abandon all solid dielectrics, and hence all conventional transmission lines. It is primarily because of dielectric losses that parallel-wire or coaxial line ceases to be effective at frequencies much above 3 GHz, although certain types of coaxial cable may be used up through 18 GHz for short runs.

As an example of the sort of attenuation encountered with coaxial cable, one source[1] specifies the attenuation of an RG-8/U (new equivalent RG-213/U) cable as approximately 2 dB per 100 feet at 100 MHz. At 1 GHz, however, the attenuation has quadrupled to 8 dB per 100 feet. Another way of appreciating this statistic is to consider that a 1 GHz signal will lose 84% of its power in traversing 100 feet of RG-8/U cable.

SIGNAL PROPAGATION IN WAVEGUIDES

It is precisely because of the loss factors outlined above that waveguides came into being. By definition, a waveguide is simply a pipe of virtually any consistent cross-sectional shape through which an E-M wave travels by reflection, not conduction. It is for this reason that we will have occasion to speak of electric and magnetic fields rather than voltage and current as was the case with transmission lines. A few typical waveguide shapes are shown in figure 4–3.

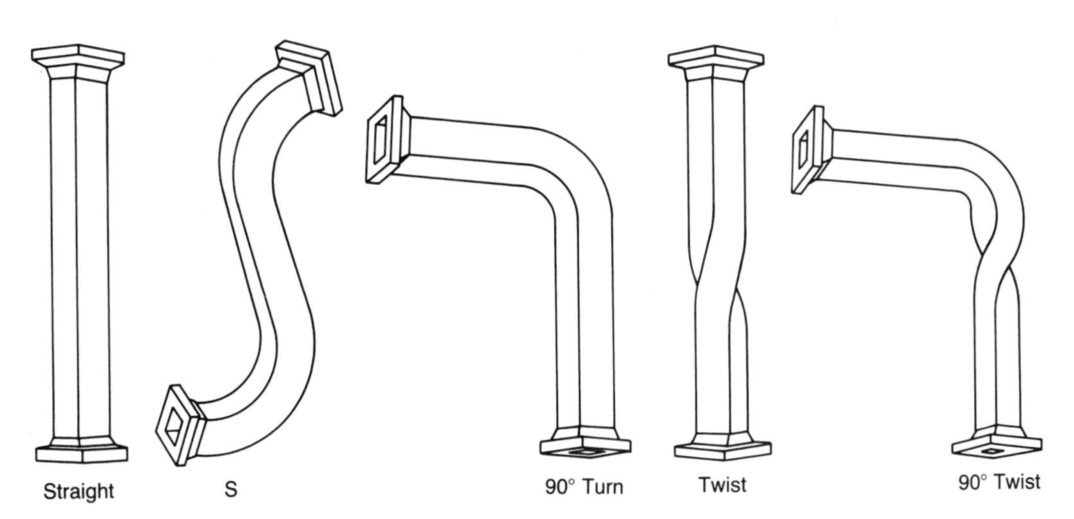

FIGURE 4–3
Typical waveguide shapes.

[1]Saad, T., and Hansen, eds., *Microwave Engineer's Handbook,* vol. 1. Dedham, Mass: Artech House, 1971.

Since the dielectric most commonly encountered in waveguides is air, the dielectric attenuation is virtually nil. For example, one data handbook[2] specifies the average attenuation of a particular waveguide as 0.16 dB/100 feet at a frequency of about 1 GHz. This is considerably less than that specified above at the same frequency for RG-8/U coaxial cable.

It is important to understand, however, that a TEM wave *cannot* traverse a waveguide as it could a conventional transmission line. As will be seen later, the electric and magnetic fields associated with the propagation of a signal in a waveguide do not array themselves in any simple mutually-perpendicular fashion as was the case with the TEM wave illustrated earlier in connection with figure 2–1.

Since the method of propagation in a waveguide is by means of reflection, it is vital to recognize that the interior surfaces be smooth, often silvered, free of debris or moisture, and unencumbered by abrupt changes in shape or direction that will cause reflections to occur back toward the source.

With the above as a prologue, we are now ready to undertake a more sophisticated analysis of the mechanism of wave propagation in a waveguide.

ANALYSIS OF WAVE PROPAGATION

It was mentioned in chapter 3 that a quarter-wavelength section of transmission line exhibited a certain distribution of voltage standing waves. Let us take a closer look at this distribution as shown in figure 4–4.

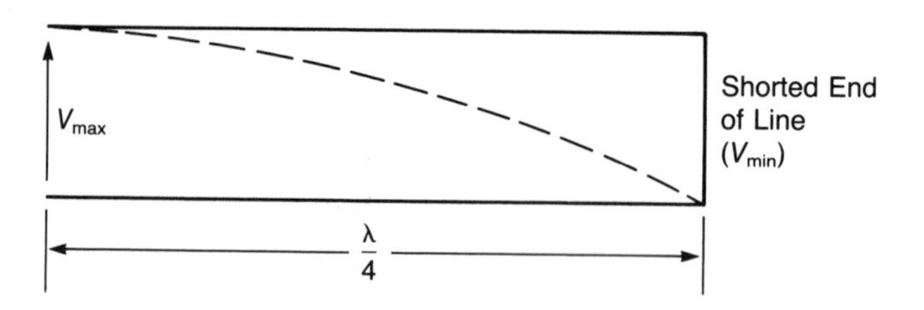

FIGURE 4–4
Distribution of voltage standing wave on short-circuited, quarter-wave transmission line.

Figure 4–4 shows that λ/4 away from the shorted end of the line is a voltage maximum. Consider, now, the effect of splicing another λ/4 section of line onto the first, as shown in figure 4–5.

Figure 4–5 illustrates that the voltage maximum occurs in the center of the rectangle, and V_{min} occurs at the shorted ends. For the time being, let us not trouble ourselves with *how* such a voltage distribution could have been established in the first place. For the present suffice it to say that it *was* somehow established.

---

[2]Waveguide is RG-69/U at f_c = 0.9 GHz, $TE_{1,0}$. Johnson, R., and Jasik, H., eds. *Antenna Engineering Handbook,* 2nd ed. New York: McGraw-Hill, 1984.

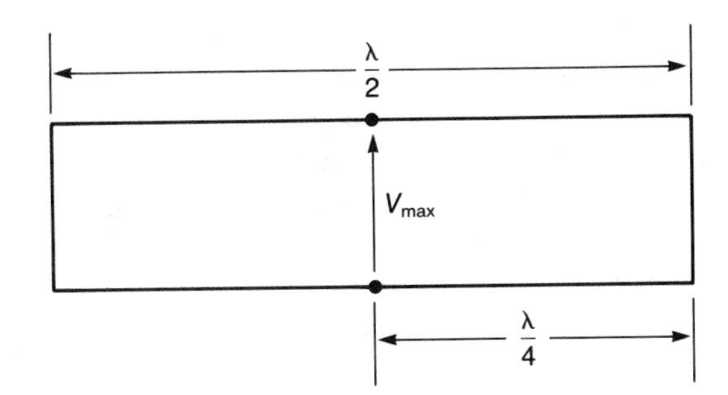

FIGURE 4–5
Two λ/4 shorted sections spliced together.

Consider the effect of adding an infinite number of such rectangular shapes one on top of the other to form the solid shape depicted in figure 4–6, which we will call a *rectangular waveguide*. Note that the basic electric field distribution has not been altered as a result of the translation of this two-dimensional shape into the third physical dimension.

FIGURE 4–6
A rectangular waveguide.

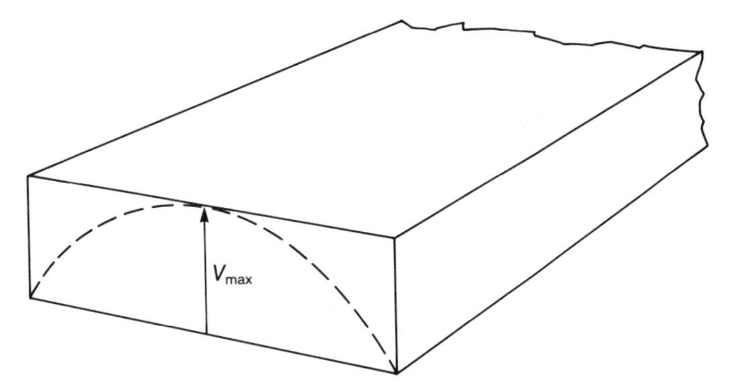

We have now begun to understand how a voltage maximum may be sustained between two short-circuited walls λ/2 apart. There are, however, two slight problems with what we think we know. First, we still have no idea how this voltage got there or how it moves down the guide. Secondly, if we were to measure the actual width of a waveguide section designed to work at a particular frequency, we would be surprised (and perhaps a little disappointed) to find that it was somewhat wider than λ/2. We will now attempt to resolve these two disparities.

In chapter 2, we mentioned how certain types of waves, called "sky waves," could be propagated great distances by alternately bouncing back and forth between the earth's surface and the ionosphere. Such waves were "launched" by means of an antenna. In a similar way, if one end of our waveguide were closed off and a small antenna, called a probe, was located a certain distance (to be determined later) from this end, the waves leaving the probe would travel down the guide by bouncing back and forth between the walls.

The location of the probe, as will be seen, is paramount if the waves are to move through the guide properly. It is fundamental to recognize that the wave must be launched in such a way that the maximum bunching of E-field lines of force occurs at the *center* of the guide and not near what, for convenience, we have called the B walls of the waveguide (see figure 4–7). This condition must be maintained because whenever an electric line of force is adjacent (and parallel) to a conductive surface, it induces a voltage in that surface and shorts itself out. Recall from chapter 3 that this is exactly how a receiving antenna acquired its induced voltage. This is precisely the reason waves cannot be sent straight down the guide, but rather must propagate by reflection in a zigzag fashion. Figure 4–7 shows the orientation of the electric lines of force within a properly excited waveguide. These lines are perpendicular to the other two walls which, again for convenience, we will call the A walls.

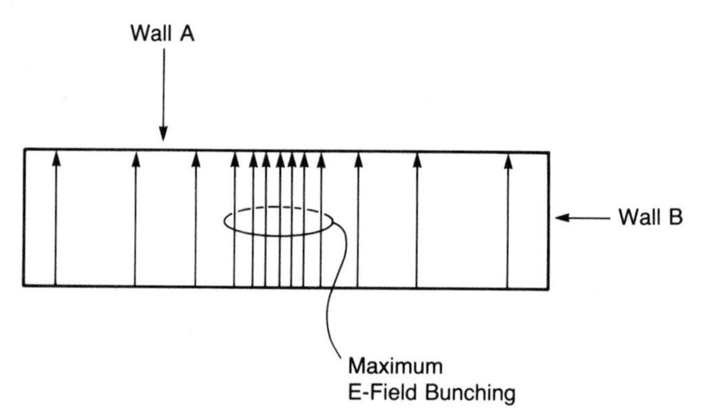

FIGURE 4–7
Distribution of electric lines of force in a waveguide.

Thus, we see that a standing wave of voltage exists between the two B walls, while traveling waves are propagated down the guide by bouncing off the B walls.

In theory, the actual A walls are not required, but in practice they maintain the correct separation between the B walls and confine the wave. The actual B dimension determines the power handling capability of the guide, but does not actually enter into theoretical calculations.

It should be pointed out that the electric lines are always accompanied by magnetic lines of force which orient themselves at right angles to the electric (E) vector, as shown in figure 4–8. Note that the magnetic (H) lines form closed loops, and that both the E and H vectors reverse direction periodically as might be expected from a cyclically varying excitation source. Note, too, that there is a component of the magnetic field in the direction of propagation that is quite unlike that of the ordinary TEM wave as diagramed vectorally in figure 2–1 or shown in figure 4–2 in connection with transmission lines. In general, the simultaneous representation of both the E field and the H field is difficult to show. Accordingly, in the interest of clarity only the E field will be shown in most of our simple drawings.

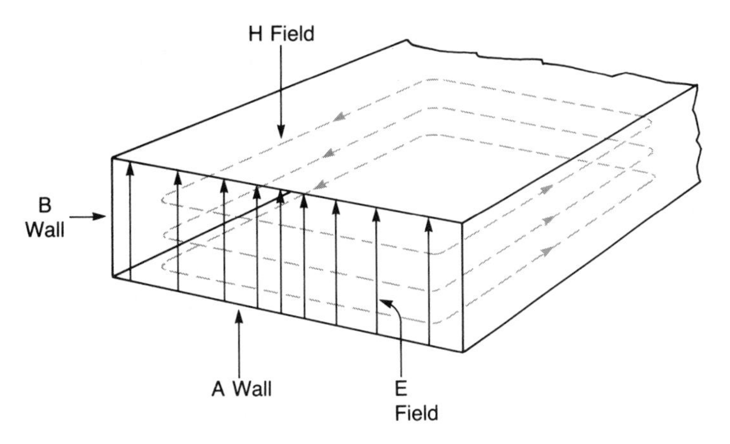

FIGURE 4–8
Electromagnetic fields in a rectangular waveguide.

Guide Wavelength

We will now attempt to resolve the fact that the $\lambda/2$ dimension of the guide is actually somewhat longer than its free-space dimension. To this end, we will consider the geometry of the situation of an incident wave as viewed in a plane perpendicular to the E-field vector, as shown in figure 4–9.

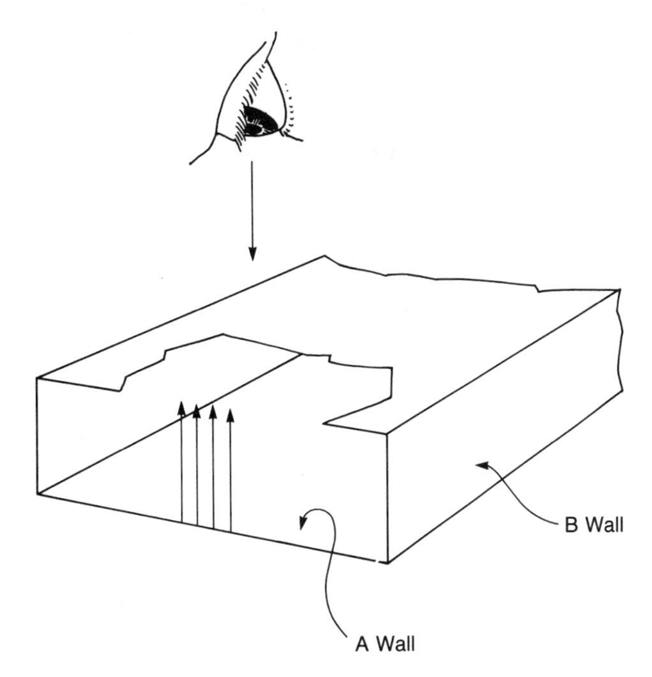

FIGURE 4–9
View of the electric wave as visualized in a rectangular waveguide.

The incident wave, which we may visualize if viewed from the vantage point of figure 4–9, is shown in figure 4–10. Note that we are looking down on the lower A wall, having removed part of the upper A wall.

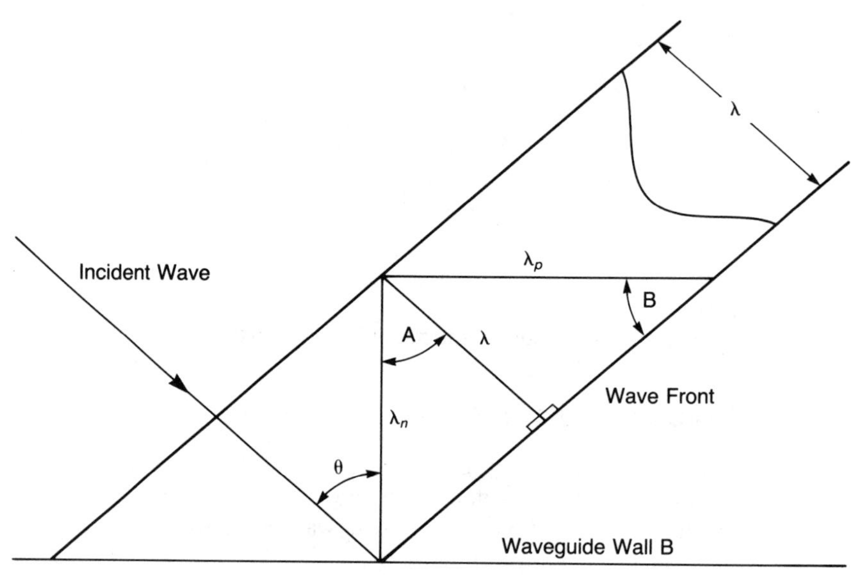

FIGURE 4–10
Incident-wave geometry.

The first thing we notice from figure 4–10 is that the wavelength in the direction of propagation (called the guide wavelength, λ_p) is not the same as the free-space wavelength (λ) which, in turn, is not the same as the wavelength (λ_n) normal to the B walls. This is essentially the same result one would obtain by cutting through a piece of corrugated plastic sheet at successively larger angles. A cut directly across the corrugations yields one crest-to-crest dimension, while an oblique (diagonal) cut yields a greater dimension, finally resulting in an infinite dimension when the cut is orthogonal. In other words, depending on how one views the situation, there are actually three wavelengths that one might define and talk about in connection with waveguides.

From a knowledge of the simple relationships existing among the sides of a right triangle, it is apparent from figure 4–10 that λ_n is, indeed, longer than λ and, therefore, $\lambda_n/2 > \lambda/2$, as we stated earlier. Let us now take a look at the simple mathematics of the situation.

We will need a convenient reference angle from which to specify the various measurements. For convenience, we will use the angle (θ) which the incident ray makes with the normal to the B wall. From the principles of plane geometry, we observe that

$$\angle\theta = \angle A = \angle B$$

It is apparent, then, that

$$\lambda_n = \frac{\lambda}{\cos \theta} \tag{4–1}$$

Therefore, the actual physical dimension of the waveguide as measured along the A wall is not $\lambda/2$, but rather

$$A = \frac{\lambda_n}{2} \tag{4–2}$$

Equation (4–2) may also be written as

$$A = \frac{\lambda}{2} \cos \theta \tag{4–3}$$

It may have occurred to you that dimension A given by equation (4–2) is only valid at one particular frequency. As the frequency increases, more and more half-wavelength bunches may be accommodated between the B walls separated by a given A dimension. Consequently, it is possible for more than one E-field bunch to fit between the walls. In other words, equation (4–2) does not quite tell the whole truth. However, if we now define a positive integer *m,* which represents the number of such E-field bunches existing between the B walls, we obtain,

$$A = m\left(\frac{\lambda_n}{2}\right) \tag{4–4}$$

When $m = 1$, we say that the wave is propagated in the *dominant mode*. This is the mode for which the lowest possible frequency (longest wavelength) may be propagated for a given A dimension, and is called the *cutoff frequency* (λ_0). We will derive an expression for finding λ_0 in the next section.

It must be recognized that it is the wave's own attempt to traverse the guide that is responsible for establishing the proper number of integral multiples of E-field bunches. No human intervention is required to underwrite this condition.

Group Velocity

It is probably obvious to you that the forward (linear) progress of the wave down the longitudinal axis of the guide is much slower than that of the free-space velocity (c) due to the zigzag path followed by the wavefront. Figure 4–11 shows that the velocity of the forward progress of the wave down the guide (called the group velocity, v_g) is given by

$$v_g = c \sin \theta \tag{4–5}$$

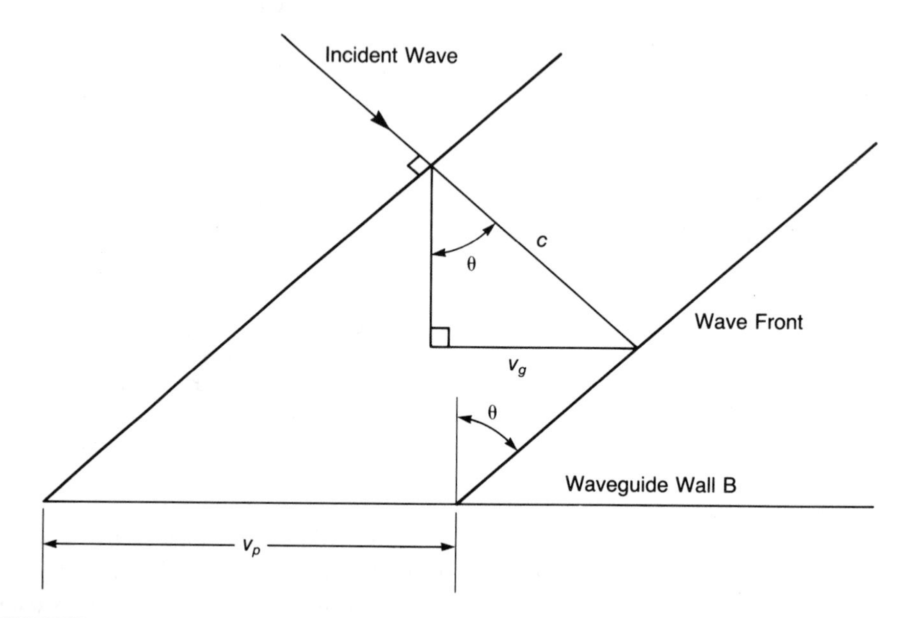

FIGURE 4–11
Group and phase velocities.

Note that the smaller the incident angle (θ), the slower the group velocity will be.

Figure 4–11 also shows that the wave undergoes a change in its phase along the B wall. Note that the vector representing this rate of change in phase is considerably longer than that representing the velocity of light (c). This does not mean, however, that the velocity exceeds that of light. Rather, the phase velocity (v_p) is simply an *apparent* velocity, not an actual velocity. Nothing exceeds the speed of light.

In chapter 1, we defined the free-space wavelength by equation (1–8) as $\lambda = v_p/f$, where v_p was the phase velocity. From figures 4–10 and 4–11, it is apparent that a rearrangement of equation (1–8) will give us an expression for the frequency of a signal within the waveguide as

$$f = \frac{v_p}{\lambda_p} \tag{4–6}$$

It should be noted that this result is consistent with the free-space equation $f = c/\lambda$. This is expected since the guide wavelength (λ_p) increases by the same amount that the phase velocity increases, so that their quotient remains the same.

CUTOFF FREQUENCY

From figure 4–10, it is apparent that

$$\lambda_p = \frac{\lambda}{\sin \theta} \tag{4–7}$$

where $\angle\theta = \angle B$.

Moreover, if we solve equation (4–4) for λ_n, and set the result equal to equation (4–1), we have

$$\frac{2A}{m} = \frac{\lambda}{\cos \theta} \tag{4–8}$$

which may be solved for cos θ giving

$$\cos \theta = \frac{m\lambda}{2A} \tag{4–9}$$

We may now use the Pythagorean identity

$$1 = \sin^2 \theta + \cos^2 \theta$$

and substitute $m\lambda/2A$ from equation (4–9) for cos θ to obtain

$$\lambda_p = \frac{\lambda}{\sqrt{1 - \left(\dfrac{m\lambda}{2A}\right)^2}} \tag{4–10}$$

Note that the radicand of this last equation can never be negative. This translates into the fact that there is some greatest value of the free-space wavelength (λ) such that

$$1 - \left(\frac{m\lambda}{2A}\right)^2 = 0$$

from which,

$$\lambda = \lambda_0 = \frac{2A}{m} \tag{4–11}$$

where λ_0 is called the *cutoff wavelength* defined as the longest wavelength (lowest frequency) that is just barely *unable* to propagate in a waveguide of given "A" and *m*.

UNIVERSAL WAVEGUIDE EQUATIONS

If we now substitute $1/\lambda_0 = m/2A$ from equation (4–11) into equation (4–10) we obtain the new equation

$$\lambda_p = \frac{\lambda}{\sqrt{1 - \left(\dfrac{\lambda}{\lambda_0}\right)^2}} \tag{4–12}$$

Equation (4–12) gives the guide wavelength (λ_p) in terms of the cutoff wavelength (λ_0) and free-space wavelength (λ).

Equation (4–12) is but one of several waveguide equations that expresses important guide parameters in terms of λ and λ_0 using the same radical term $\sqrt{1 - (\lambda/\lambda_0)^2}$ as will be shown below. These various equations are collectively called *universal waveguide equations* and assume much importance in engineering calculations.

Phase Velocity

In equation (4–6), it was shown that $f = v_p/\lambda_p$. This equation may be rewritten as $v_p = f\lambda_p$ for the phase velocity. Now, by substituting equation (4–12) for λ_p, we obtain,

$$v_p = \frac{f\lambda}{\sqrt{1 - \left(\frac{\lambda}{\lambda_0}\right)^2}} \tag{4–13}$$

Now, from the fact that $\lambda = c/f$, we substitute $c = f\lambda$ into equation (4–13) to obtain

$$v_p = \frac{c}{\sqrt{1 - \left(\frac{\lambda}{\lambda_0}\right)^2}} \tag{4–14}$$

Note that the denominator is the same as that used in equation (4–12).

It is also interesting to observe that as long as $\lambda_0 > \lambda$, the denominator of equation (4–14) is less than unity, which means $v_p > c$, as noted earlier. Remember, however, this is only an *apparent* velocity.

Group Velocity

From figure 4–10, we observe that

$$\sin\theta = \frac{\lambda}{\lambda_p} \tag{4–15}$$

from which,

$$\lambda_p = \frac{\lambda}{\sin\theta} \tag{4–16}$$

Also, from equation (4–6), $f = v_p/\lambda_p$, we may write

$$v_p = f\lambda_p \tag{4–17}$$

Combining equations (4–16) and (4–17) we write

$$v_p = \frac{f\lambda}{\sin\theta} = \frac{c}{\sin\theta} \tag{4–18}$$

Moreover, from equation (4–5), $v_g = c \sin\theta$, we may write the product $v_g v_p$ as,

$$v_g v_p = \frac{c \sin\theta}{1} \times \frac{c}{\sin\theta} = c^2 \tag{4–19}$$

That is,

$$v_g = \frac{c^2}{v_p} \tag{4–20}$$

On substituting from equation (4–14) we get

$$v_g = c\sqrt{1 - \left(\frac{\lambda}{\lambda_0}\right)^2} \tag{4–21}$$

Note the appearance of the same radical as used in connection with equation (4–12). Note, too, that equation (4–21) provides a mathematical reaffirmation of the earlier statement that $v_g < c$.

Before leaving this section, we should mention that the division of equation (4–14) by equation (4–12) gives the fraction c/λ, which is the frequency obtained in connection with equation (4–6). This further points up the fact that the frequency of the signal does not change as it traverses the waveguide.

MODES OF PROPAGATION

Depending on how the probe is arranged in the waveguide, there may either be an electric or magnetic component of the wave in the direction of propagation. We will discuss probe placement in the next section. In the meantime, we must make clear the distinction between the two possible *modes* of propagation.

The word *mode* refers to the manner in which the E and H fields arrange themselves in the waveguide. In the United States, the mode designation standard adopted for use with rectangular waveguides is shown below.

1. Transverse electric (TE) A mode is said to be transverse electric if there is *not* an E-field component of the wave in the direction of propagation.
2. Transverse magnetic (TM) A mode is said to be transverse magnetic if there is *not* an H-field component of the wave in the direction of propagation.

The two designations (TE and TM) are always followed by integral subscripts (m,n) which designate the number of half-wavelength bunches of intensity (electric for TE; magnetic for TM) between *each pair* of walls. The m specifies the number of such bunches of intensity between the A walls and the n the number between the B walls.

For example, in figure 4–8 there is *not* an electric component of the wave in the direction of propagation and, accordingly, this mode is said to be transverse electric. Moreover, there is only one bunch of half-wavelength E-field intensity between the A walls and zero bunches between the B walls. Therefore, by our designation standard, this mode of propagation is called the $TE_{1,0}$ mode.

We must remember that for TE modes, the m and n integers refer *only* to bunches of *electric* intensities, while for TM modes, the integers refer *only* to the bunches of *magnetic* intensities.

For the $TE_{m,n}$ mode, where $n \neq 0$, we must modify equation (4–11), $\lambda_0 = 2A/m$, since we now have an E-field between the B walls as well as between the A walls. Therefore, we use

$$\lambda_0 = \frac{2}{\sqrt{\left(\frac{m}{A}\right)^2 + \left(\frac{n}{B}\right)^2}} \qquad \textbf{(4–22)}$$

Note that since the magnetic (H) fields always form closed loops (unlike electric fields), it is *not* possible to have a $TM_{1,0}$ mode. Therefore, the $TM_{1,1}$ mode is the dominant transverse magnetic mode just as the $TE_{1,0}$ is the dominant transverse electric mode of propagation in rectangular waveguides. The $TM_{1,1}$ mode is shown in figure 4–12.[3]

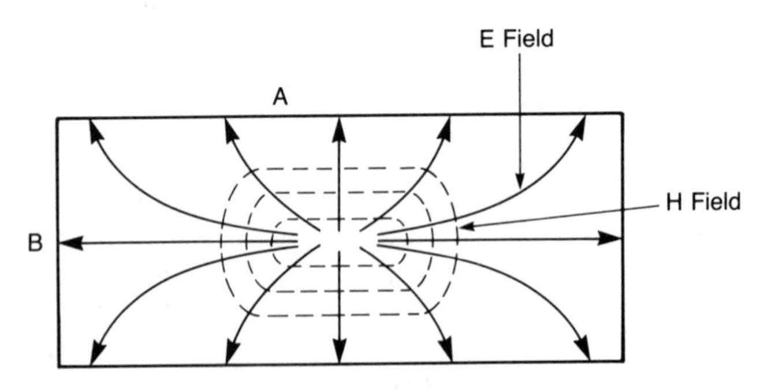

FIGURE 4–12
The $TM_{1,1}$ dominant transverse magnetic mode of propagation.

---

[3]Diagrams and representations of various other TE and TM modes of propagation may be found in standard microwave engineering data books. For example, Saad and Hansen, eds. *Microwave Engineer's Handbook*, vol. 1. Dedham, Mass: Artech House, 1971.

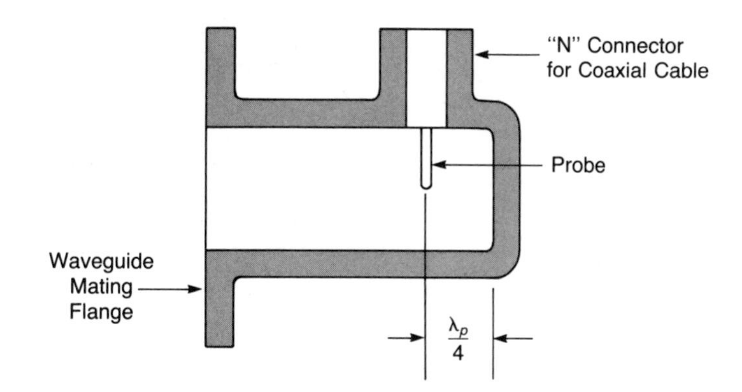

FIGURE 4–13
Waveguide adapter for $TE_{1,0}$ mode.

METHODS OF EXCITING WAVEGUIDES

Figure 4–13 shows the position of the probe inside a waveguide adapter that is designed to "launch" the $TE_{1,0}$ mode of propagation in a waveguide.

As mentioned earlier, the probe is basically an antenna that is capable of both transmitting as well as receiving. Inferentially, then, this same adapter may be used to remove energy from the waveguide as well as introduce it. Note that the probe is positioned along the center line of the guide, $\lambda_p/4$ from the closed end of the adapter, and in the plane of the E field. Placement of the probe in any other position will fail to properly launch the desired mode.

Figure 4–14 shows the probe arrangement required to excite a waveguide to the $TE_{2,0}$ mode. One method of exciting the $TM_{1,1}$ mode is shown in figure 4–15. Figure

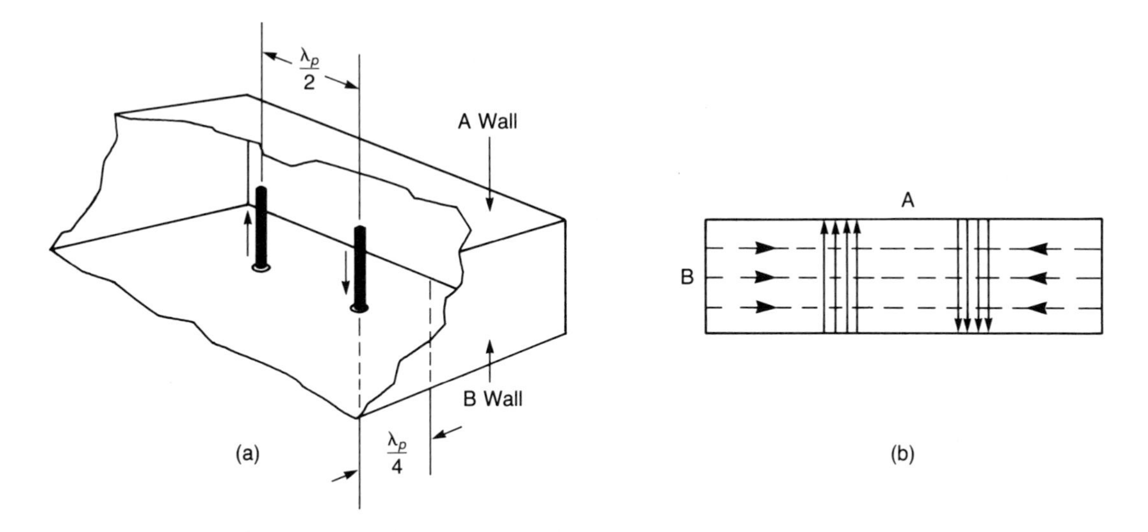

FIGURE 4–14
(a) Probe arrangement for $TE_{2,0}$ mode. (b) E-M field pattern of $TE_{2,0}$ mode.

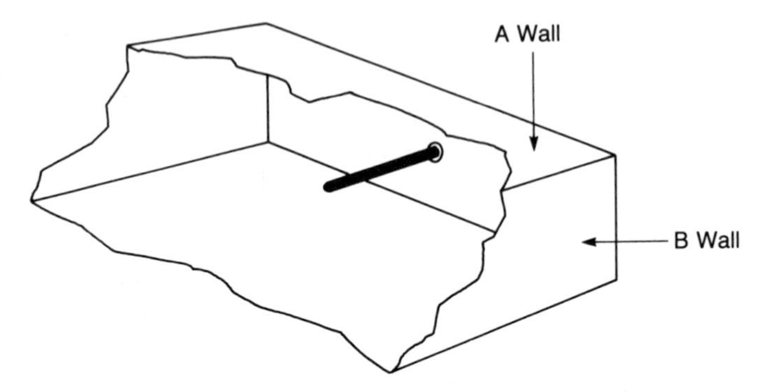

FIGURE 4–15
Exciting the $TM_{1,1}$ mode.

4–16 shows a method of exciting the $TE_{1,0}$ mode using loop coupling. With probe coupling, the axis of the probe lies entirely within the plane of the magnetic field. With loop coupling, the plane of the loop is at right angles to the plane of the magnetic field, but may be positioned at any point convenient along the closed path of the magnetic flux. The degree of coupling may be adjusted by rotating the loop axis. Probes are used primarily to couple to electric fields, while loops are used mostly for coupling to the magnetic field. Both fields are inextricably linked, however, and any moving field of one kind is always accompanied by the other.

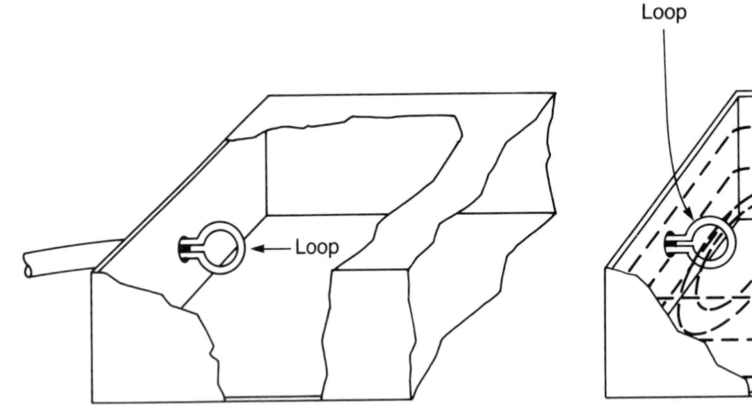

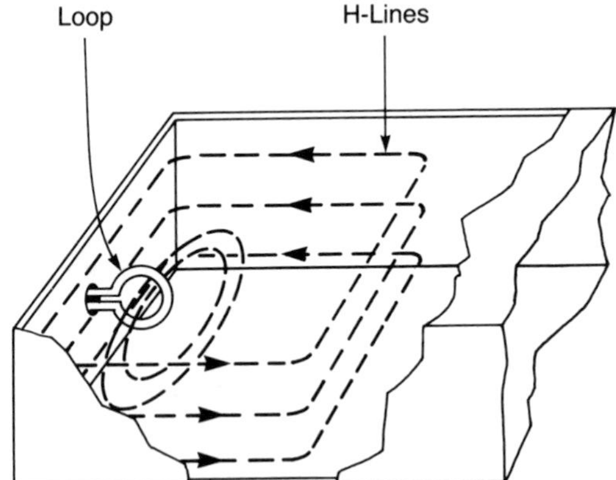

FIGURE 4–16
Loop used to excite the $TE_{1,0}$ mode.

In addition to launching the proper mode, the adapter also serves to match the impedance of the coaxial cable to that of the waveguide to prevent reflections which would result in unacceptably high values of VSWR. For example, the Hewlett-Packard model X281A waveguide adapter is designed to match the 50-ohm RG-8/U coaxial cable with an N connector to an X-band waveguide with a UG-135U flange. As will be shown later, waveguides have characteristic impedances just like transmission lines. However, since the guide impedance is frequency dependent, it is important to specify the operating range of waveguide components. The X in the H-P model number specifies that this adapter is for operation in the X-band (8.2–12.4 GHz). Incidentally, waveguide adapters are precision-machined parts designed to yield the lowest possible SWR. (The X281A has an SWR of 1.25.) As a result, these adapters, though physically small and simple in appearance, may cost over $100 each.[4]

WAVEGUIDE IMPEDANCE

In chapter 1, we developed equation (1–12) which showed that the ratio of the electric field intensity (E) to the magnetic field intensity (H) of a wave in free space always resulted in the same numerical value of 377 ohms. That is, because of the permittivity (ϵ) and permeability (μ) inherent in free space, any TEM wave traversing this medium is subject to a wave impedance (Z_w) of 377 ohms.

A similar situation exists for E-M waves (not TEM waves) in guide structures where, because of wave orientation and guide geometry, the guide impedance (Z_0) is never quite equal to 377 ohms. Moreover, the guide impedance is frequency dependent since it is a function of both λ and λ_0.

For rectangular waveguides propagating TE modes, the waveguide impedance is given by

$$Z_0 = \frac{377}{\sqrt{1 - \left(\frac{\lambda}{\lambda_0}\right)^2}} \tag{4–23}$$

Similarly, for guides propagating TM modes, the impedance is found by

$$Z_0 = 377\sqrt{1 - \left(\frac{\lambda}{\lambda_0}\right)^2} \tag{4–24}$$

Note that in each case as the aperture area of the guide increases beyond all bounds, λ_0 also becomes infinite, as seen from equation (4–11). This situation causes the radicands of equations (4–23) and (4–24) to approach unity, and Z_0 in each case approaches 377 ohms. Recall from chapter 3 that this was the reason given for flared openings of horn antennas. That is, by flaring the opening, the waveguide's impedance (Z_0) could be matched to that of the wave impedance (Z_w) of free space.

OTHER WAVEGUIDE SHAPES

The principles of waveguide operation are easiest to understand in connection with rectangular shapes, but these are not the only possible configurations. Ridged, circular, and flexible geometries also exist which have wide application and these will now be discussed briefly.

---

[4]The 1987 Hewlett-Packard catalog lists the price of an X281A waveguide adapter at $120.

Ridged Waveguides

Ridged waveguide has the property of allowing the $TE_{1,0}$ mode to propagate over a broader range of frequencies. That is, the cutoff frequency is lowered while the $TE_{2,0}$ and higher modes have their cutoff frequencies elevated. Thus, the ridged guide functions as a broadband device. However, the addition of the ridge increases the linear attenuation and lessens the power handling capability of the guide. Nonetheless, for a given size, the ridged guide possesses greater bandwidth than a simple rectangular guide of comparable dimensions.

Since the cutoff frequency is lowered (λ_0 increases), the net effect of the ridge must be to extend the range of its impedance. From equation (4–23) it is apparent that as λ_0 approaches λ, Z_0 becomes infinite. One major application, then, of ridged waveguide is to act as an impedance matching device since its impedance is easily changed by tapering its cross-section gradually. Single and double-ridged waveguide sections are illustrated in figure 4–17.

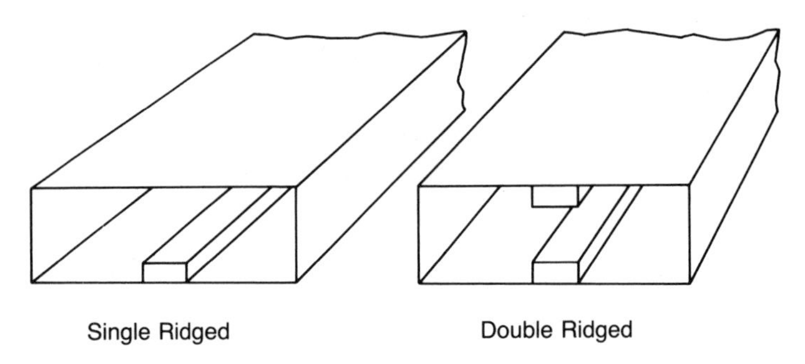

FIGURE 4–17
Ridged waveguide.

Circular Waveguide

Figure 4–18 illustrates one obvious disadvantage of the circular waveguide. For a given frequency, the waveguide is physically larger and heavier than its rectangular counterpart.

Despite the size limitation, the circular guide has the advantage of being the only practical shape in those situations (e.g., radar antennas) where rotation of the guide structure is required. Moreover, the circular guide offers greater power handling capability than a comparably-sized rectangular guide. The circular guide also exhibits less attenuation for a given cutoff frequency.

Since there are no A or B walls per se, the mode designation system must be changed somewhat. To this end, the integer m will now represent the number of *full-wave* intensity variations around the circumference. The integer n will denote the number of *half-wave* intensity changes *radially*. As an example, the dominant mode $TE_{1,1}$ and the symmetrical mode $TM_{0,1}$ are shown in figure 4–19.

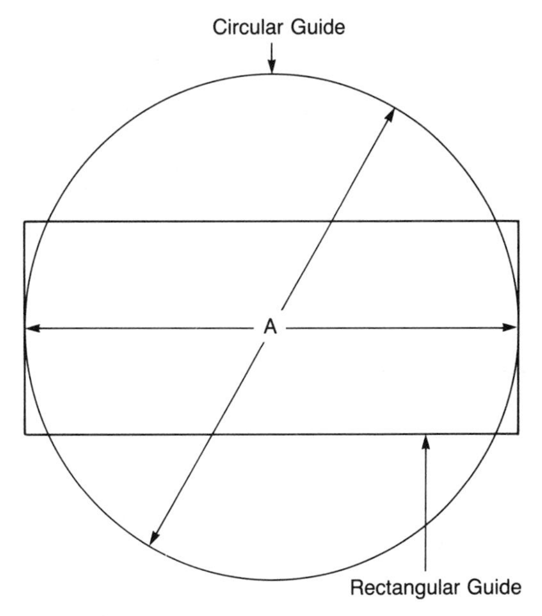

FIGURE 4–18
Comparison of circular and rectangular waveguide sizes.

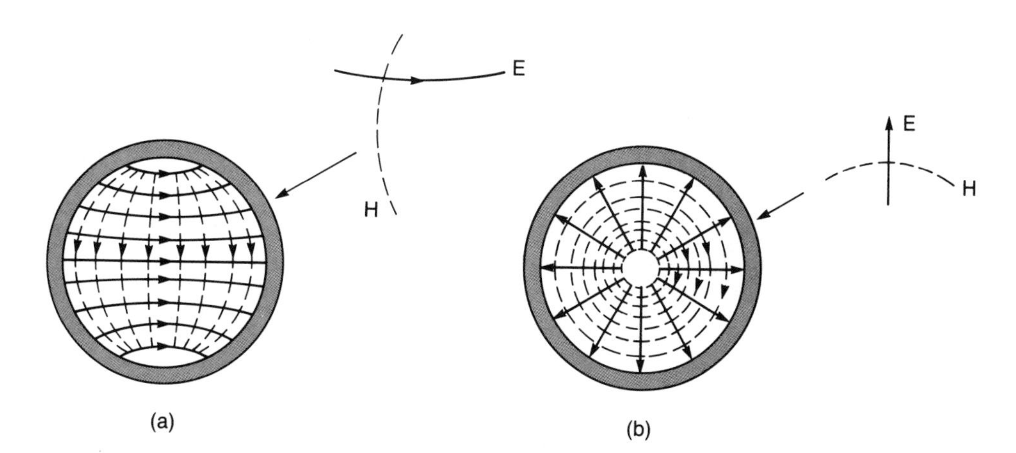

FIGURE 4–19
(a) Dominant mode $TE_{1,1}$. (b) Symmetrical mode $TM_{0,1}$.

Because of interior surface irregularities of the guide, the intended polarization of a particular signal may change. An example of the consequence of such a change involves a receiving antenna that is only minimally excited (or fails to be excited) due to inconsistent polarization. In this case, the symmetrical mode may, indeed, be the only answer to the problem.

All of the previous equations governing waveguide behavior apply equally to the circular waveguide with the exception of equation (4–11) for the cutoff wavelength. A new equation involving the zeros of Bessel functions[5] is required. The equation for the cutoff wavelength in circular waveguides is given by

$$\lambda_0 = \frac{2\pi r}{B_{(m,n)}} \qquad \textbf{(4–25)}$$

where r is the guide radius, and $B_{(m,n)}$ is the appropriate Bessel function solution corresponding to the particular m,n mode being propagated. For example, the Bessel solution for mode $TM_{0,2}$ is $B_{(m,n)} = 5.52$. If a circular guide has a radius of, say, 4 cm, then the cutoff wavelength for the $TM_{0,2}$ mode is

$$\lambda_0 = \frac{2\pi 4}{5.52} = 4.6 \text{ cm}$$

Values of Bessel function solutions $B_{(m,n)}$ for selected modes are shown in table 4–1.

TABLE 4–1
Partial listing of Bessel function solutions.

Mode	$B_{(m,n)}$
$TE_{0,1}$	3.83
(Dominant) $TE_{1,1}$	1.84
$TE_{2,1}$	3.05
$TE_{0,2}$	7.02
$TE_{1,2}$	5.33
$TE_{2,2}$	6.71
(Symmetrical) $TM_{0,1}$	2.40
$TM_{1,1}$	3.83
$TM_{2,1}$	5.14
$TM_{0,2}$	5.52
$TM_{1,2}$	7.02
$TM_{2,2}$	8.42

Flexible Waveguide

Flexible waveguide is used where vibration between adjacent fixed sections is a part of normal equipment operation. Such pliant waveguide sections are also used to take up lateral slack and compensate for misalignment between sections of rigid guide. Flexible waveguide has an essentially elliptical cross-section, which is due to the corrugations in

[5]Bessel functions of the first kind are the solutions to certain differential equations. For a discussion of Bessel functions, see Wylie, C. *Advanced Engineering Mathematics*, 3rd ed. New York: McGraw-Hill, 1966. Also see, Kustner and Kastner, Eds. *The VNR Concise Encyclopedia of Mathematics*. New York: Van Nostrand Reinhold, 1975.

the material (usually brass or aluminum) which allows the flexing to occur. The ends of the pliant section are fitted with appropriate flanges for mating with rectangular waveguide. Signal attenuation is considerably greater in flexible guide. A section through a flexible waveguide is shown in figure 4–20(a).

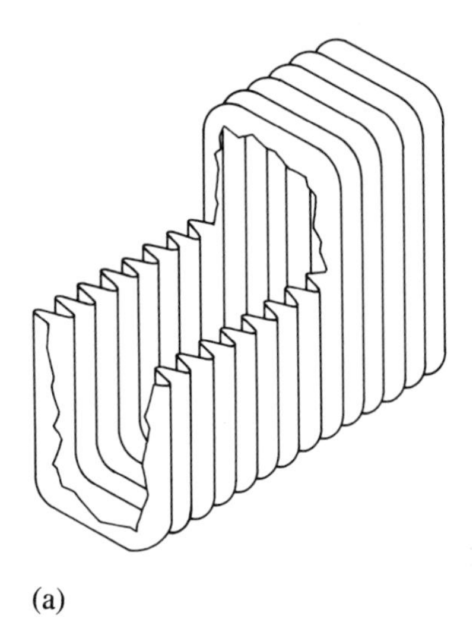

(a)

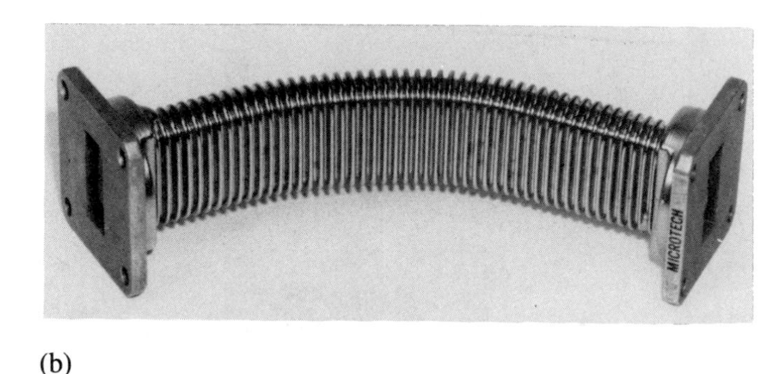

(b)

FIGURE 4–20
(a) Flexible waveguide section.
(b) Photo of flexible waveguide from *Microwave Journal* 29, no. 9 (September 1986): 165. Used with permission of Microtech, Inc.

WAVEGUIDE-RELATED STRUCTURES

Now that we have some feeling for waveguide principles, we may put our knowledge to further use in examining the properties of useful auxiliary components and structures. Such structures are usually formed by the combination of two or more waveguide sections interconnected in a particular way. We shall begin our study of related structures by examining a simple modified waveguide flange.

The Choke Flange

It was shown in chapter 1 that unintentional reflected microwave energy along a transmission line represented a loss of energy. The higher the attendant SWR value, the greater the loss. The same is true in waveguides where reflections are even more likely to occur since, with higher frequencies, even a modest physical misalignment between guide sections presents comparatively large physical discontinuities from which reflections might occur.

The point at which reflections are most likely to occur is at the mating surfaces where flanges join two sections of guide together. One possible solution to this problem would be to machine the surfaces so precisely that no physical misalignment is possible.

In practical terms, however, such labor-intensive operations are not cost-effective. Even small sections of waveguide, machined within standard production-line tolerances, are expensive. For example, a simple Hewlett-Packard waveguide adapter (such as model MX292B) of only a few inches in length is priced at \$270.[6] It is apparent, then, that an alternate solution must be pursued. Happily, the answer to the problem adds little to the cost of the device, and actually makes precision alignment unnecessary.

Consider the two mating flanges shown in figure 4–21. The flange of the lower section of waveguide is a plain flange, but the upper flange has a circular, L-shaped slot cut into its surface whose *total* length is $\lambda_p/2$. The significance of this slot dimension may be appreciated by looking at the effect of such a stretched-out slot length, as shown in figure 4–22.

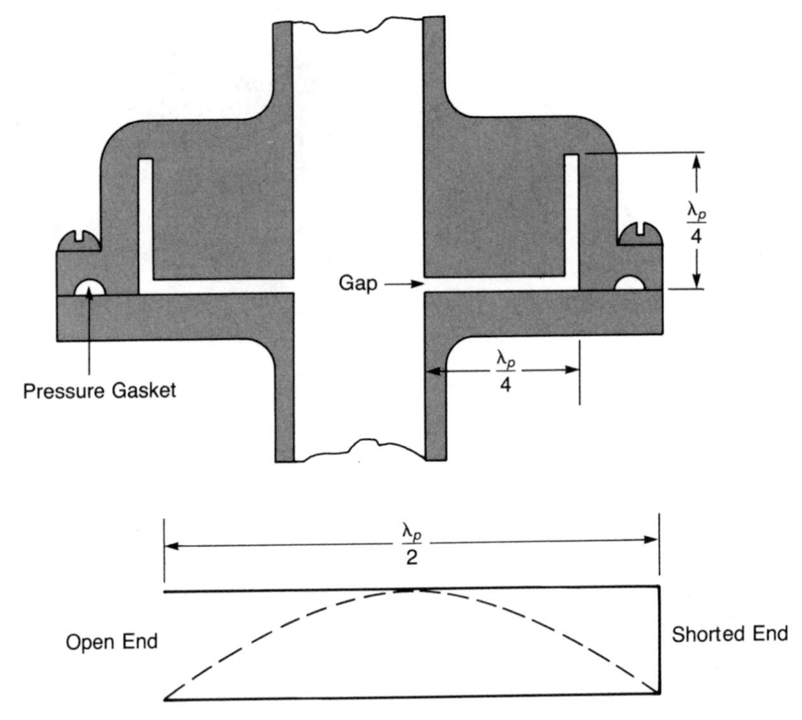

FIGURE 4–21
The choke flange.

FIGURE 4–22
Reflected short of the choke flange.

As may be seen from figure 4–22, the voltage distribution along the total slot length is such that an effective short circuit appears across the flange gap, which provides an electrical continuity that would otherwise be missing. Observe that the gap is L-shaped in the interest of cutting down the total flange dimension in a given direction, but has no other particular functional significance. The choke flange in end view is shown in figure 4–23.

[6]From Hewlett-Packard 1987 catalog, p. 564.

FIGURE 4–23
End view of choke flange.

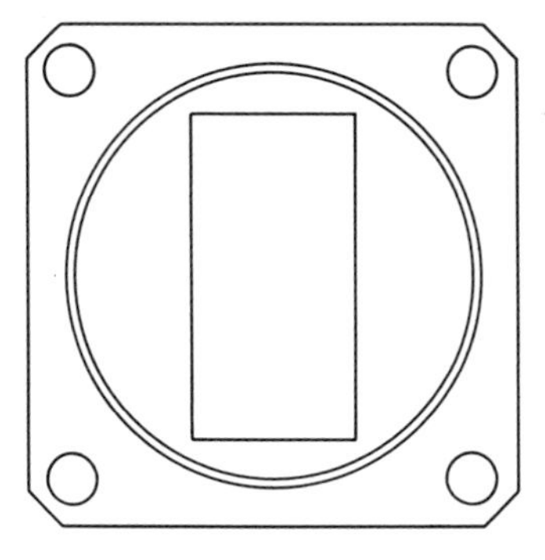

The Rotary Joint

As mentioned earlier, there are situations wherein waveguide sections are required to rotate relative to one another. One type of radar antenna, for example, must rotate in order to scan. Obviously, then, the circular waveguide cross-section is the only possible shape capable of fulfilling this requirement. As shown in figure 4–24, the rotary joint is basically one length of circular guide slipped within another circular section and terminated in a rectangular waveguide at either end. Section A rotates relative to section B in the illustration.

FIGURE 4–24
The rotary joint.

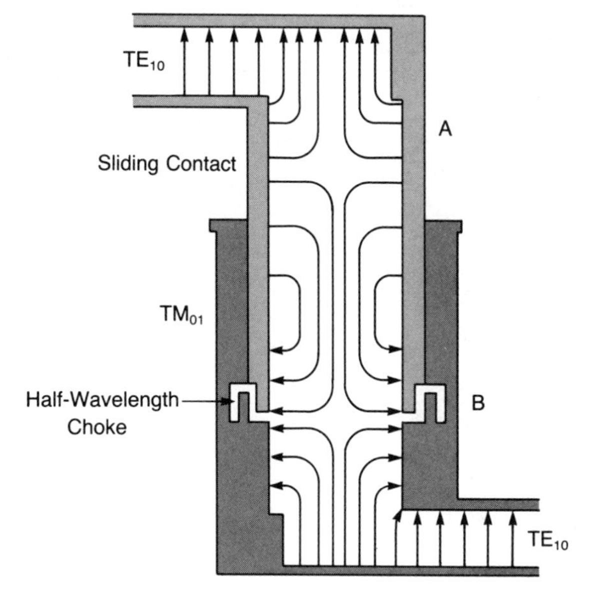

As may be seen from figure 4–24, the $TE_{1,0}$ dominant mode within the rectangular guide changes to the symmetrical $TM_{0,1}$ within the rotary joint. Consequently, the initial wave polarization is maintained between input and output sections of the rotary joint.

FIGURE 4–25
In-line rotary joint. (Courtesy Maury Microwave Corp.)

An in-line rotary joint is shown in figure 4–25. The joint is designed for use at 18–26.5 GHz and has an insertion loss of 0.6 dB, and a VSWR of 1.35 maximum.

The Hybrid Tee

Before we may gain an appreciation of the hybrid tee, we must have an understanding of the distribution of electric field intensities of the two conventional tee sections. Such distributions are shown in figure 4–26 where the $TE_{1,0}$ mode of wave propagation is assumed. Part (a) of figure 4–26 shows a cross-section in the E-field plane, while part (b) is a section in the H-field plane (orthogonal to the E-field plane).

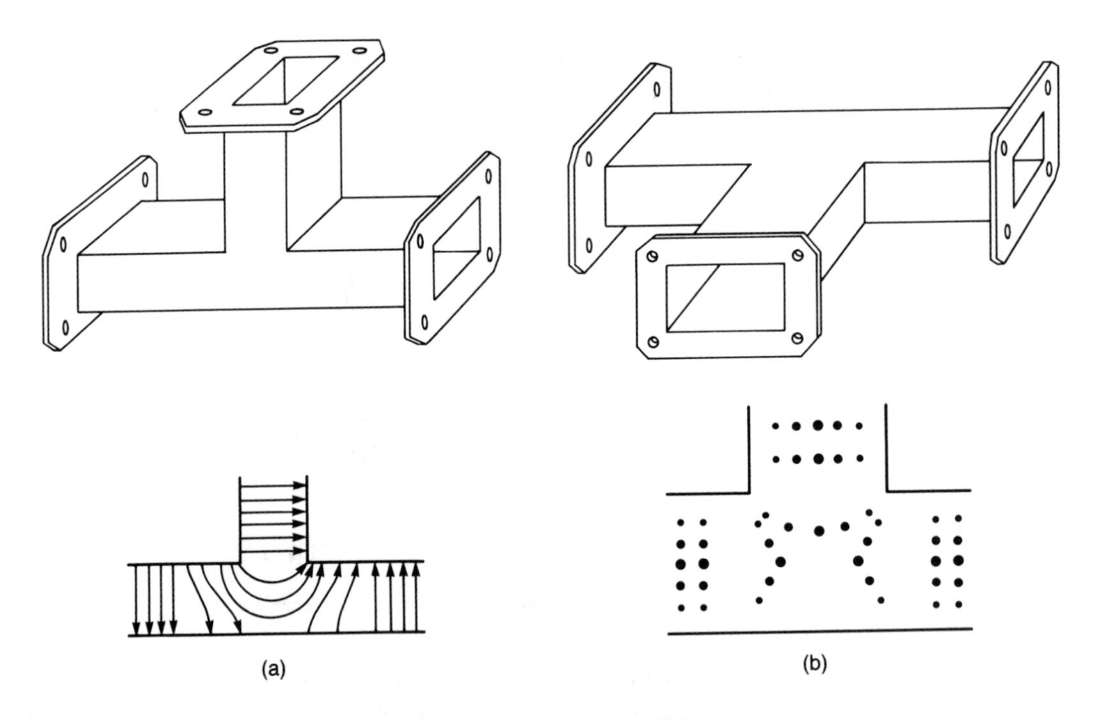

FIGURE 4–26
(a) The E-plane or "series" tee. (b) The H-plane or "shunt" tee.

FIGURE 4–27
The hybrid tee.

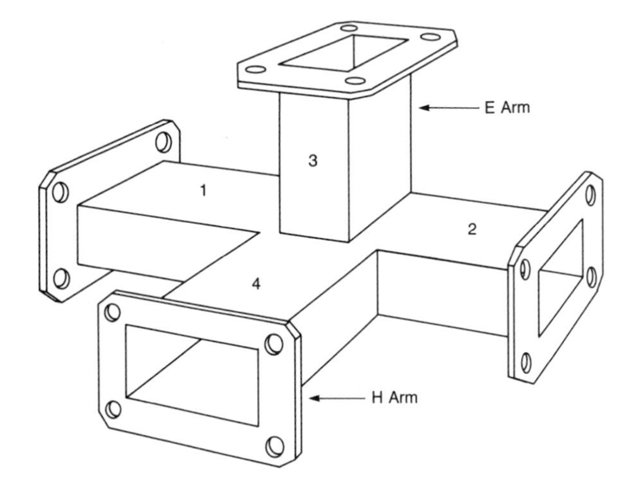

Figure 4–26 (b) shows a cross-section through the electric lines in a plane parallel to the H-plane. The diameter of the dots represents the magnitude of the electric intensity at that point.

The hybrid tee shown in figure 4–27 combines an E-plane tee and an H-plane tee in such a way that arm 3 will communicate with arms 1 and 2, but not 4. Conversely, arm 4 will communicate with arms 1 and 2, but not 3.

An extensive application of the magic tee is in microwave receivers where, as shown in figure 4–28, two separate signal sources are required to feed a mixer without interfering with each other. Careful impedance termination and matching are required to prevent unwanted reflections from occurring. Such matching is often effected with the use of iris and post intrusions into the tee itself.

An iris is simply a thin partition which reduces the waveguide cross-section geometry by some fixed amount at a given point. The iris may act capacitively or induc-

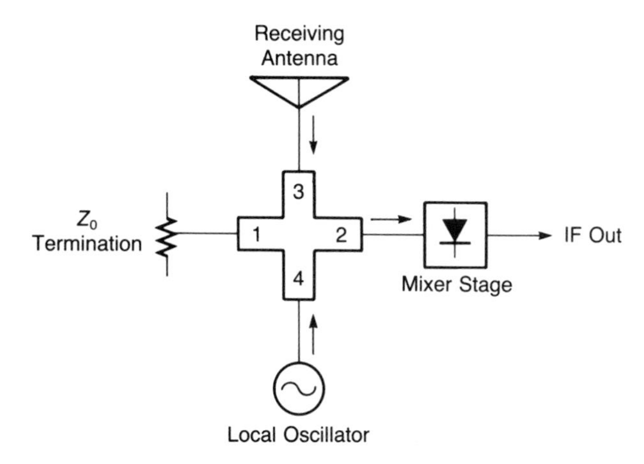

FIGURE 4–28
The hybrid tee used in a microwave receiver.

tively. A capacitive iris is shown in figure 4–29(a) and an inductive iris is illustrated in figure 4–29(b). In part (a) of the figure, the A walls are much closer together than at any other point, and so the capacitance is greater at this point. In part (b), the partition is in the middle of the maximum E field and will have a current induced in it causing a magnetic field at that point which will act inductively.

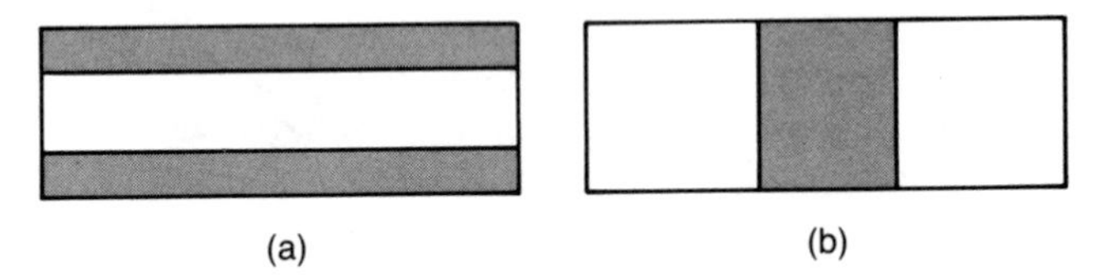

FIGURE 4–29
(a) A capacitive iris. (b) An inductive iris.

An iris is fixed in position and size. Consequently, its effect is fixed. However, if provision is made for lowering or retracting a screw into the guide's cross-section, variable reactance effects are possible. Figure 4–30 shows an adjustable post mounted on a waveguide section in such a way that it will intrude into the E field of a properly-launched $TE_{1,0}$ mode signal.

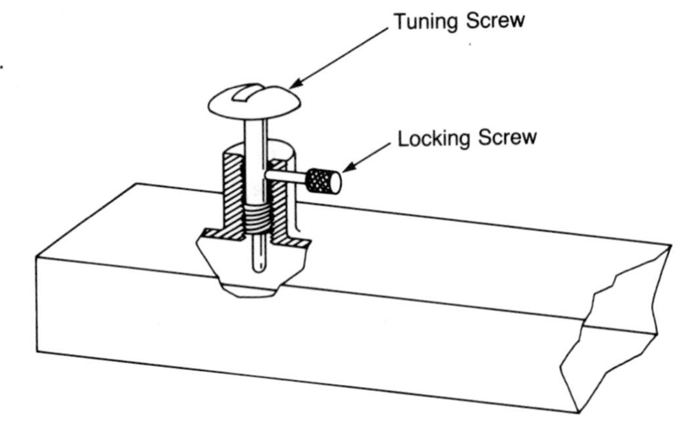

FIGURE 4–30
An adjustable post.

The Hybrid Ring

A device similar in function to the hybrid tee is the hybrid ring (not to be confused with Wagner's ring), sometimes called the "rat race." Basically, the hybrid ring is a rectangular waveguide bent in the E plane to form a complete ring. Four arms project from the ring, which are spaced as shown in figure 4–31.

The median circumference of the ring is $3\lambda_p/2$ and the arms are spaced such that either an in-phase or out-of-phase component from a given arm will either aid or oppose the signal from another arm. Consequently, any one arm will couple to the other two, but not to the fourth. For example, arm 1 will couple with arms 2 and 4, but not arm 3. Also, arm 3 will communicate with arms 2 and 4, but not arm 1.

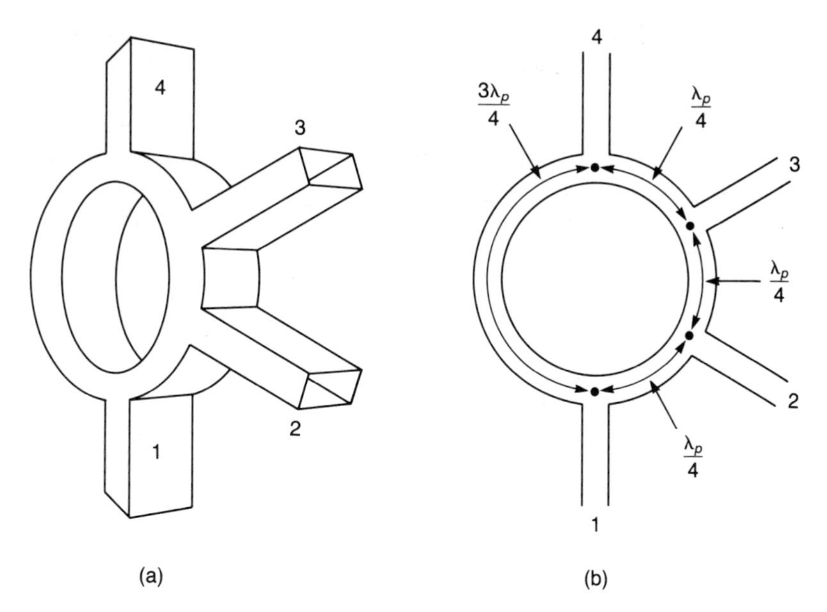

FIGURE 4–31
The hybrid ring: (a) physical appearance and (b) arm spacing.

The logic of this operation will become apparent if we consider an example. If a signal is introduced into arm 1, it will divide equally with half moving clockwise $5\lambda_p/4$ to arm 2, while the other part travels counterclockwise $\lambda_p/4$ to arm 2. The result is that the two signal halves will arrive at arm 2 in phase, as shown in figure 4–32, reinforcing each other and producing an output signal at port 2.

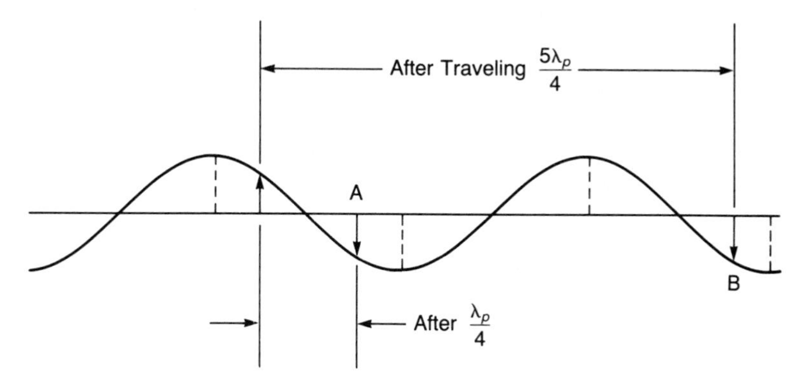

FIGURE 4–32
Reinforcement of signals 360 degrees apart. A and B are in phase.

On the other hand, any signal entering arm 1 will split equally, with half traveling clockwise a distance of λ_p to arm 3, and the other half moving $\lambda_p/2$ counterclockwise to this same arm. Therefore, as shown in figure 4–33, the two signals will arrive 180 degrees out of phase and will cancel each other. There will be no output from port 3. It remains as an exercise for the student to show that arm 3 will communicate with arms 2 and 4, but not 1.

The hybrid ring may be used interchangeably with the hybrid tee, but does not require impedance matching.

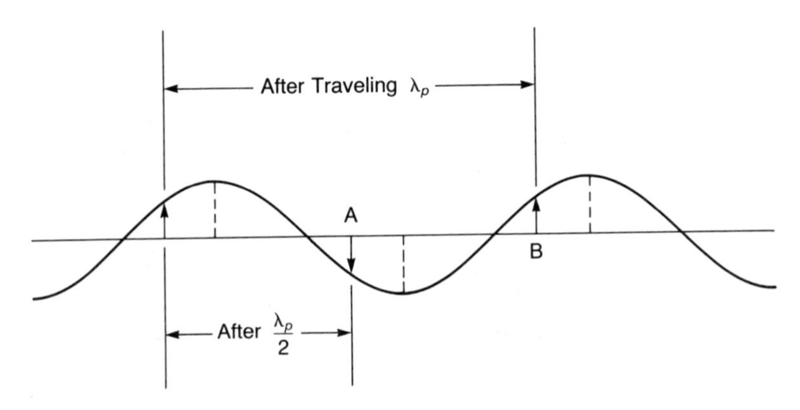

FIGURE 4–33
Cancellation of signals 180 degrees apart. A and B are 180 degrees out of phase.

FIGURE 4–34
Hewlett-Packard model X752 X-band directional coupler.

Directional Couplers

A directional coupler, shown in figure 4–34,[7] is actually two waveguides joined along their A walls and coupled with exactly-spaced holes through their connected sides only.

The action of the coupler is such that a fraction of the forward energy passing through the *main arm* will be coupled to the *auxiliary arm*. On the other hand, any reflected energy traveling back along the main arm will not be coupled to the auxiliary arm. The result is a highly-directional means of coupling signal energy between a primary system and a secondary microwave circuit. The theory of operation of the coupler is illustrated in figure 4–35.

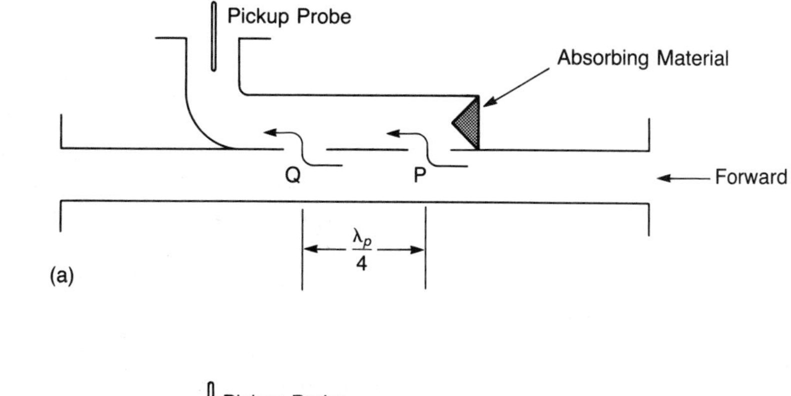

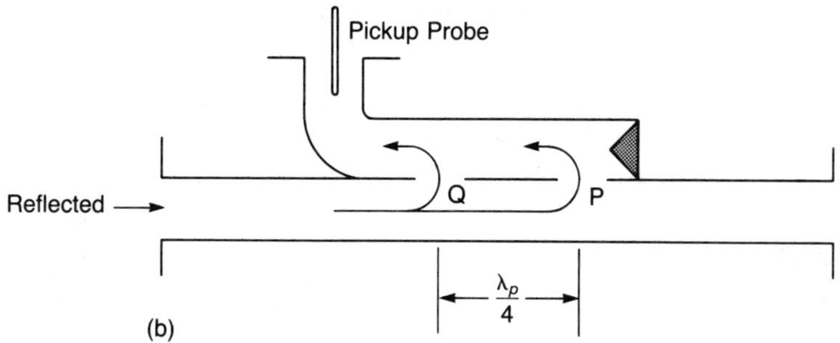

FIGURE 4–35
Operation of the directional coupler: (a) forward operation and (b) reverse operation.

As illustrated in figure 4–35(a), forward energy traveling to the right passes through holes P and Q, and then travels the same distance to the probe located in the auxiliary arm. These signals are in phase and reinforce each other. In figure 4–35(b), the reflected energy also passes through holes P and Q, but that part of the energy

[7]The student may find it interesting to note that the 1987 Hewlett-Packard catalog lists the price of the X752 coupler at $680.

passing through hole P must turn around and travel back again a distance of $\lambda_p/4$. Consequently, the total distance traveled by this part of the reflected signal is $\lambda_p/2$. The result is that cancellation will occur, and no signal will reach the probe.

Note that any energy traveling to the left in the auxiliary arm is dissipated by a microwave-absorbing material, and thus reflections from that end of the arm are prevented. The actual shape of the absorbing material, which is aquadag,[8] has a pyramidal configuration to prevent reflection, which would certainly occur had the shape been a flat surface incapable of gradually dissipating the incident energy. The pyramidal shape is a common geometry employed in many microwave devices, such as adjustable shorts and certain attenuators. Since microwave signals possess the same properties as light, flat surfaces act much in the same way as mirrors, and must be avoided.

Directional couplers have many important ratings, some of which will now be discussed.

Coupling The coupling rating refers to the amount of energy coupled from the main arm to the auxiliary arm. For example, the Hewlett-Packard model X752 shown in figure 4–34 is available with coupling specifications of 3, 10, and 20 dB. A 10 dB coupler, for example, would couple 1/10th the forward energy to the auxiliary arm.

As a further example, suppose 3 mW of power was detected and measured at the output of the auxiliary arm of a coupler rated at 10 dB. We may find the input power to the main arm by solving

$$\text{dB (coupling)} = -10 \log \frac{P_{\text{aux}}}{P_i} \qquad \textbf{(4–26)}$$

where P_{aux} = power out of auxiliary arm
P_i = power in to main arm

Therefore, $-10 \text{ dB} = 10 \log 3 \text{ mW}/P_i$. On solving the equation, $P_i = 30$ mW.

Directivity Directivity is a measure of the degree of isolation between the main and auxiliary arms when the direction of energy flow is reversed. Directivity (d) is defined as

$$d(\text{dB}) = 10 \log \frac{P_{\text{aux(fwd)}}}{P_{\text{aux(rev)}}} \qquad \textbf{(4–27)}$$

where $P_{\text{aux(fwd)}}$ = power detected in auxiliary arm when a given power is sent along the main arm in the forward direction.

$P_{\text{aux(rev)}}$ = power detected in auxiliary arm when the same power is sent along the main arm in the reverse direction.

---

[8]Aquadag is a trademark of Acheson Colloids Co., and is a dispersion of colloidal graphite in water. Aquadag is highly conductive.

The higher the directivity (d) in dBs, the greater the isolation and the better the coupler. For example, the Hewlett-Packard X752 series couplers have a directivity of 40 dB which indicates that the ratio of forward to reverse power is 10,000 to 1.

Insertion Loss Insertion loss is the power lost in the main arm due to dissipation. Expressed in dBs, the lower this value, the better the coupler. For example, an insertion loss of -0.1 dB means that the ratio of P_o to P_i in the main arm is 0.98 to 1.0. In other words, 2% of the input power is lost in the main arm through dissipation.

Frequency The frequency is the range over which the coupler is intended to operate. The Hewlett-Packard X752 is an X-band coupler intended to operate over the range of 8.2 to 12.4 GHz. As the frequency departs from nominal, the P-to-Q hole spacing of $\lambda_p/4$ shown in figure 4–35 is no longer valid, and measurements tend to be less accurate.

Impedance The characteristic impedance of the coupler must match that of the other devices used in the system in order to insure maximum power transfer. The most common impedance in microwave work is 50 ohms.

Input VSWR Input VSWR is a measure of the degree of match offered by the input port of the coupler. Typically, this value is close to 1.05.

Power Power is the power handling capability of the coupler expressed in watts CW or kW peak. Exceeding the power rating will irreparably damage the coupler.

Cavity Resonator

A tank circuit using conventional discrete inductors and capacitors is not a viable idea at microwave frequencies due to the errant reactive properties of these devices. Moreover, the diminutive wavelengths involved at these frequencies often equal the component's physical size. These factors create all sorts of problems of reflection and radiation which tend to degenerate intended circuit performance.

Consider, now, the single turn of wire with its distributed capacitance, as shown in figure 4–36. At microwave frequencies, it is entirely possible for the single loop, together with its inherent capacitance, to exhibit the property of resonance at some frequency.

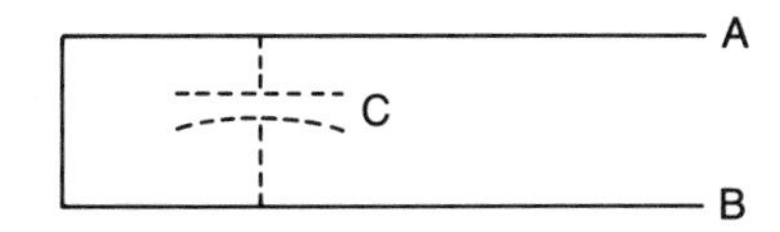

FIGURE 4–36
L-C tank circuit formed from a single loop of wire.

If we were to rotate the L-C tank circuit shown in figure 4–36 about an imaginary axis through A-B, we will have generated a surface of revolution called a cylinder. This cylinder would be made up of an infinite number of elemental single-wire loops connected in parallel, and a similar amount of distributed capacitance which is also shunt wired.

Note the effect of such parallel connections on the value of the resonant frequency as given by

$$f_r = \frac{1}{2\pi\sqrt{LC}} \tag{4–28}$$

As more and more turns of wire are connected, the value of the total inductance (L) decreases while the total capacitance (C) increases proportionately. Consequently, the resonant frequency (f_r) remains the same as it was originally for a single loop. The only new condition created, however, is that the total resistance (R) of the parallel circuit has decreased. Therefore, the Q of the circuit has increased sharply. A direct consequence of this is that the bandwidth of our "resonant tank cylinder" has become extremely narrow and, therefore, highly selective.[9]

If we impose the further size constraint on our resonant cylinder that its diameter be exactly $\lambda/2$ at the value of f_r, then, by the same reasoning used earlier in connection with rectangular waveguides, a dense concentration of the electric field will be situated in the exact center of the cylinder, as shown in figure 4–37. Note that the resonant cavity in cross-section looks very much like the cross-section of a rectangular waveguide, including the voltage distribution.

FIGURE 4–37
E-field distribution in the resonant tank cylinder.

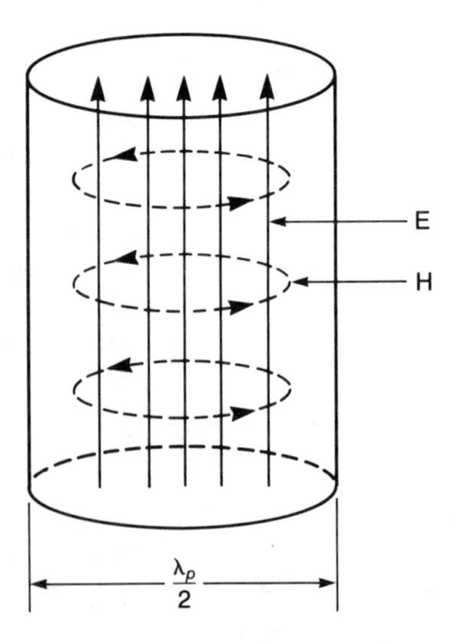

What we have been calling a resonant tank cylinder up to this point is, in fact, technically known as a *resonant cavity*. Its essential purpose is to replace the conventional L-C tank circuits made from discrete components.

[9]See Appendix F for a discussion of bandwidth and circuit Q.

It may be apparent to you that any such cavity will resonate at an infinite number of frequencies, all harmonically related to the fundamental. This is due to the fact that the λ/2 diameter of the cavity at the fundamental frequency (f) is λ at $2f$. This will have the effect of creating two bunches of E-field intensity along the cavity's diameter, but resonance will still occur. The frequency of this new harmonic-resonance value will depend on which harmonically-related stimulus frequency excites the cavity.

Unfortunately, in those situations where the cavity is excited by pulses rich in harmonics, the cavity output will also be in the form of pulses rather than a pure sinusoid. This is because the cavity produces oscillations by virtue of the flywheel effect. All harmonics of the pulse will thus compete in an attempt to excite the cavity, with the result that the output will still be in the form of pulses.

To overcome this problem, the cavity is not configured as a regular, symmetrical shape. Instead, the practical cavity has an irregular geometry which insures that the various oscillating frequencies are not harmonically related. One such shape shown in figure 4–38 is called a *reentrant* cavity because one of its surfaces reenters the cavity itself.

By inserting an adjustable plunger into the cavity, it is possible to vary the cavity size and hence the frequency of resonance. A plunger-tuned cavity is shown in figure 4–39. As the plunger enters the cavity, its size is reduced and the frequency increases. The frequency is lowered as the plunger is withdrawn.

One typical application of the adjustable cavity is as an absorption-type frequency meter. The Hewlett-Packard model P532A is shown in figure 4–40.

FIGURE 4–38
Cross-sectional shape of one type of reentrant cavity.

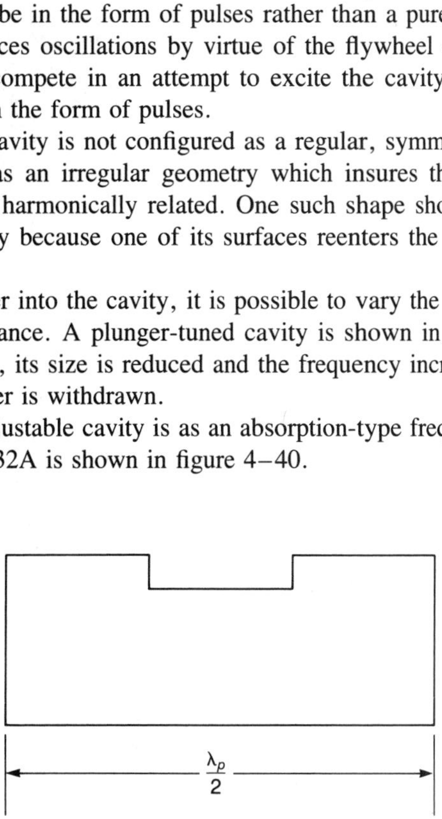

$\frac{\lambda_p}{2}$

FIGURE 4–39
A plunger-tuned resonant cavity.

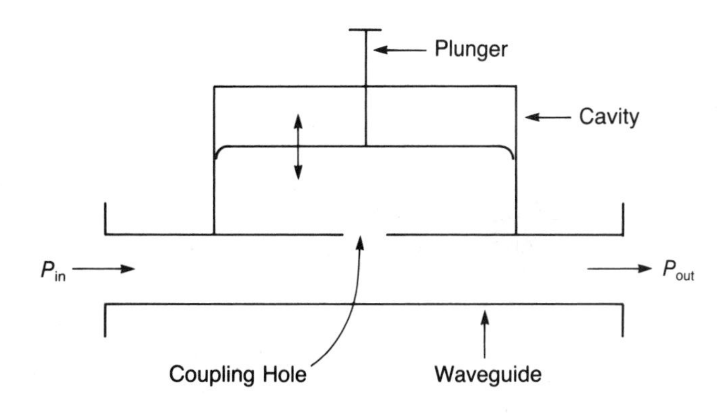

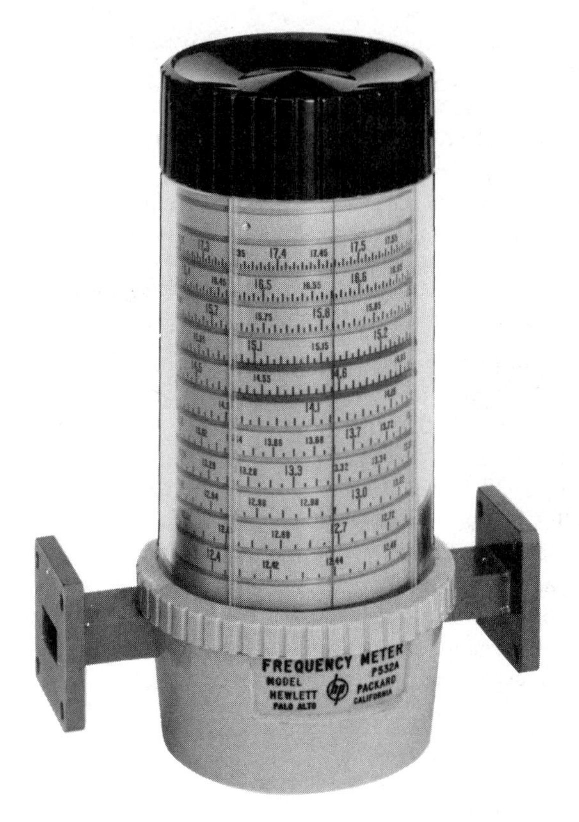

FIGURE 4–40
The Hewlett-Packard model P532A frequency meter (12.4–18.0 GHz).

A calibrated plunger allows for the direct reading of the resonant frequency to which the cavity has been adjusted. A 1 dB (20%) drop in power between input and output ports will be detected at the resonant frequency set by the meter scale. At off-resonance frequency settings, nearly 100% of the input power is transmitted. The 1987 Hewlett-Packard catalog price of the P532A is $1,155.

FERRITE DEVICES

In microwave work, it is often necessary to isolate a signal source from a load whose impedance variations may cause changes in the generator's output power or frequency (frequency pulling). Such load variations cause unwanted signal reflections to travel back to the generator output at each instant that $Z_0 \neq Z_L$. The result is an overall system degradation with significant fluctuations in output power from the microwave source.

To prevent unwanted reflections from entering the signal source, a number of non-reciprocal (one-way) devices have been developed that make use of a variety of materials called ferrites. Before discussing specific devices, however, we must pause to discuss ferrite behavior and an electromagnetic wave phenomenon known as Faraday rotation.

Faraday Rotation

Ferrites are multiple-oxide compounds of ferric oxide (Fe_2O_3) and another metallic oxide, commonly zinc, manganese, nickel, aluminum, or comparable ferromagnetic substances such as yttrium-iron-garnet ("YIG"), $Y_3Fe_2(FeO_4)_3$.[10] These materials are formed by sintering a mixture of the various compounds and pressing them into the desired shape. When completed, ferrites exhibit interesting magnetic dipole-moment properties particularly useful in microwave work since they are transparent to microwave energy and are electrical insulators. Yet their magnetic properties allow them to interact with the magnetic fields of a properly polarized wave.

The magnetic properties of ferrites manifest themselves as a consequence of the magnetic dipole moment associated with a spinning electron. Such a spinning electron has essentially the same gyroscopic properties as a spinning top subject to unbalanced forces, and may be seen to precess[11] about its axis in much the same way. The action of a spinning electron within a ferrite material subjected to a uniform axial DC (static) magnetic field (B_0) is shown in figure 4–41.

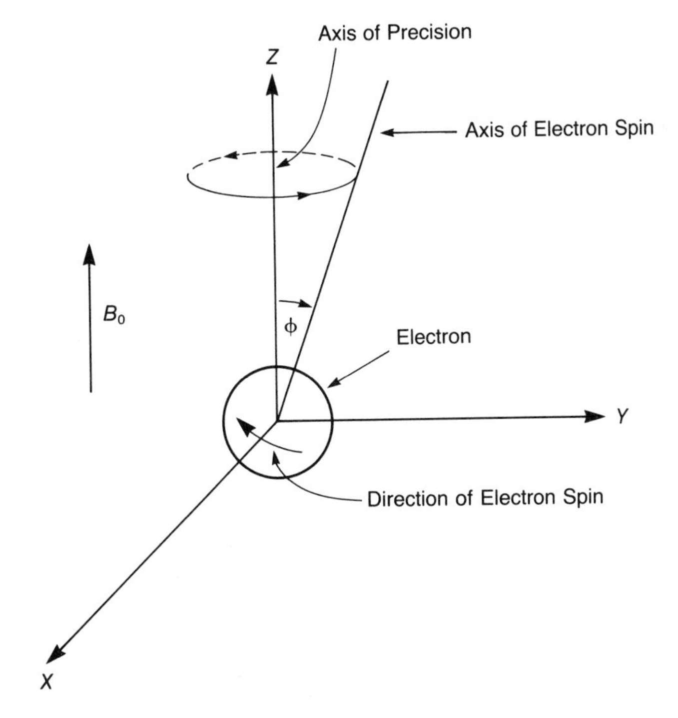

FIGURE 4–41
Free precession of a spinning electron in static DC magnetic field.

---

[10] Hawley, G. *The Condensed Chemical Dictionary*, 9th Ed. New York: Van Nostrand, 1977. See also Gray & Isaacs, eds. *New Dictionary of Physics*. London: Longman Group Ltd., 1975.

[11] When the axis of spin of a rotating body shifts, the resultant rotation is called precession, and the axis about which the precession occurs is called the precession axis. See Morgan, J. *Introduction to University Physics*, vol. I. Boston: Allyn & Bacon, 1963.

If a circularly-polarized AC magnetic field (B_1) traveling along the Z-axis in a plane normal to B_0 is superimposed on the DC field, the electron will undergo a forced precession which will incline its spin axis to an amount exceeding θ in figure 4–41. The result will be a rotation of the plane of polarization of the wave's E field propagating through the ferrite. This phenomenon is referred to as Faraday rotation.[12]

The magnitude (in degrees) of rotation of the polarization plane is a function of the size and shape of the ferrite as well as the strength of the static axial field.

Faraday Isolator

An isolator (or "uniline") is a two-port, unilateral microwave device which makes use of ferrite magnetic properties. As shown in figure 4–42, one version of the isolator consists of a ferrite rod, tapered at both ends to prevent wave reflection, located along

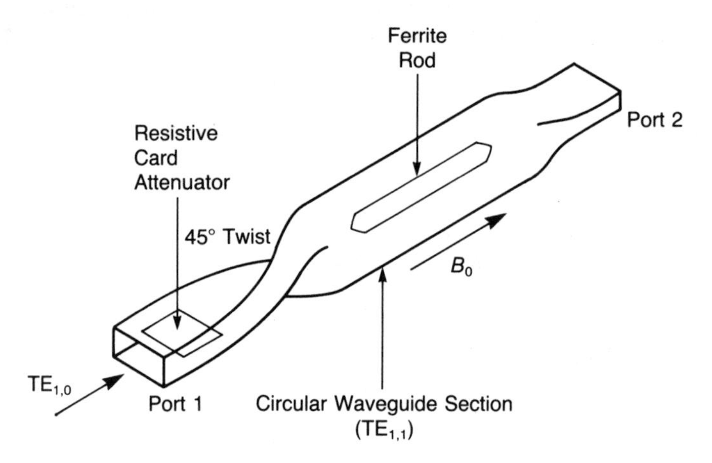

FIGURE 4–42
A Faraday rotation isolator.

the center line of a circular section of waveguide. Port 1 of the isolator is a 45-degree twist transition waveguide section and port 2 is another rectangular-to-circular guide section. The axial magnetic field (B_0) is supplied by an external magnet, which is not shown in the illustration.

A wave entering port 1 in the $TE_{1,0}$ mode has its plane of polarization rotated counterclockwise by 45 degrees in passing through the twisted section. Upon entering the circular section, the required circularly-polarized AC magnetic field is achieved as the wave mode is changed to the dominant mode $TE_{1,1}$. This field, now properly polarized as required, interacts with the ferrite in such a way that its polarization plane is

[12]Collin, R. *Foundations of Microwave Engineering*. New York: McGraw-Hill, 1966.

rotated clockwise, as shown in figure 4–43 by 45 degrees. The plane is now what it was when it entered port 1.

As a reflected wave enters port 2, it will have its plane of polarization rotated 45 degrees clockwise by the Faraday rotator action of the ferrite. At the circular-to-rectangular guide section of port 1, the wave will experience another 45 degree clockwise rotation. Consequently, the E field will be parallel to the resistance card attenuator and will be absorbed. The situation is illustrated in figure 4–44.

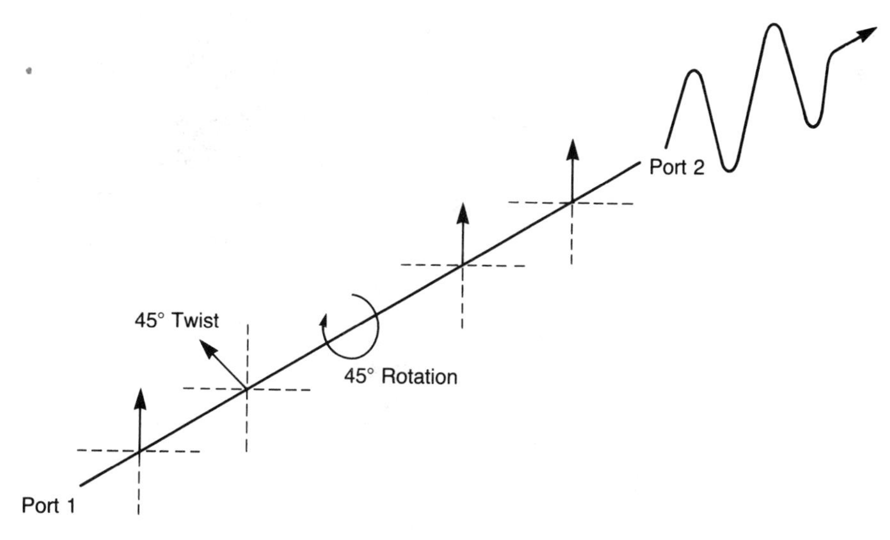

FIGURE 4–43
Port 1-to-port 2 wave propagation.

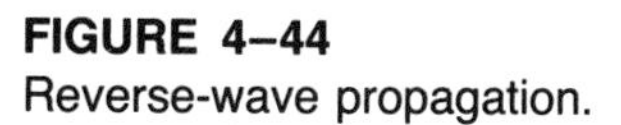
FIGURE 4–44
Reverse-wave propagation.

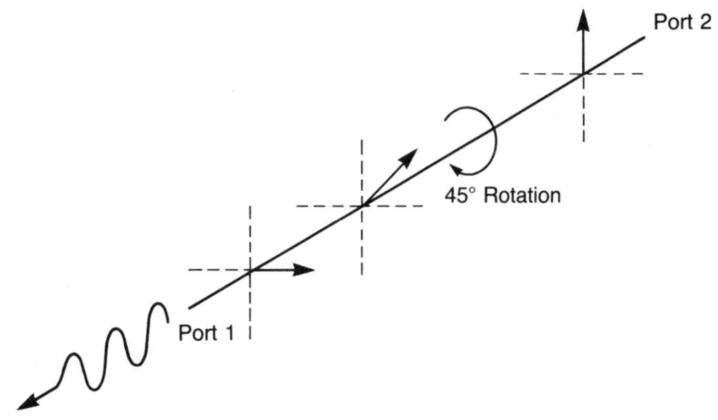

Attenuation in the reverse direction is on the order of 20 to 30 dB, while forward reductions are typically less than 2 dB. For example, the Hewlett-Packard waveguide isolator model number Q365A shown in figure 4–45 is used in the 33–50 GHz range and has a minimum isolation of 25 dB with an insertion (forward) loss of 1.5 dB maximum. The power handling capability of the Q365A is 1.5 watts. The 1987 Hewlett-Packard catalog lists this isolator at $990.

FIGURE 4–45
The Hewlett-Packard Q365A isolator.

Resonant Absorption Isolator

Once again, consider the situation shown in figure 4–41. If a circularly polarized wave (i.e., rotating magnetic field) is caused to travel along the Z axis in a plane perpendicular to the DC axial magnetic field (B_0), it will give up a certain amount of its energy if it is in the same direction as the electron's precession and at a frequency close to the electron's orbital (precessional) velocity. If the frequency of the signal coincides exactly with that of the precessional frequency, a condition of resonance exists wherein great amounts of energy will be removed from the wave and dissipated as heat in the crystalline lattice of the ferrite. The amount of energy absorbed may be controlled by the strength of the DC magnetic field. Rotation of the field in the opposite direction does not fulfill the requirements of resonance, hence no net energy is absorbed. This accounts for the unidirectionality of a device called the resonant isolator. One version of the isolator is shown in figure 4–46, which shows a longitudinal section of ferrite placed about $\lambda_p/8$ from one wall of the waveguide in a region of maximum magnetic flux density. The mode of wave propagation is $TE_{1,0}$.

Circulator

A ferrite circulator is a multi-port device having the property that only rotationally-adjacent ports are connected. Schematically, the circulator is shown in figure 4–47 wherein port 1 connects to port 2, port 2 connects to port 3, port 3 to port 4, and port 4 to port 1. Circulators find wide application in radar sets (see chapter 7) where isolation must be provided between the transmitter and receiver connected to the same antenna.

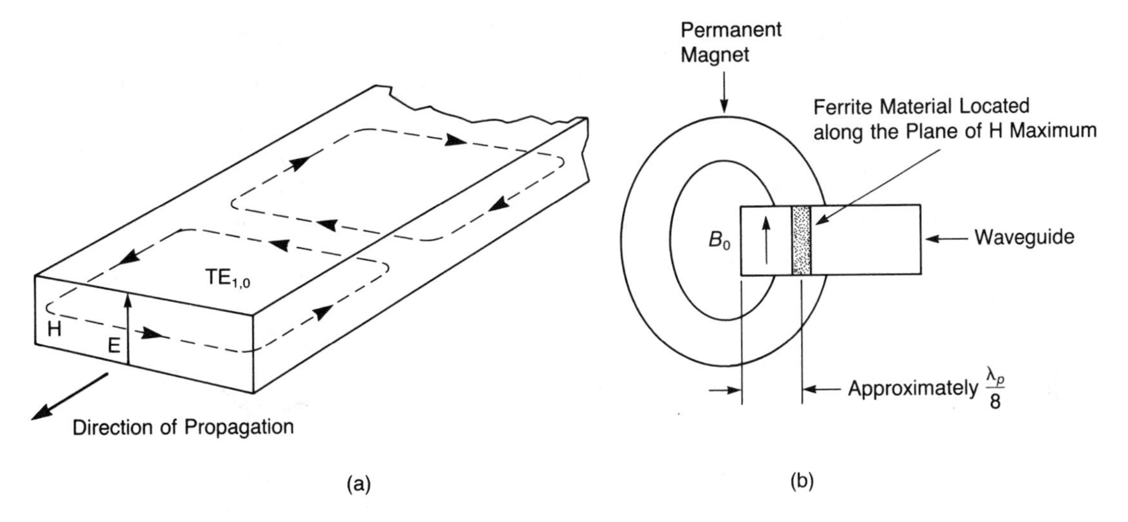

FIGURE 4–46
(a) Location of maximum flux density in rectangular waveguide operating in $TE_{1,0}$ mode.
(b) Cross-section of absorption isolator.

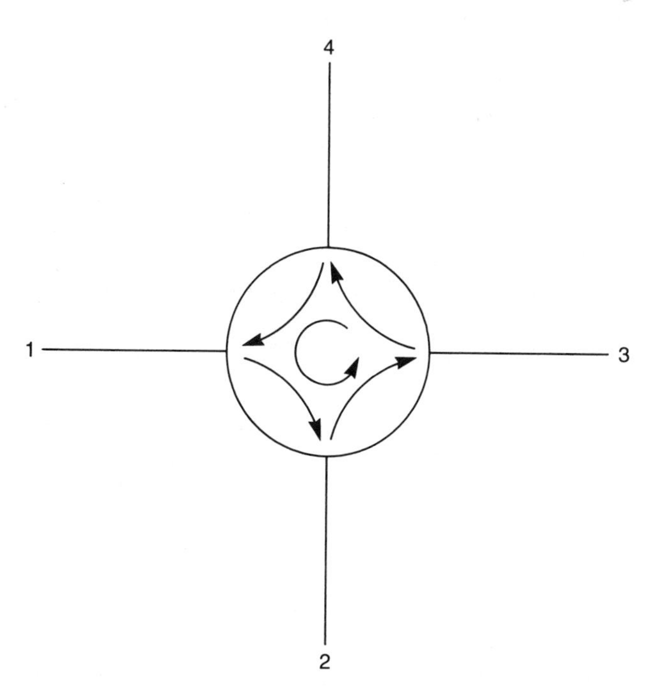

FIGURE 4–47
Schematic diagram of 4-port circulator.

A Faraday-rotator circulator is shown in figure 4–48. The action is similar to the Faraday isolator, and the resistive absorption cards are removed.

Energy entering port 1 in the $TE_{1,0}$ mode is transformed to the circular $TE_{1,1}$ mode and rotated clockwise 45 degrees by the ferrite action, and leaves port 2 which is the first suitably aligned termination. Similarly, energy into port 2 will be rotated so that only port 3 presents an appropriate exit port. In the same way, port 3 exits port 4, and port 4 is coupled only to port 1.

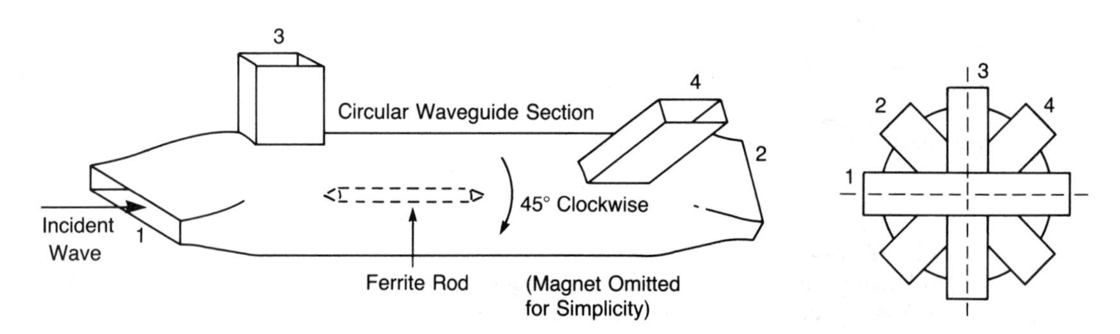

FIGURE 4–48
A 4-port Faraday-rotator circulator.

SUMMARY

The chapter opened with a discussion of the three principal losses encountered with coaxial and parallel-wire transmission lines. The losses were stated as being due to conduction, radiation, and dielectric heating. As a result of such losses, a more efficient method of carrying E-M waves had to be developed. This led to the design of waveguides which were defined basically as pipes through which waves propagated by reflection instead of conduction.

The rectangular waveguide was the first simple guide shape to be discussed. It was shown that such a guide was, in principle, an extension of quarter-wave transmission line geometry. The shape was such that a dense concentration of E-field lines of force was positioned near the center of the guide so that they would not be shorted out by the conductive walls of the waveguide.

It was recognized that there were three primary velocities associated with the wave inside the waveguide: the free-space velocity (c); the group velocity (v_g); and the phase velocity (v_p). The group velocity (v_g) was slower than the free-space velocity (c) due to the zigzag motion with which the wave moved along the axis of the guide. The phase velocity (v_p) was shown to be an apparent velocity only, and was defined as the rate with which the wave changed phase along a boundary.

Two principal wavelengths were also associated with waveguides. The free-space wavelength (λ) was shown to be shorter than the guide wavelength (λ_p) since the wave slows down upon entering the guide. The frequency inside or outside the guide, however, remained the same, and was given as either $f = v_p/\lambda_p$ or $f = c/\lambda$. It was shown that for a given mode of propagation, there is some longest wavelength (lowest frequency) at which the wave would not propagate in the waveguide. This wavelength was called the cutoff wavelength (λ_0), and was defined by equation (4–11).

Universal waveguide equations were developed for λ_p (equation 4–12), v_p (equation 4–14), and v_g (equation 4–21) which employed a universal factor stated only in terms of the free-space wavelength and the cutoff wavelength.

The various TE and TM modes of propagation were discussed, and a system of designating these modes was presented. A different equation (4–22) for finding λ_0 was developed for those situations wherein field intensities were maintained between both walls of the waveguide. Various methods of exciting the different modes within waveguides were presented, and loop and probe methods were mentioned.

The concept of characteristic waveguide impedance (Z_0) was discussed, and it was determined that Z_0 for waveguides, unlike transmission lines, was frequency dependent. Moreover, different equations were developed for the TE and TM modes of propagation.

Wider bandwidth ridged and higher power circular waveguides were discussed. A different formula for finding λ_0 in circular waveguides was developed using the Bessel functions. Flexible waveguide was also mentioned.

The choke flange was shown to be an effective and less expensive method of mating waveguide flanges while reducing SWR at coupling surfaces. The rotary joint and the hybrid tee were shown as examples of combination shapes and guide structures which had specific properties and advantages over simple guide geometries. The iris and post method of introducing inductance or capacitance into waveguide structures was discussed. The hybrid ring was introduced as a selective port coupler, and the directional coupler was mentioned as a means of sampling power flow along a main guide channel.

The cavity resonator was discussed as an extension of waveguide geometry, and the frequency meter was presented as an application of cylindrical resonant tank circuits.

Ferrite devices were shown to be of value as isolators and circulators where no moving mechanical parts were present. The principle of such devices was shown to be the Faraday rotation effect which resulted in a specified rotation of the plane of polarization of the wave.

PROBLEMS

1. A technician makes the following sketch of a waveguide which she measured. If the guide is to propagate the $TE_{1,0}$ mode, what is the nominal frequency of operation? [10 GHz]

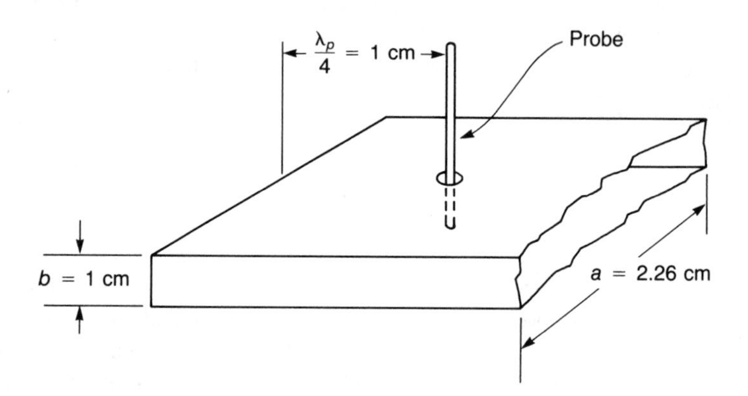

2. What is the cutoff frequency of the waveguide in problem 1? [6.64 GHz]
3. What is the maximum frequency this waveguide can propagate in the $TE_{1,0}$ mode? [13.27 GHz]
4. How much above and below the X-band of frequencies does the $TE_{1,0}$ bandwidth of the guide in problem 3 extend? [1.56 GHz below; 0.87 GHz above]
5. What is the theoretical $TE_{1,0}$ bandwidth of a WR19 waveguide whose internal dimensions are 0.188 inch by 0.094 inch? [31.38–62.76 GHz]
6. What is the cutoff frequency of a circular waveguide 5.25 cm in diameter operating in the $TM_{1,1}$ mode? [1.82 GHz]
7. What is the A dimension of a rectangular waveguide whose characteristic impedance (Z_0) is 408 ohms, and which is required to propagate a 9.6 GHz signal in the $TE_{1,0}$ mode? [4.09 cm]
8. A power of 15 mW is detected at the output of the auxiliary arm of a directional coupler rated at 20 dB. How much power appears at the output of the main arm? [1.485 watts]
9. A directional coupler has a directivity of 28 dB. If the power in the auxiliary arm was 2 mW for the forward main-arm power test condition, what power was detected in the auxiliary arm when the main-arm power was reversed? [3.17 μW]

QUESTIONS

1. Why is skin effect directly proportional to frequency? Why does increasing the wire's diameter tend to offset the skin effect?
2. Why is radiation loss from coaxial cable less than that of parallel-wire transmission line?
3. Explain the nature of energy loss in the dielectric.
4. Why is it necessary to keep electric lines of force away from waveguide walls?
5. Draw a diagram that illustrates why the guide wavelength (λ_p) is greater than the free-space wavelength (λ).
6. Why is $v_g < c$ in the waveguide?
7. What is "universal" about the universal waveguide equations? Why is this important?
8. What is meant by the cutoff wavelength?
9. Describe the mode-labeling system used with rectangular and circular waveguides.
10. What is the function of a waveguide adapter?
11. What evidence can you find for the assertion that characteristic waveguide impedance is frequency dependent?
12. How can the cross-section of a waveguide be varied to provide a gradual impedance match with free space?
13. What is a common problem with the $TE_{1,1}$ mode in circular waveguides?
14. Explain the operation of the choke flange. Can the L-shaped slot be other than $\lambda_p/2$ in length? Why or why not?
15. What is a primary application of the magic tee?
16. Explain the principle behind the iris and the post.
17. In the hybrid ring, explain why arm 3 will communicate with arms 2 and 4, but not arm 1 (see figure 4–31).
18. Explain the principle of operation of a directional coupler.
19. Why does the absorbing material in the auxiliary arm of a directional coupler have a pyramidal shape?
20. Explain the basic principle of the cavity resonator.
21. What is a ferrite? What is Faraday rotation?
22. Explain the operation of the Faraday isolator.

5

MICROWAVE THERMIONIC DEVICES

In the early 1950s, the invention of the transistor heralded the demise of the vacuum tube. Indeed, in only a very short time, electron tubes began to disappear from virtually all electronic equipment, and nearly every industry felt the impact.

In the microwave industry, however, a few special-purpose tubes clung tenaciously to their place in a burgeoning technology and refused to be unseated by the new solid-state contender. These tubes, of which the magnetron and klystron are examples, have survived the thirty-odd years of semiconductor competition and show no signs of losing ground. In fact, a recent market survey[1] has projected a 27% increase in sales over a five-year period representing $152 million.

Clearly, then, the thermionic device is far from gone, and an understanding of its principles is important to anyone wishing to function successfully in the microwave industry.

REVIEW OF THERMIONIC EMISSION DEVICES

The word *thermionic* pertains to the release of electrons by thermal means. Credit for first having noticed this effect goes to Thomas Edison, who in 1883 observed that electrons were released from the filaments of his modified incandescent lamps. Hence, the production of electrons by thermal means now bears his name, and is appropriately called the *Edison effect*.

At the heart of every thermionic device, then, is a specially treated *cathode* which emits electrons when heated electrically to incandescence by a filament. There are several types of cathodes, all treated with various exotic combinations of materials designed to enhance their electron emission. While whole industries are devoted to the design of

[1]Bierman, H. "Microwave Tubes Reach New Power and Efficiency Levels," *Microwave Journal* 30, February, 1987.

better cathode structures, we will not dwell on this area of the device. Instead, we will look at what happens to these electrons after they leave the cathode. Specifically, we will explore how electrons are actually put to use in active devices.

The Vacuum Tube Diode

The simplest thermionic device is the two-element *diode* consisting of an electron source (the filament-cathode assembly) and an anode for collecting the free electrons. These two electrodes are placed in an envelope from which all the air has been evacuated.[2] Most often, the cathode and anode are coaxially arranged, and the whole tube is cylindrical, as shown in figure 5–1.

FIGURE 5–1
Typical vacuum tubes.

Schematically, the diode is symbolized as shown in figure 5–2. Occasionally, the entire symbol is placed inside a circle which represents the tube's envelope. However, it has become common practice to omit the circle. The filament also is often left out of the schematic representation since it is understood that the device cannot function properly unless the filament is operating.

The electrons emitted from the heated cathode accumulate in sort of a cloud around the cathode. This "space charge," as it is called, acts as a reservoir since it

[2]One reason the air is removed is to prevent the heater and cathode from burning up in the presence of any oxygen. Furthermore, trace amounts of gas will result in plasma conduction which interferes with proper tube operation. Any residual gasses are removed from the envelope by inductively heating a magnesium ribbon (called the "getter") to its flash point. The magnesium then combines chemically with any remaining gas and is deposited harmlessly as a silvery film on the inner surface of the tube envelope.

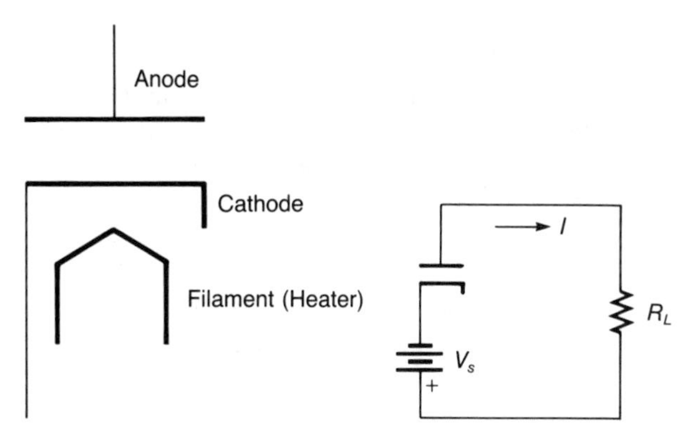

FIGURE 5–2
Diode schematic symbol.

FIGURE 5–3
Diode connected to a DC voltage source.

usually holds more free electrons than necessary for proper tube operation. If a positive charge is now placed on the anode, there will be a flow of these free electrons from the cathode region to the anode (also called the "plate") through any external load, through the power supply, and back to the cathode. The circuit thus described is shown in figure 5–3.

The voltage drop across the load is given by Ohm's law as $V_L = IR_L$. Note that if the positive voltage on the anode was high enough, the diode would act as a short circuit and the entire supply voltage, V_s, would appear across the load.

If an AC voltage source was placed in the series circuit of figure 5–3 instead of the DC supply, electrons would flow during each positive alternation of the supply voltage. The result would be a rectified voltage output, as shown in figure 5–4.

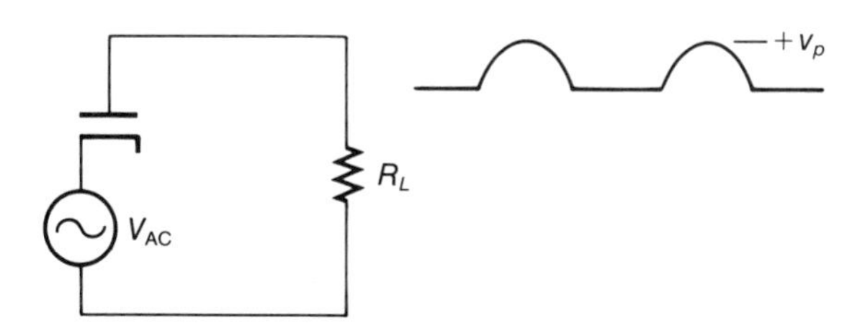

FIGURE 5–4
Ideal diode vacuum tube used as a rectifier.

One cannot help but notice the similarity between the rectified voltage from the vacuum tube diode and its solid-state diode counterpart. In point of fact, if the vacuum tube were removed and a semiconductor put in its place, the output voltage would be essentially the same. Why, then, has the vacuum tube been supplanted by silicon technology? The main reasons are that solid-state diodes require no delicate filament-cathode assembly and consequent drain on the power supply; there is no "warm-up" period; silicon diodes of the same ampacity may be only 1/100th the size and less than 1/1,000th the cost; and semiconductor diodes run much cooler and are more reliable.

The Vacuum Tube Triode

The diode cannot provide voltage gain since there is no way of controlling the large space-charge current with a small signal voltage. However, if a screenlike cylindrical structure was inserted between the cathode and anode so that electrons on their way to the plate could be influenced by its electric field, then control of the current would be possible.

The screenlike structure referred to is called the *control grid,* and the resulting vacuum tube is known as a *triode*. The triode's schematic symbols are shown in figure 5–5.

Note that if the control grid is made more negative than the cathode, electrons will still be attracted by the large positive anode voltage, but very few electrons will actually strike the grid itself. Consequently, if the grid is biased with a negative voltage, then any suitable AC signal voltage will cause this bias voltage to increase or decrease in step with the signal frequency. Therefore, the tube current (called the plate current, I_p) will also vary in step with the signal variations. If this plate current is now allowed to flow in an external circuit, then a large voltage drop will be developed across any external resistance. This voltage drop being greater than the input signal, we say we have voltage gain or that amplification has occurred. A simple triode amplifier circuit might look like that shown in figure 5–6. Note the 180-degree phase reversal between

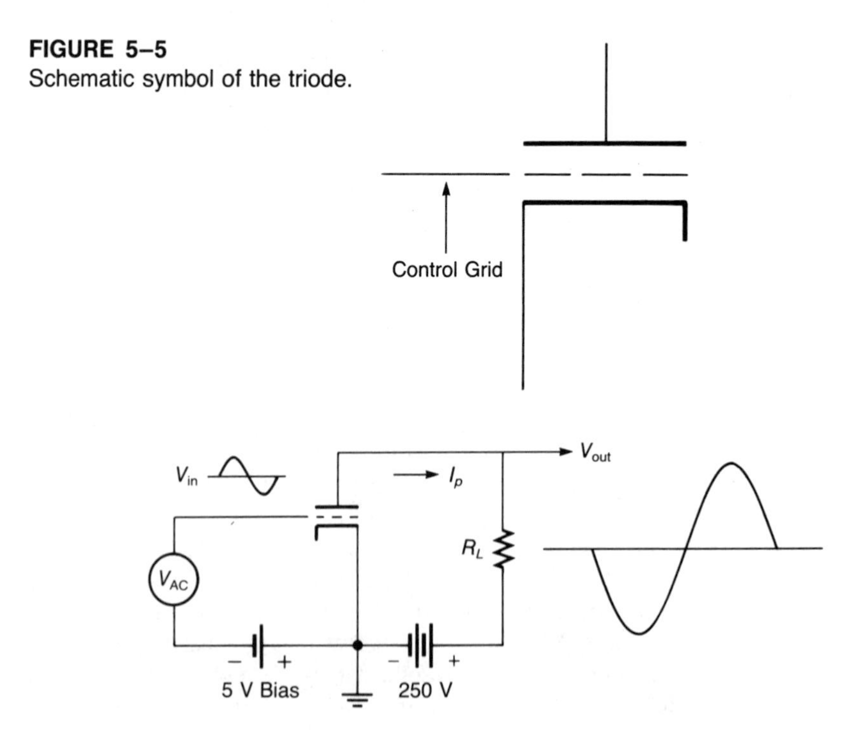

FIGURE 5–5
Schematic symbol of the triode.

FIGURE 5–6
Triode amplifier circuit.

V_{in} and V_{out}. This is the same phase relationship one gets with a common-emitter transistor amplifier. Note, too, the similarity between the tube and transistor elements, as shown in table 5–1.

TABLE 5–1
Comparison of tube and transistor elements and functions.

Tube	Transistor	Function
Plate	Collector	Collects electrons
Grid	Base	Controls electrons
Cathode	Emitter	Emits electrons

Triode Characteristics

The following are a few of the more important triode characteristics which will give the student some insight, albeit limited, as to what tube parameters mean and how they interrelate. The discerning reader will notice many similarities between vacuum tube triode and bipolar transistor characteristics. Unfortunately, a complete discussion of all the various tube types and performance nuances is beyond the scope of this text, but there are several good references on the subject which the interested student may consult.[3]

Amplification Factor Amplification factor (μ) is defined as:

$$\mu = \frac{\Delta V_p}{\Delta V_g}$$

where ΔV_p = change in plate voltage
ΔV_g = change in grid voltage

Suppose V_p was changed by 90 volts and this change caused I_p to change by 3 mA. And, suppose V_g was changed by 3 volts and I_p changed by this *same* amount (3 mA). Then 90/3 = 30.

The tube μ is a figure of merit for the triode and tells how much more effective the grid is than the plate in controlling the plate current. In this example, the grid is 30 times more effective than the plate. Obviously, then, voltage amplifier tubes need to have very large μ values.

---

[3]Bureau of Ships, Navy Dept. *Radar Electronic Fundamentals*. Navships 900,016. Washington, D.C., June, 1944. Section IV: Vacuum Tubes and Applications.

Depts. of Army and Air Force. *Basic Theory and Application of Electron Tubes*. TM 11–662/TO 16–1–255. Washington, D.C., February, 1952.

Marcus, A. *Electronics for Technicians*. Englewood Cliffs, NJ: Prentice-Hall, 1969.

Note that μ has no units since it is the ratio of two voltages. Note, too, that amplification factor (μ) is not the same as stage gain (A). However, there is a relationship.

To the load (R_L), a vacuum tube looks like a resistance (R_p) in series, as shown in figure 5–7.

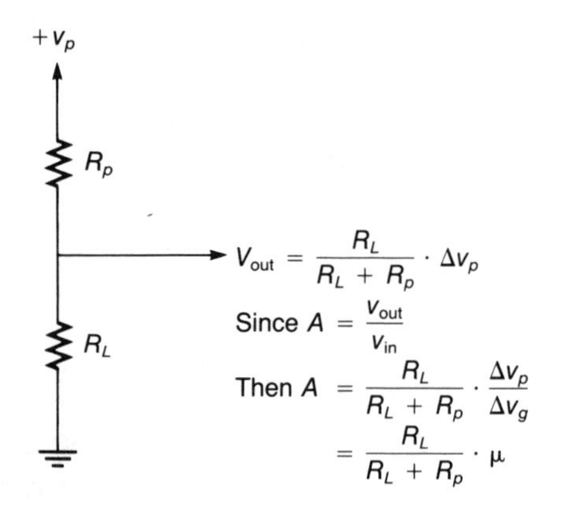

FIGURE 5–7
Load and plate resistance in series.

Transconductance Transconductance (G_m) is an indication of how much grid voltage affects control over plate current. Transconductance is another performance figure of merit.

$$G_m = \frac{\Delta I_p}{\Delta V_g}$$

G_m is measured in Siemens (formerly mhos) and is derived simply from Ohm's law: $G = I/V$.

Plate Resistance Plate resistance (R_p) is a fictitious quantity whose existence can only be inferred, but not measured directly.

$$R_p = \frac{\Delta V_p}{\Delta I_p} \qquad \text{(with } V_g \text{ held constant)}$$

Specific values of R_p may be determined by using the plate-family of characteristic curves. Thus, the relationship between μ, G_m, and R_p is

$$\mu = G_m R_p$$

since $(\Delta I_p/\Delta V_g)(\Delta V_p/\Delta I_p) = \Delta V_p/\Delta V_g = \mu$.

The Triode at UHF

Unintentional capacitance and inductance exist in triode vacuum tubes simply by virtue of mechanical construction. Any time there is a separation between conductors, a certain amount of capacitance will exist. Moreover, leads that allow the various signals to move into or out of the tube have a small value of inductance. At lower frequencies, these stray reactances may be all but ignored and seldom impair tube operation. At or above the UHF region (300 MHz–3 GHz) interelectrode capacitance can no longer be ignored, and lead inductance becomes a critical problem. At these frequencies, the equivalent triode circuit looks like that shown in figure 5–8.

FIGURE 5–8
Equivalent triode at UHF.

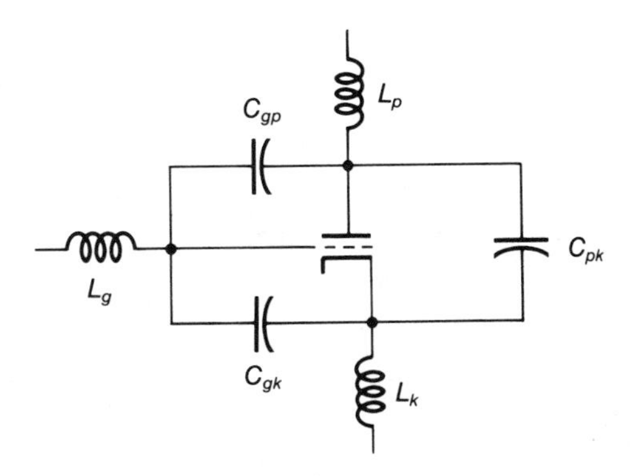

Since $X_c = 1/(2\pi fC)$ and $X_L = 2\pi fL$, increasing the frequency decreases X_c and increases X_L. Therefore, if these values become critical (which they do at UHF), then the signal attempts to bypass the tube action entirely.

Moreover, transit time (time required for the electron to travel from cathode to anode) becomes significant. This is due to the fact that an electron has mass and cannot follow the rapid changes in the voltage applied to the control grid. So, for example, by the time an electron leaves the vicinity of the cathode when the grid is, say, positive, the polarity on the grid may change before the electron reaches the plate. As a result, the output signal is distorted in phase as well as shape.

Up to a certain point, changes in tube geometry can mitigate the problems of both transit time as well as interelectrode reactance. For example, if the tube elements are brought closer together, transit time is reduced, but capacitance increases. To solve this problem, surface areas of tube elements are reduced, but so is power handling ability. To get around this problem, radiating fins may be placed on the anode connection to dissipate more heat.

The means employed to extend the useful range of the ordinary vacuum triode into the UHF region and beyond are many and devious. But there are limits. Around the late 1930s, Russel and Sigurd Varian[4] abandoned further efforts to extend the useful

---

[4]Russel and Sigurd Varian built the first klystron in 1937 using the calculations of W.W. Hansen of Stanford University.

range of conventional triodes and embarked on an entirely novel approach. Their efforts culminated in the invention of the *klystron*.[5] The word *klystron* is derived from the Greek word "klyzein" meaning "sea waves." The significance of this will be illustrated in the next section.

KLYSTRONS

Since the transit time problem inherent in the vacuum tube triode cannot be resolved by ordinary (and often extraordinary) means, the transit effect was turned to an advantage in a device called the multicavity klystron. In the conventional triode, electrical energy is converted to RF energy by exercising control over cathode emission to vary the current density of the electron beam moving toward the anode. This process is referred to as *current density modulation*. In the klystron, the velocity of electrons in a constant-density beam is varied in step with the input RF signal, and their energy is delivered to an output circuit. This process is called *velocity modulation*.[6] The construction of the single-transit, two-cavity klystron is shown in figure 5–9.

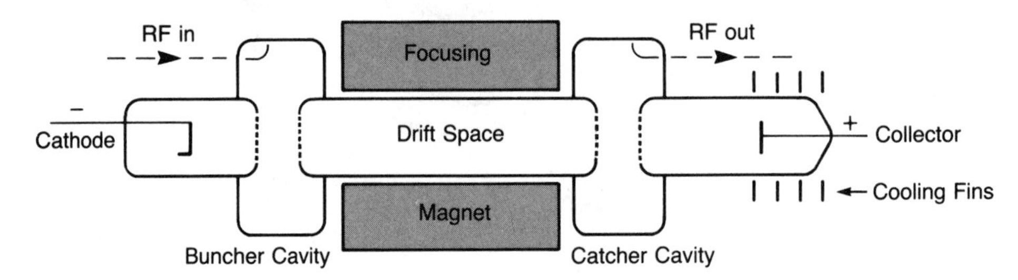

FIGURE 5–9
The single-transit (two-cavity) klystron.

Theory of Operation

Consider three electrons e_1, e_2, and e_3 that leave the cathode at times t_1, t_2, and t_3 of the input RF sinewave voltage, as shown in figure 5–10. The position of each electron at its respective time is shown in figure 5–11.

Electron e_1 left the cathode earlier (at t_1 in figure 5–10), but traveled slower since the grid was negative at that time. Electron e_2 left the cathode later than e_1, but traveled somewhat faster since the grid was at zero potential. Finally, e_3 left the cathode last (at t_3), but its velocity was the highest since the grid was at a positive value. The net result is that all three electrons arrive at the buncher grid at about the same time in a single "bunch." These bunches arrive at the buncher grid once per cycle of the RF signal

[5]Varian, R., et al. "A High-Frequency Amplifier and Oscillator." *Journal of Applied Physics* 10, May, 1939.
[6]For a complete discussion, see Harrison, A. *Klystron Tubes*. New York: McGraw-Hill, 1947.

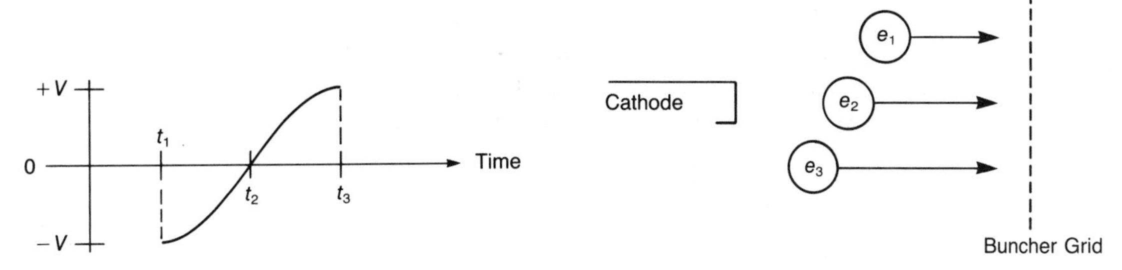

FIGURE 5–10
An RF sinewave timing diagram.

FIGURE 5–11
Electron bunching effect.

voltage. They then continue through the drift space, the length of which is no longer a problem since the entire concept of transit time is irrelevant with this particular tube design. Upon passing the catcher grid, the electron bunches give up considerable energy to the output cavity. The output cavity is excited to oscillation at the bunching frequency, which is the same as the input signal frequency.[7] The result is considerable power gain which, for larger devices, may be on the order of 90 dB.

The purpose of the external focusing magnet is to keep the electron beam converged and thus assure more uniform bunching as the electrons continue toward the collector. At the collector, the electrons are still quite kinetic, and much heat is given off as they give up their residual energy. Consequently, some provision for cooling the device must be made, which may range from simply freely circulating air by natural convection for smaller devices to combinations of forced air and water jackets for larger tubes.

Since bunching is never quite complete in the two-cavity device, multiple cavities are often used to insure a more homogeneous beam for greater efficiency[8] and less noise produced by errant electrons that do not participate fully in the bunching process. As many as seven total cavities may complement a single klystron. Power outputs tend to exceed actual requirements in most cases. For example, the Varian model 4K3SL-3, four-cavity S- and L-band klystron amplifier for troposcatter systems has a CW output of 1 kW and a gain of 38.6 dB. This means that less than 140 mW of input power is required to achieve maximum output of the device.

The principal drawbacks to high-power, multicavity klystrons are the often complicated cooling systems needed to keep temperatures within operating limits, the heavy (often over 400 pounds) external beam-focusing magnet, and elaborate regulated high-voltage power supply.

If the klystron is to be used merely as a low-power oscillator and not an amplifier, the size-weight problem may be lessened by the use of a *reflex klystron* such as the one

---

[7]A resonant cavity is the microwave equivalent of a tuned circuit and will respond most strongly at one particular frequency. See chapter 4 for a detailed discussion of resonant cavities.

[8]Efficiencies are typically less than 45%.

shown in figure 5–12. Such low-power oscillators have efficiencies typically less than 10%, but require no external focusing magnet. Since the device is physically short, bunching is achieved by the out-and-back "reflex" action due to a negative repeller voltage at the end of the drift space. The electron beam simply does not have the opportunity to diverge sufficiently, and so the focusing requirement is eliminated.

Adjustments in frequency within the range of the device are possible by physically deforming the resonant cavity *slightly* by means of a micrometer-controlled lever arrangement, as shown in figure 5–12. The construction of the reflex klystron is shown in figure 5–13.

Initiating oscillations is simple enough with switching transients or other noise. However, in order to sustain oscillations, the returning bunch of electrons must pass through the gap at the precise moment to give up most of its energy to the cavity. This will occur at a time in the oscillation voltage cycle that will present the strongest deceleration force on the electron bunch. This occurs when the grid nearest the repeller is positive and the grid closest to the cathode is negative; that is, it will occur when the oscillator voltage is at its maximum value across the gap, as shown in figure 5–14.

An Applegate diagram for a reflex klystron shows that optimum bunching occurs if an electron bunch forms and returns to the gap 1¾ cycles of the oscillator frequency after it left the gap, as shown in figure 5–15.

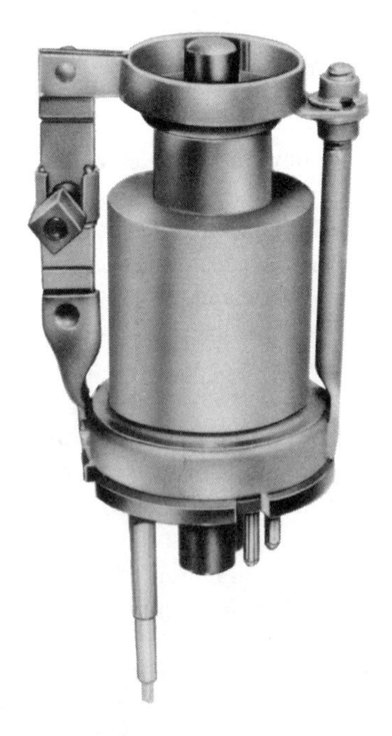

FIGURE 5–12
25 mW reflex klystron.

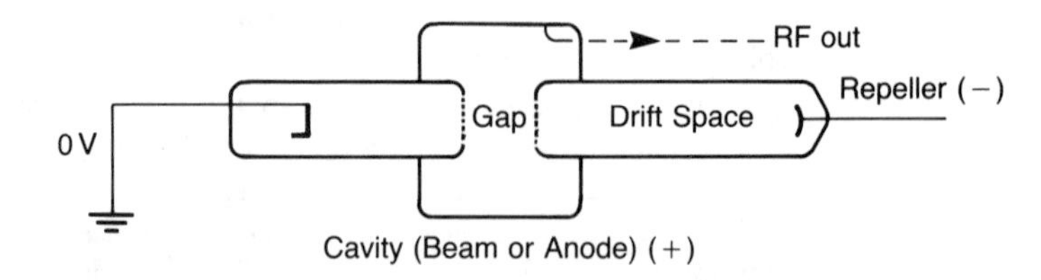

FIGURE 5–13
Reflex klystron.

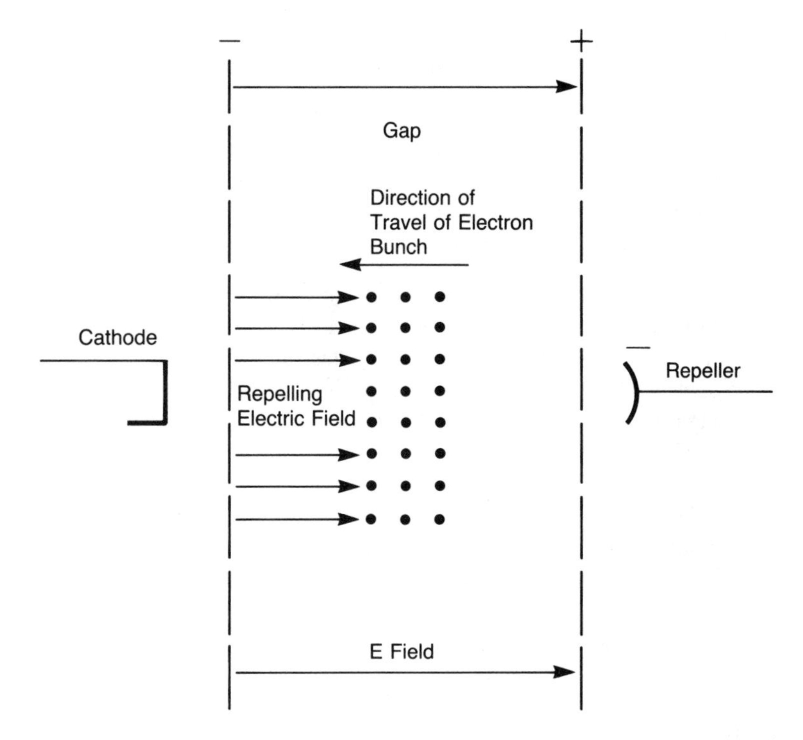

FIGURE 5–14
Returning electron bunch decelerated by cavity gap.

FIGURE 5–15
Applegate diagram for reflex klystron.

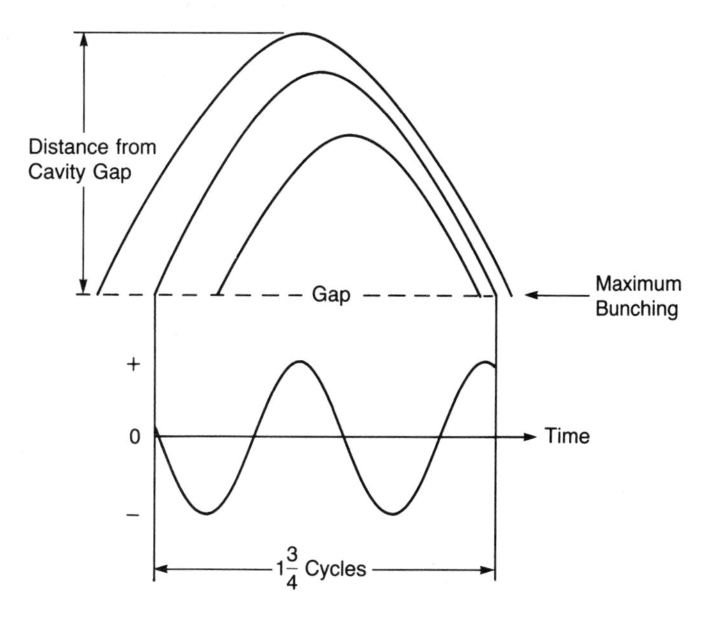

In practice, the electron bunches may return on *any* positive half-cycle maximum. Accordingly, the time for the return of an electron bunch is given by

$$T = n + \frac{3}{4}$$

where T = transit time in cycles for out-and-back trip of electrons
n = a positive integer

For each value of n, we say we have established a *mode* of operation for the klystron. In practice, we may have modes of $\frac{3}{4}$, $1\frac{3}{4}$, $2\frac{3}{4}$, up to about $6\frac{3}{4}$ cycles of the gap voltage. It is these variations in the mode that substantially determine the power output of the device. Consequently, once the cavity voltage is set as recommended by the manufacturer, variations in the repeller voltage may be made to effect minor variations not only in frequency, but also in power.

In closing, it should be mentioned that great care must always be exercised when operating the reflex klystron. Since the repeller is not designed to actually receive any emitted electrons, it is a somewhat flimsy affair, subject to almost instantaneous destruction if accidentally bombarded by electrons. Therefore, to insure against damage, never apply the positive cavity voltage until the tube has had time to warm up and unless the negative repeller voltage is present.

MAGNETRONS

The magnetron is a microwave oscillator vacuum tube capable of high peak powers in short-duration pulses. The pulsed power may range from 10 mW in the UHF band to

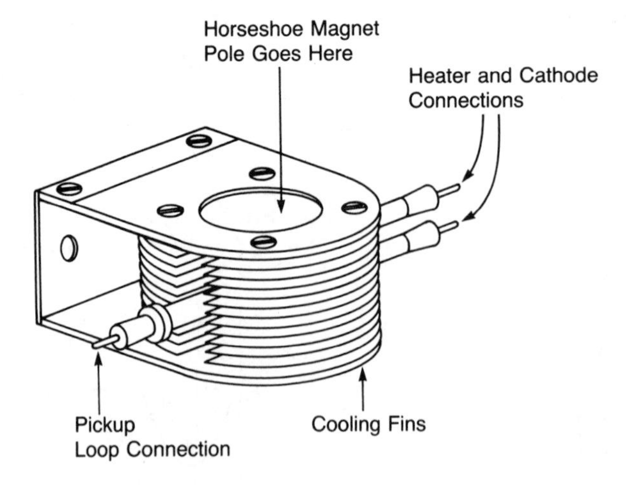

FIGURE 5–16
Cavity magnetron.

10 kW in the millimeter waveband (30–300 GHz). Lacking frequency stability, the magnetron finds wide application in radar (chapter 7) as a pulsed device or in industrial heating as CW oscillators where frequency drift is unimportant. Home appliance microwave ovens also employ a CW version of the magnetron which usually operates at 2.45 GHz. One type of magnetron is shown in figure 5–16.

Theory of Operation

The magnetron is often referred to as a diode, but only in the sense that it has two elements. It has nothing whatsoever to do with the rectification process. The anode consists of a massive cylindrical copper block into which resonant cavities have been cut. The cavities connect with the cathode region (interaction space) by means of slots cut radially into the anode. The heater-cathode assembly is located coaxially along the major axis of the anode, as shown in figure 5–17.

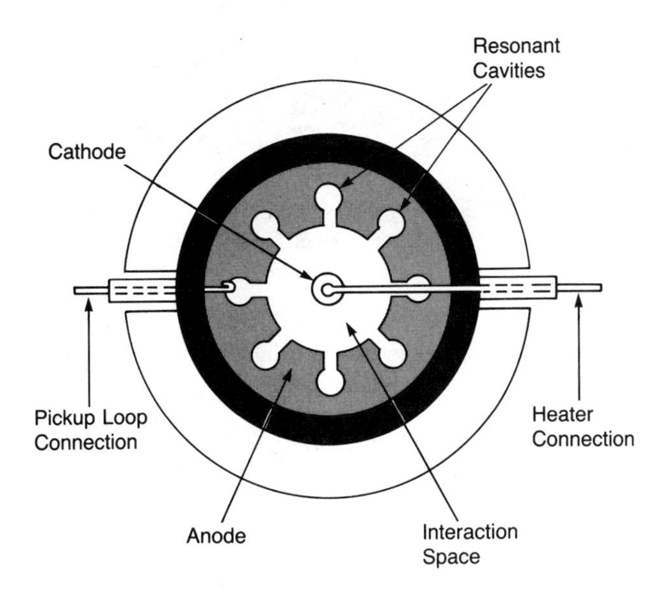

FIGURE 5–17
Cross-sectional view of a magnetron.

Figure 5–18 (p. 178) clearly shows the cavities and slots that are cut into the copper anode. The magnetron shown has a peak power of 240 kW and a duty ratio of 0.001.

The magnetron uses the cross-field principle of operation in which electrons interact with both a magnetic field and an electric field at right angles to one another; hence, the name "cross-field" device. The magnetic field is supplied by a powerful permanent

FIGURE 5–18
Magnetron with cover machined away to show cavities in copper anode.

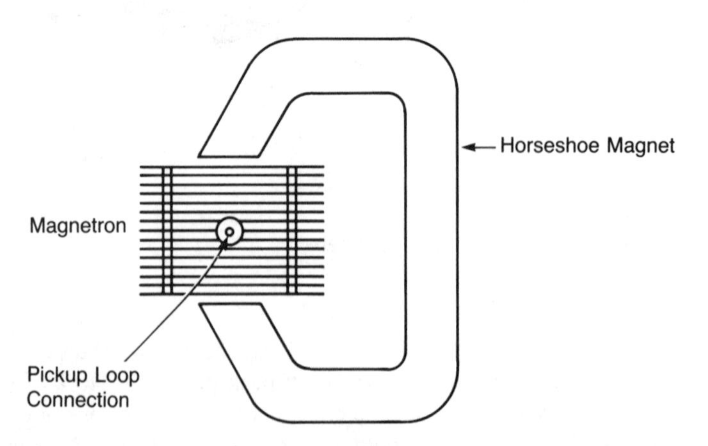

FIGURE 5–19
Mounting of external permanent magnet.

magnet mounted externally to the magnetron and having its lines of force parallel to the major axis of the cathode, as illustrated in figure 5–19.

Suppose the magnetic lines of force extend out of the page toward the reader. Then, according to Flemming's rule, an electron attempting to move along an electric line of force from cathode to anode in figure 5–20 will be deflected to the left *provided no other forces are acting*. If the magnetic field were powerful enough, any electron escaping the cathode region (with fixed anode potential) would be returned to the cathode.

If the cavities are now shock excited to resonance, an RF electric field[9] will be established between adjacent anode sections which will alternately aid or oppose the DC electric field between cathode and anode. Such a field will alternately aid or oppose any electron attempting to move toward the left, as pictured in figure 5–21, where the interaction space is shown straightened out.

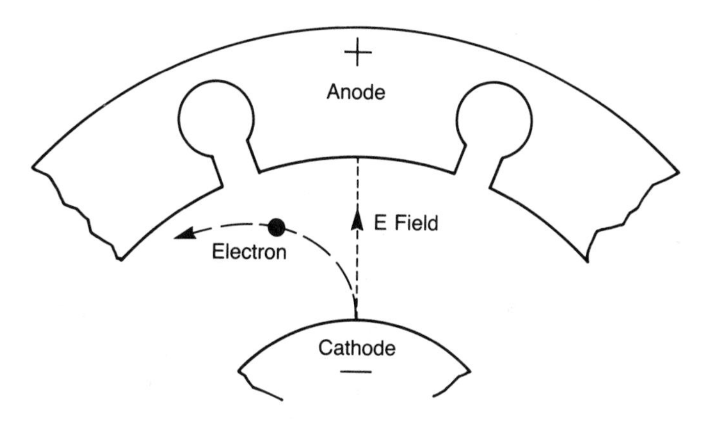

FIGURE 5–20
Deflection of electron path in magnetic field.

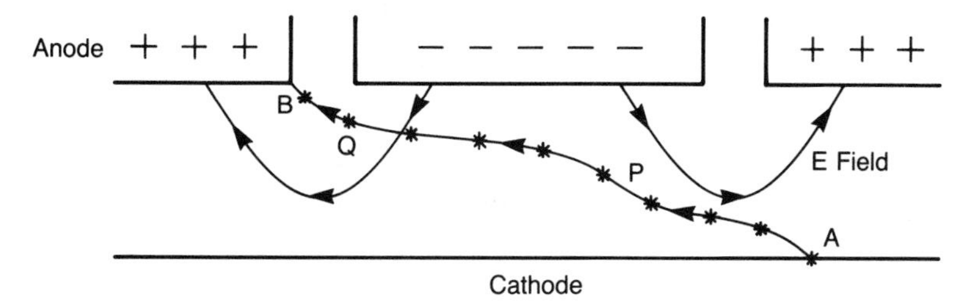

FIGURE 5–21
Path of single electron (*) under the influence of three fields: E, H, and RF.

---

[9]Each cavity behaves like a quarter-wave section of transmission line with a high-voltage across each gap.

The odd-shaped path of the electron moving from A to B is important because the longer the electron can remain in the interaction space before reaching the anode, the more energy it can give up to the RF field as it decelerates, and the easier (and stronger) will be the oscillations. In figure 5–21, the electron will be decelerated at P, thus giving its energy to the RF field, and accelerated at Q, thereby extracting energy from the field. The question then arises, "With this alternate give and take of energy, how can there be any net energy gain?" The answer is that at P, since the electron slows down, it can remain in the interaction space for a *longer time,* thereby giving up more of its energy than can be taken away at Q by its *shorter time* in the field. Consequently, an overall net energy gain is possible.

In actual practice, individual electrons form bunches similar to those in a klystron and sweep past the cavity gaps, imparting their energy in the process. Such bunched electron "clouds" or "spokes," as they are sometimes called, are shown in figure 5–22.

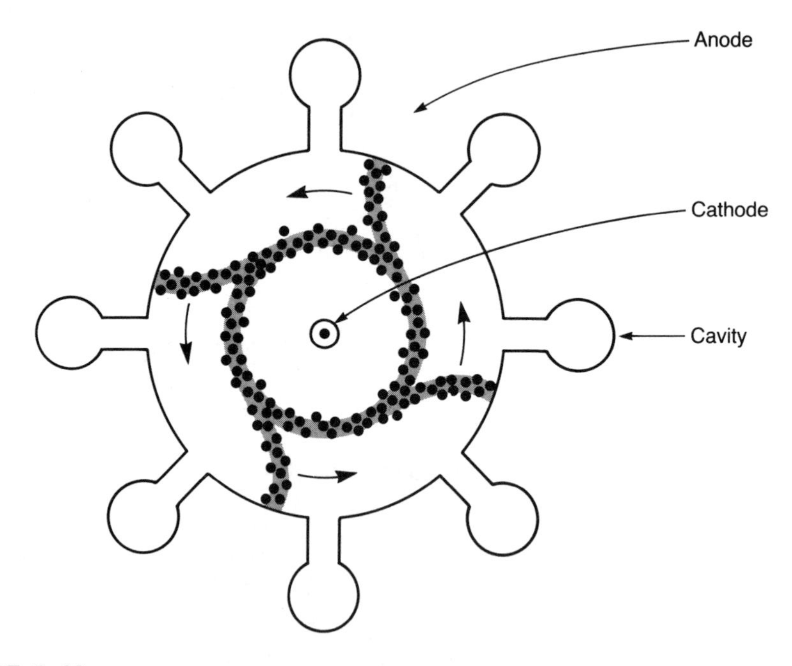

FIGURE 5–22
Bunched electron clouds sweeping past cavity gaps.

FIGURE 5–23
Duty cycle timing diagram.

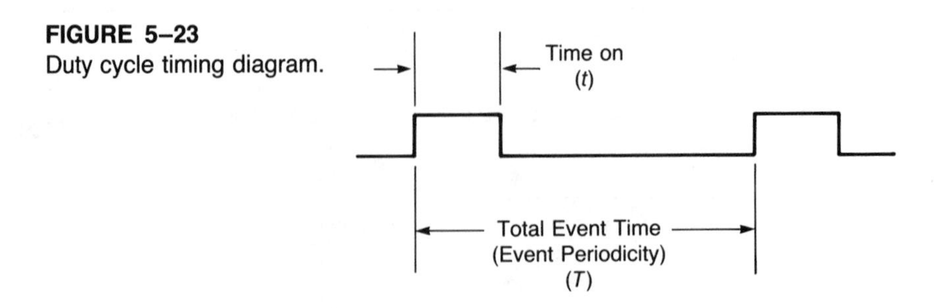

The operation of the magnetron described above is for the most popular type of "mode" since it delivers the most power. It is called the π mode because the RF voltage phase difference between adjacent anode segments is 180 degrees (π radians).

As was the case with the klystron, energy is removed from the magnetron by means of a coupling loop, as shown in figure 5–17 (also see chapter 4). While it may appear that the energy is being extracted from only one cavity, all cavities are, in fact, coupled together and their combined energies may be removed at any convenient cavity.

As mentioned in the opening of this section, magnetrons are used for very high-power, pulsed operation where they may deliver far more power for a brief interval than possible to achieve on a continuous basis. This may be explained by looking at the timing diagram shown in figure 5–23. From the figure we will define duty cycle as the ratio $t/T \times 100$. That is,

$$\text{duty cycle} = \frac{\text{time on}}{\text{total event time}} \times 100 \tag{5–1}$$

Suppose, for example, an *average* of 18 watts is delivered to the load over a total of 900 1-microsecond pulses. Then the amount of *peak* power in each pulse is

$$\frac{\text{power}}{\text{pulse}} = \frac{18\text{ W}}{1} \times \frac{1\text{ S}}{900\text{ pulses}} \times \frac{\text{pulse}}{1\ \mu\text{S}} = 20{,}000\text{ W}$$

Obviously, this is considerably more power than might be handled continuously. In this case, the duty *ratio* is given by

$$\text{duty ratio} = \frac{\text{pulse width}}{\text{pulse periodicity}} = \frac{t}{T} = \frac{1\ \mu\text{S}}{\frac{1}{900}}$$

Putting all these facts together leads us to conclude that

$$\text{peak power} = \frac{\text{average power}}{\text{duty ratio}} \tag{5–2}$$

TRAVELING-WAVE TUBES

The traveling-wave tube (TWT) was invented in the mid-1940s independently by Kompfner in Great Britain[10] and Pierce at Bell Telephone Laboratories in America.[11] Essentially, the device is a high-gain, low-noise microwave amplifier with extremely broad band characteristics. This is because the tube does not rely on resonant phenomena for its operation. Current operating frequencies exceed 50 GHz, and gains above 40 dB are common. Basic structural details may be seen in figure 5–24.

---

[10]Kompfner, R. "The Traveling-Wave Valve." *Wireless World* 52, 1946.

[11]Pierce, J. *Traveling-Wave Tubes*. New York: Van Nostrand, 1950.

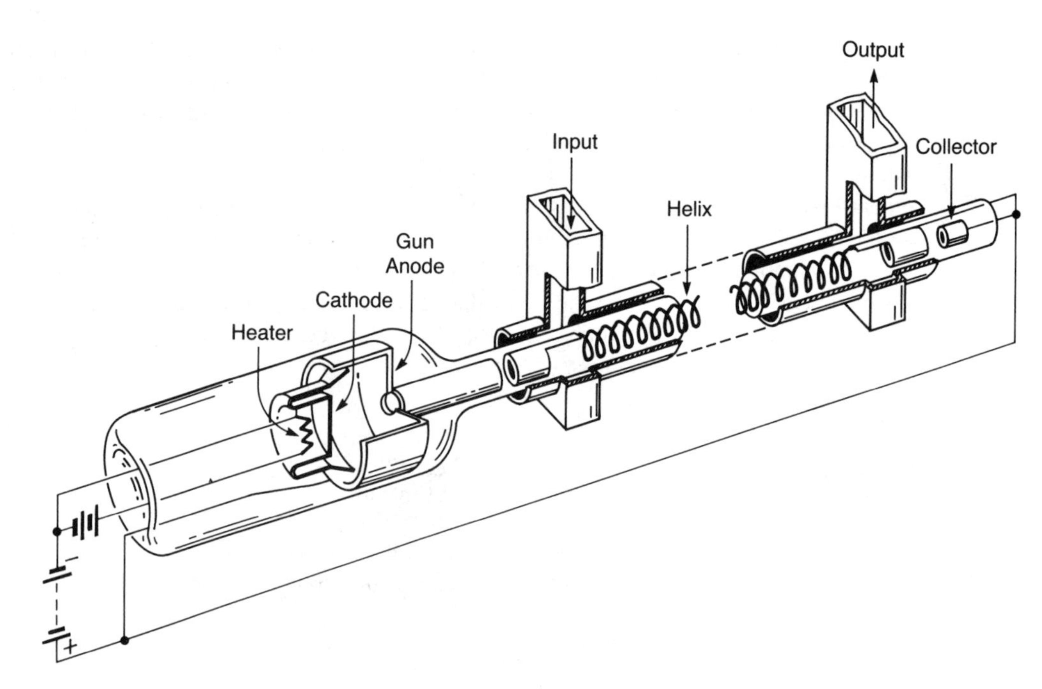

FIGURE 5–24
Traveling-wave tube (redrawn from J. Pierce, *Traveling-Wave Tubes.* New York: D. Van Nostrand, 1950. P. 7, Fig. 2.1).

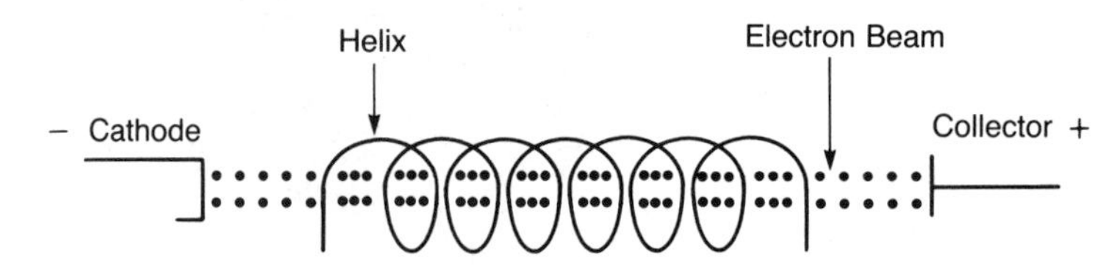

FIGURE 5–25
TWT helix structure.

Resonant cavities are conspicuously absent from the illustration and, indeed, their absence is responsible for the broad band qualities of the traveling-wave amplifier (TWA). Amplification is achieved by the interaction of an electron beam and an RF signal applied to the helix or ''slow-wave'' structure. The input signal is coupled to the helix inductively from the waveguide on the left, and the amplified signal is taken out of the device by the waveguide at the right.

The principle of operation of the TWT is based upon the sustained interaction of the electron beam with the signal wave. Since the interaction is continuous, there is a constant energy exchange occurring between the beam and signal with a consequent increase in the strength of the RF signal.

The heart of the TWT principle is the slow-wave helix structure shown in figure 5–25. An external focusing magnet keeps the electron beam from physically contacting the helix, which would destroy the tube.

Since the RF signal travels at the speed of light, but the electron beam can obtain only perhaps 10% of that value, there must be some provision to slow the wave down if beam-field interaction is to be sustained. This is accomplished by forcing the signal to travel a circuitous path around the helix, while allowing the electron beam to move along the linear axis of the helix. In point of fact, the electron beam actually travels slightly faster than the wave. This is necessary in order to allow the electron bunches a longer interaction time during deceleration to impart more energy to the wave than is removed during acceleration. It should be apparent that the forward velocity of the signal wave along the helix axis is in the ratio of helix pitch to circumference. That is,

$$v_p = c\left(\frac{p}{C}\right)$$

where $c = 300 \times 10^6$ m/S; p = helix pitch (distance between successive turns); and C = helix circumference. Obviously, then, if the pitch is small and the circumference is large, the axial velocity of the signal will be greatly retarded.

Bunching of the electron beam begins to occur at the cathode end as emitted electrons first encounter a weak RF field. As the beam moves further downstream, the signal, and hence the bunching, increases because of continued interaction. The resulting loss of energy to the RF wave further increases the bunching, which further increases the strength of the signal.

As seen in figure 5–26, the electron beam density is greatest just after the maximum negative electric field has occurred. This is consistent with what one might have expected since the greatest deceleration of the beam is occurring at this time, and the electrons will tend to bunch up. Conversely, as the electric vector becomes positive, the electron beam speeds up, and there is a thinning out of the density. Note that each time the beam slows down, energy is given up to the signal.

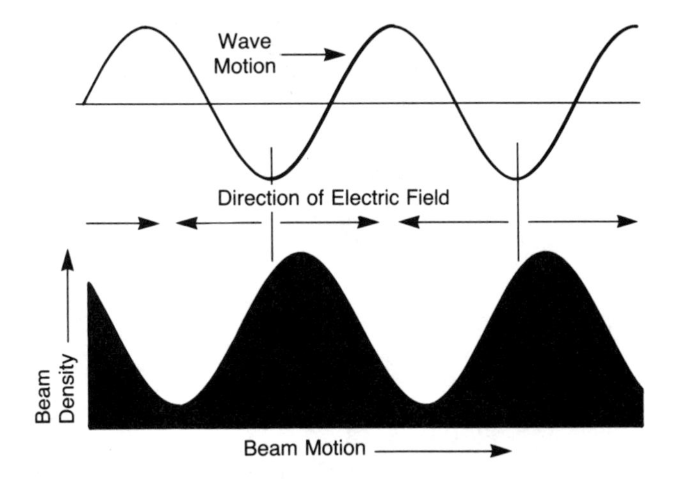

FIGURE 5–26
Beam density versus wave amplitude.

Because the signal build-up is so great, it is possible for reflected waves to travel back along the helix and initiate sustained oscillations. To prevent this from happening, a lossy material (Aquadag, for example) is applied to the outside surface of the tube. This acts as an attenuator of both the forward and reverse signals, but will not interfere with the bunching process. Consequently, any loss of forward signal strength is of no real detriment.

BACKWARD-WAVE OSCILLATORS

A backward-wave oscillator (BWO) is essentially a TWT with the attenuator removed. As a result, any amplified signal reflections from the unmatched collector end will traverse the helix in a reverse direction, and may be removed from the cathode end of the

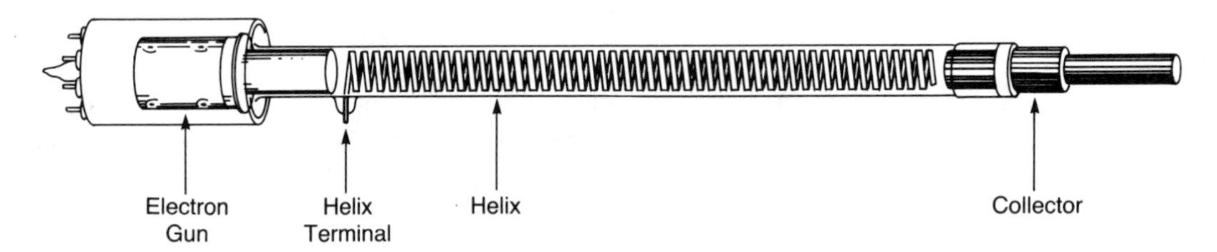

FIGURE 5–27
BWO construction.

FIGURE 5–28
Typical BWO packaging configurations (photos courtesy of Raytheon Company).

device at the helix terminal. Typically, oscillations are initiated by shot noise from the electron beam which induces noise voltages in the helix. The tube action is then similar to that of the TWT, with amplification reinforcing oscillation and vice-versa. Basic structural details of the BWO are shown in figure 5–27.

As with the TWT, one will notice the absence of any resonant cavities normally associated with other types of microwave oscillator devices (e.g., klystrons and magnetrons). Two complete BWO assemblies with integral focusing magnets are shown in figure 5–28.

A significant difference between the TWT and BWO is that a ring cathode structure within the BWO emits a hollow, cylindrical electron beam which is focused magnetically and kept much closer to the helix than was the case with the TWT. This situation allows for a stronger beam-signal interaction with subsequent broader frequency tuning capabilities.

Oscillation frequency is a function of beam velocity; hence, tuning is accomplished by means of voltage variations in the helix-collector network, and have typical ranges of slightly better than one octave. An octave is defined as the interval between any two frequencies whose ratio is 2:1. Therefore, the interval, in octaves, between any two frequencies is given by

$$\text{octave interval} = \log_2 \left(\frac{f_{hi}}{f_{lo}}\right)$$

where f_{hi} is the upper frequency and f_{lo} is the lower frequency.

A more useful formula for the octave interval makes use of the base-transformation relationship[12] and the common log. This gives us the formula

$$\text{octave interval} = 3.322 \log_{10} \left(\frac{f_{hi}}{f_{lo}}\right) \qquad \textbf{(5–3)}$$

For example, the Raytheon BWO #QKB-931 has a tunable frequency within the X-band between 7.2 GHz and 12.4 GHz. Therefore, its tuning range in octaves is

$$\text{octave interval} = 3.322 \log_{10} \left(\frac{12.4}{7.2}\right) = 0.8$$

Lower frequency devices seem to favor higher octave intervals. For example, the ITT BWO #F-2513 (1.3–4 GHz) has a much higher octave interval of 1.6.

Low-power BWO devices used as instrument signal sources are called O-BWO (from the French *ordinaire*) tubes, and have full-octave tuning capability. The maximum power attainable from the outputs of such oscillators is typically on the order of a few-hundred milliwatts. One serious drawback to these broad frequency range instruments, however, is that power output is unequal across the swept-frequency range. This is due

[12]From Rice, H., et al. *Technical Mathematics with Calculus,* 2nd ed. New York: McGraw-Hill, 1957.

$$\log_b N = \frac{\log_a N}{\log_a b} \qquad \log_2 N = \frac{\log_{10} N}{\log_{10} 2}$$

to the fact that RF power output is a function of beam current, which is controlled by helix-anode voltages. However, frequency is also controlled by these voltages. Consequently, as frequency increases, so does output power. For example, the Hewlett-Packard X-band sweep oscillator Model 8690B shown in figure 5–29 has a power-versus-frequency response curve similar to that shown in figure 5–30.

Obviously, then, something must be done to "level" the power output across the entire band of frequencies being swept. This is essential if one wishes to compare two or more device performances (e.g., microwave horn antennas) across an entire range of frequencies. One way of accomplishing this is to incorporate an automatic leveling control (ALC) circuit within the sweep instrument. The general strategy employed is shown in figure 5–31.

FIGURE 5–29
Hewlett-Packard model 8690B sweep oscillator.

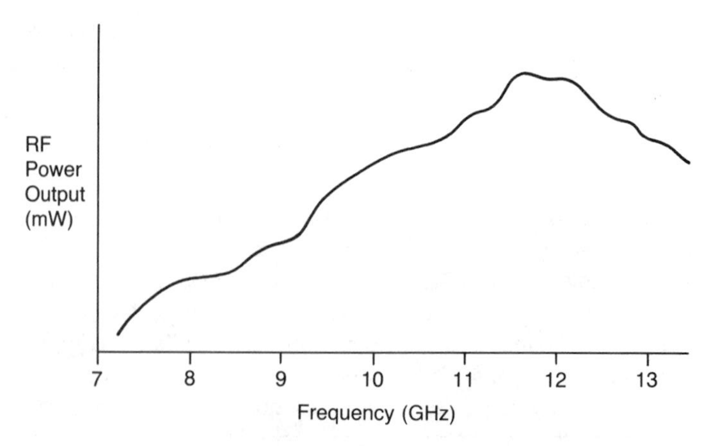

FIGURE 5–30
Typical BWO response curve showing output power as a function of frequency.

FIGURE 5–31
ALC loop.

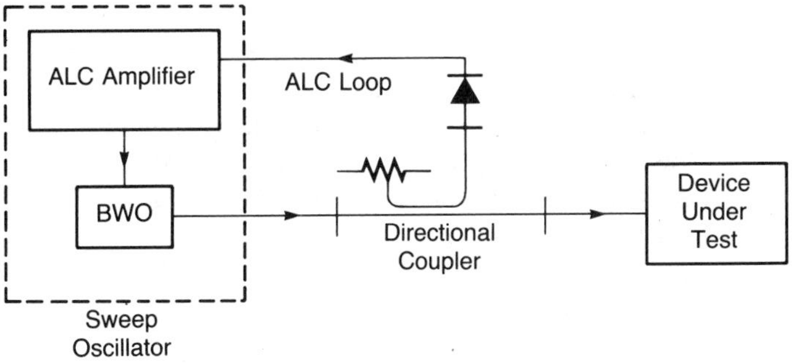

In this set-up, the directional coupler (see chapter 4) samples a small portion (e.g., 1/100th) of the output signal power from the sweep oscillator and feeds it back to the ALC amplifier where any deviation is detected and amplified. Then, a correction voltage is fed to the BWO control circuits which tend to restore power to original levels. Such closed-loop designs, called homeostatic loops, keep the power fairly even across the swept-frequency range.

Another group of microwave oscillator tubes are the M-BWO (*M* for magnetron) devices capable of high CW powers typically on the order of 300 watts, with efficiencies in the neighborhood of 30%. These are cross-field devices like the magnetron and, for convenience, have their slow-wave structures bent in the form of a loop rather than straight, as was the case for the TWT. Unlike the O-BWO device, the M type does not require beam-density modulation. Consequently, beam-wave interaction durations are longer, and more energy is extracted from the beam.

SUMMARY

At the heart of all vacuum tube devices is a cathode, which emits electrons when heated. An earlier UHF tube type called the triode was fraught with problems of stray interelectrode reactances as it struggled to perform in the microwave region. The device was further troubled by output distortion of the RF signal brought about by excessive transit time. These problems were ultimately solved by the invention of the multicavity klystron, which took advantage of long transit times in order to velocity modulate the electron beam leaving the cathode. This device was too large for many microwave oscillator applications, and so the smaller, low-power reflex klystron came into being. The invention of radar during the 1940s called for an oscillator capable of still more power at higher frequencies. This need culminated in the invention of the magnetron, a cross-field device capable of pulsed peak power outputs in the megawatt range. This was possible because the duty cycle of the magnetron is extremely small.

The call for broadband microwave amplifiers resulted in the traveling-wave tube (TWT) having high gain and relatively low noise. The TWT employed the concept of interaction between the moving input RF signal and an electron beam in order to extract the beam's energy. This was possible because the beam-signal interaction could be sustained almost continuously in a slow-wave helix structure. Previously, the RF field remained stationary and interaction times were diminutive. Further investigations into beam-signal interactions resulted in new microwave signal sources called backward-wave oscillators. These BWOs could be voltage tuned over a wide range of frequencies at modest power outputs. Continuing research has resulted in BWO types capable of relatively high power outputs.

PROBLEMS

1. A 6J5 triode has an amplification factor of 20.5 and a dynamic plate resistance of 8,360 ohms. What is the value of the mutual conductance? [2,450 μS]
2. A certain UHF triode has a grid-to-plate interelectrode capacitance (C_{gp}) of 2 pF. If the grid lead inductance is 1.63 nH, what is the resonant frequency of this tube? [2.79 GHz]
3. A klystron manufacturer states that the output of his device is 156 mW at 2.78 GHz. If this represents a power gain of 8.5 dB, what input power is required? [22 mW]
4. A ship's radar has a peak power output of 96 kW. If the pulse duration is 0.9 μS and the pulse repetition rate (PRR) is 856 pulses/sec, what is the average power? [74 W]
5. A certain TWT is to have an axial wave velocity along the helix of 25% of the speed of light. If the circumference of the helix is 4.75 mm, what is the helix pitch? [1.19 mm]
6. A certain low-power O-BWO is said to have an octave interval of 2.2 with a lower frequency of 7.5 GHz. What is the upper frequency? [34.46 GHz]

QUESTIONS

1. Explain briefly the operation of the vacuum tube triode as an amplifier. How is it similar to a bipolar transistor?
2. What is meant by the statement, "Triode plate resistance is a fictitious quantity"? How can such a concept be quantified?
3. What are three important triode characteristics and how are they related?
4. What two major problems beset the conventional triode at UHF operation? What attempts were made to solve these problems?
5. What problem, inherent in the triode, did the klystron take advantage of?
6. Briefly explain the operation of the multicavity, single-transit klystron. How does it differ in operation from the reflex klystron? Compare advantages and disadvantages of both devices.
7. Which device, the multicavity or reflex klystron, would you expect to find used as a local oscillator on an aircraft radar set? Why?
8. What is the purpose of the magnet on the multicavity klystron? Why is there no such magnet on a reflex klystron?

9. Briefly explain how bunching occurs in either klystron.
10. What is the purpose of the resonant cavities on either klystron? How do these affect bandwidth?
11. What does an Applegate diagram show?
12. What is meant by the modes of operation of a klystron oscillator?
13. How is the frequency of a reflex klystron adjusted?
14. The resonant cavities are always visible on a klystron. Why are they *not* visible on a magnetron?
15. What are the two principal advantages of the magnetron over the klystron?
16. Why is the magnetron called a cross-field device?
17. Briefly explain how the magnetron operates.
18. Explain the idea behind duty cycle.
19. How does a magnetron derive such huge powers from small average powers?
20. What is responsible for the broadband characteristics of a TWT?
21. What completely novel concept did the TWT bring to the area of active microwave devices?
22. Explain how the TWT helix is used to slow down the RF input signal. Why is it necessary to slow the signal down?
23. Why is an attenuator used on the TWT amplifier?
24. Explain the operation of the BWO.
25. Why is the power output of a BWO not constant across its frequency range? Externally, what can be done to offset this problem?

6

MICROWAVE SOLID-STATE DEVICES

Because semiconductor action occurs at the atomic level, the distances that an electron is required to traverse are considerably less than those encountered in a comparable thermionic device operating at the same frequency. Specifically, interelectrode spacing in solid-state devices is often less than the individual electrode dimensions themselves within a vacuum device. Obviously, then, the nanosecond response time of semiconductors can extend their operating frequencies by a factor of ten or more beyond similar thermionic counterparts. Moreover, semiconductor operation is achieved without wasted filament power. It is, therefore, apparent that the semiconductor is the device of choice when it comes to extending the usable frequency range of microwave devices.

Nothing is gained without cost, however. And such a cost is paid by the semiconductor in terms of its limited power handling ability brought about by those very same factors that permitted its operation at higher frequencies. For as device configurations decrease, so too will power handling capabilities since I^2R losses are an inescapable fact of diminished cross-sectional geometry. Of course, new developments in elevated-temperature superconducting materials[1] may well push the power limits of solid-state devices in the same direction and to the same degree as their frequency ceiling.

We will begin our investigation of microwave integrated circuit (MIC) devices with a brief discussion of passive components and then extend the topic to active circuit elements which employ semiconductor technology.

[1] *Science News* 132, July 4, 1987. Washington, D.C.: Science Service, Inc.

MICROWAVE PRINTED CIRCUITS

A microwave printed circuit board (MPCB) is designed using essentially the same methods used for the conventional printed circuit board (PCB).[2] However, actual structural materials become increasingly important, and low-loss laminates such as woven fiberglass, polyolefin, and high-density cross-linked polyethylene must be used.

FIGURE 6–1
(a) Stripline construction.
(b) E-H field pattern.

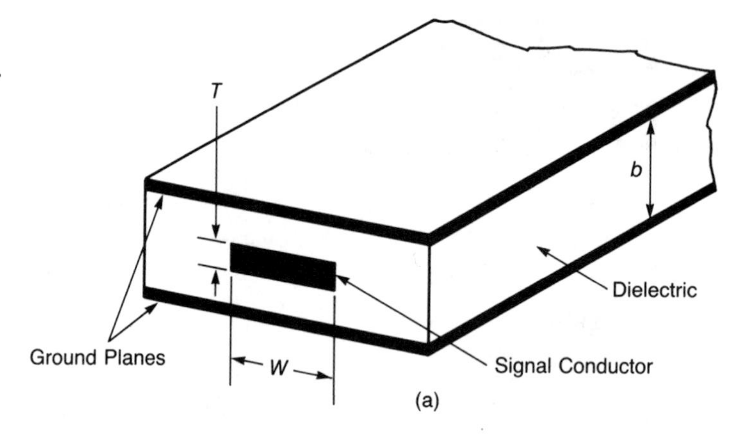

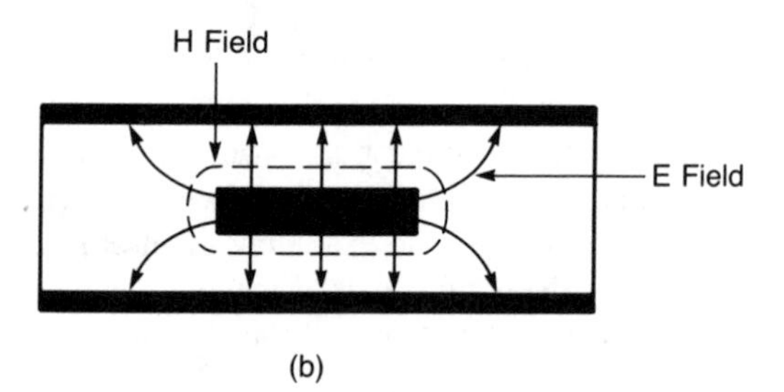

Stripline

Coaxial cable has its counterpart in a MPCB strip transmission line called *stripline*. Basic construction details and field pattern of stripline are shown in figure 6–1. Stripline MPCB construction is essentially a multilayer process using two copper-clad boards of either 1- or 2-ounce foil.[3] The lower surface of one board (the ground plane) is left

---

[2]Villanucci, R., et al. *Electronic Drafting: Printed Circuit Design*. New York: Macmillan, 1985. Chapter 6 details actual board construction.

[3]The copper thickness deposited on the bare laminate is specified in ounces per square foot. One ounce of deposited copper will produce a foil thickness of 0.0014 inch. Two ounces of copper results in a 2.8 mil foil. In other words, a one square foot sheet of copper 2.8 mils thick will weigh 2 ounces.

blank while the conductive stripline traces are etched on the other surface. The other board has copper on only one side and is laminated to the etched board by a method of dielectric bonding after all air gaps are filled with an electrically compatible material. The result is the monolithic MPCB structure shown in figure 6–1.

It is possible to have several different signals running throughout the MPCB at any given time without interference provided certain minimum spacing and trace-routing requirements are met. Note that there is no need for "side walls." This behavior was predicted from waveguide theory.

The actual characteristic impedance (Z_0) of the stripline is a function of conductor width (W) and thickness (T), substrate thickness (b), and relative permittivity (ϵ_r). The relationship between these variables is shown in figure 6–2.

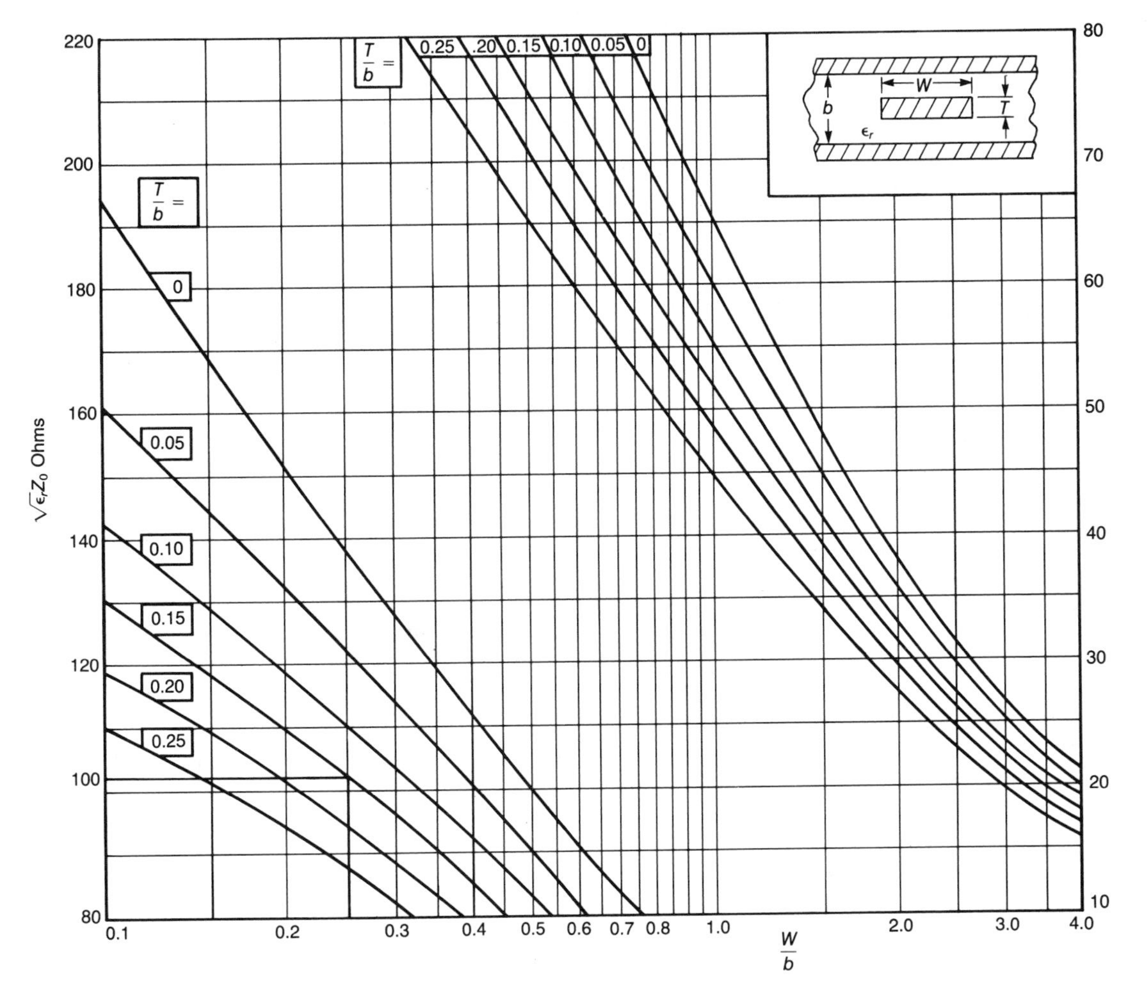

FIGURE 6–2
Graph of $\sqrt{\epsilon_r}Z_0$ versus W/b for values of T/b.

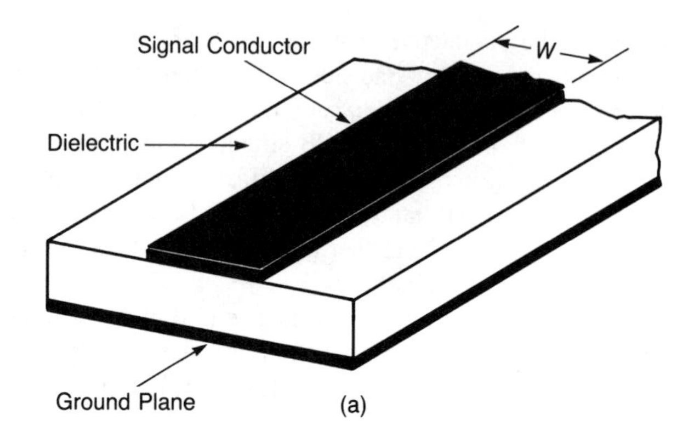

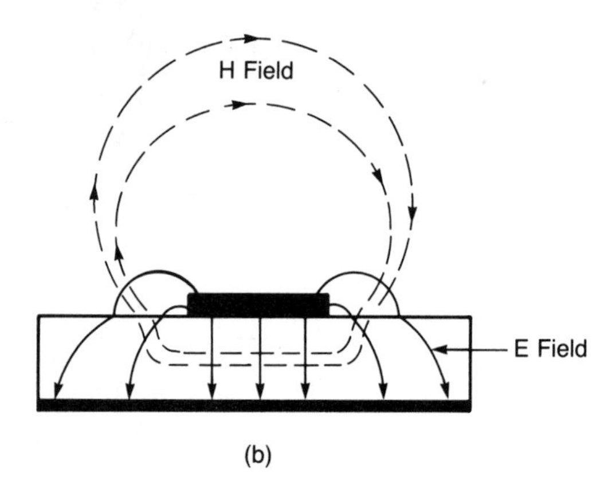

FIGURE 6–3
(a) Microstrip construction.
(b) E-H field pattern.

To use the chart in figure 6–2, suppose the ratio T/b was 0.15 and Z_0 was required to be 50 ohms with a laminate permittivity of 4.20. A simple computation shows that $\sqrt{\epsilon_r}Z_0 = 102$. Finding this value in the left-hand column shows that the ratio of W/b must be maintained at 0.25:1. The ratio of W/T for a given board thickness (b) is often determined initially on the basis of trace ampacity (i.e., current carrying ability of the conductor).

Microstrip

Parallel-wire transmission line has its counterpart in MPCBs as *microstrip*. Construction details of microstrip are shown in figure 6–3.

Microstrip has the advantage over stripline in that it is easier to fabricate and, hence, is less costly. On the other hand, its unshielded signal conductor is subject to

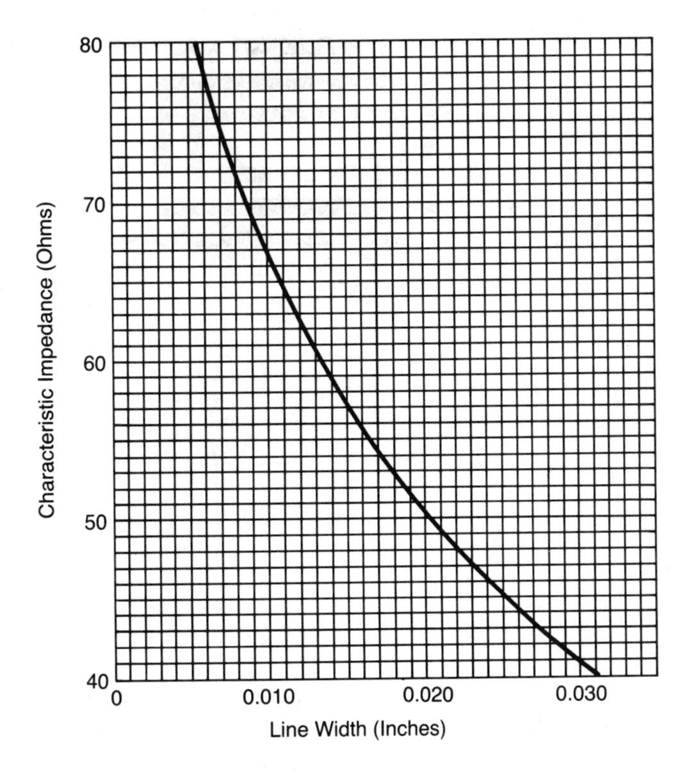

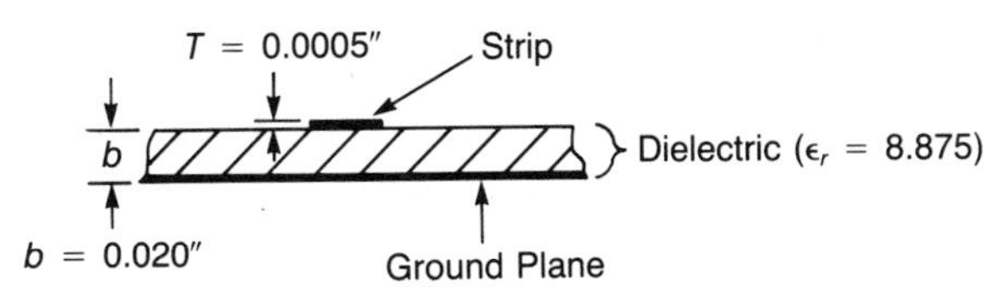

FIGURE 6–4
Z_0 versus trace width (W).

radiation and noise reception problems. As was the case with stripline, the characteristic impedance of microstrip depends on the width (W) and thickness (T) of the conductive trace; board thickness (b); and effective permittivity of the air-laminate boundary (ϵ_r). A chart showing the relationship of Z_0 to trace width (W) for a fixed board thickness (b) of 0.020 inch; conductor thickness (T) of 0.0005 inch; and dielectric constant (ϵ_r) of 8.875 is shown in figure 6–4.

Printed Microwave Components

At microwave frequencies, a significant value of inductance can be achieved with a relatively short length of conductor. Furthermore, capacitive effects are more noticeable

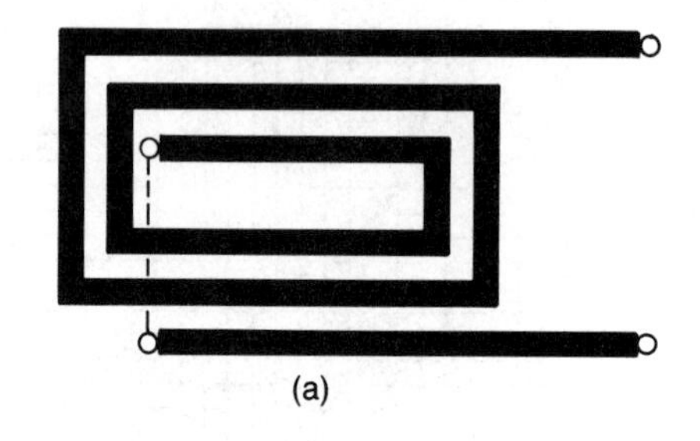

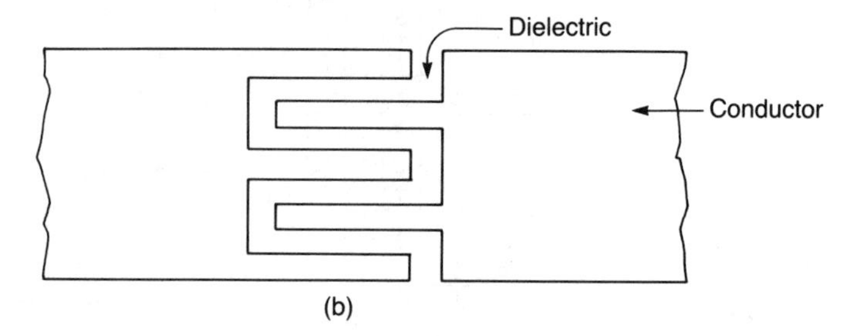

FIGURE 6–5
(a) Printed inductance.
(b) Capacitance.

at these frequencies. Therefore, it is possible to "print" (etch) L or C directly onto the MPCB, as shown in figure 6–5.

The planar (lying in one plane) inductor shown in figure 6–5(a) requires a double-sided MPCB due to the crossover path from the inner conductor. The spiral design offers the advantage of more inductance in a smaller space than comparable "S" or meander designs. Several printed spiral inductors are shown in the MMIC (microwave monolithic integrated circuit) chip illustrated in figure 6–6.

Figure 6–7 shows a Motorola #MCH 5802 thin-film inductor for use in microwave hybrid circuits. The device measures 0.22 inch square by 0.01 inch thick. The following ratings apply:

$$\begin{aligned}
\text{inductance} &= 115 \text{ nH} \\
\text{frequency} &= 0.2 \text{ GHz} \\
\text{stray capacitance} &= 0.21 \text{ pF} \\
\text{self-resonant frequency} &= 1.03 \text{ GHz} \\
\text{series R} &= 5.5 \text{ ohms} \\
I_{max} &= 250 \text{ mA at 25 degrees C}
\end{aligned}$$

The self-resonant frequency is computed by using the values of L and stray C.

Surface Acoustical Wave Devices

In order to understand the surface acoustical wave (SAW) device, we must first understand the piezoelectric effect. If a thin piece of quartz crystal, properly cut and polished, is placed between two conductive plates and a mechanical pressure applied, a voltage will appear across the plates. Releasing the pressure will cause the polarity of the voltage to reverse. This is illustrated in figure 6–8.

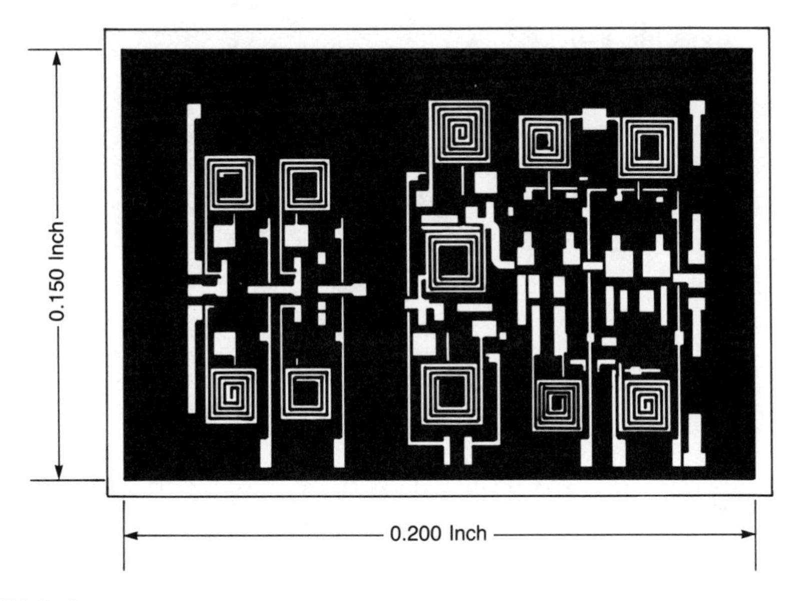

FIGURE 6–6
A Microwave monolithic integrated circuit (MMIC) chip.

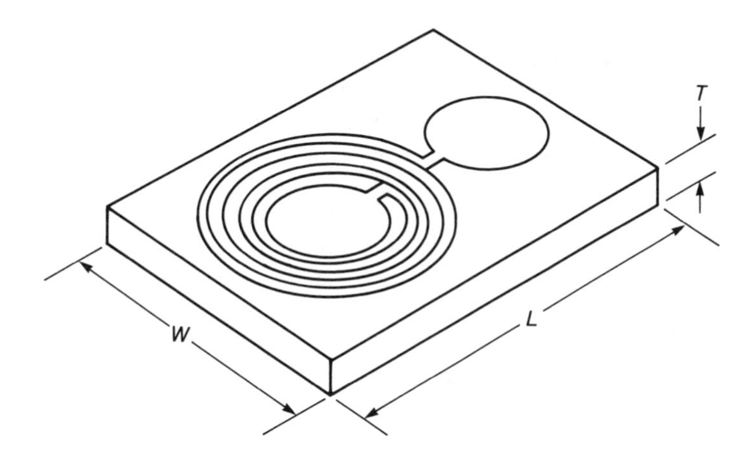

FIGURE 6–7
Then-film chip inductor. (Motorola Technical Information Center, P.O. Box 20912, Phoenix, Arizona 85036)

FIGURE 6–8
The piezoelectric effect.

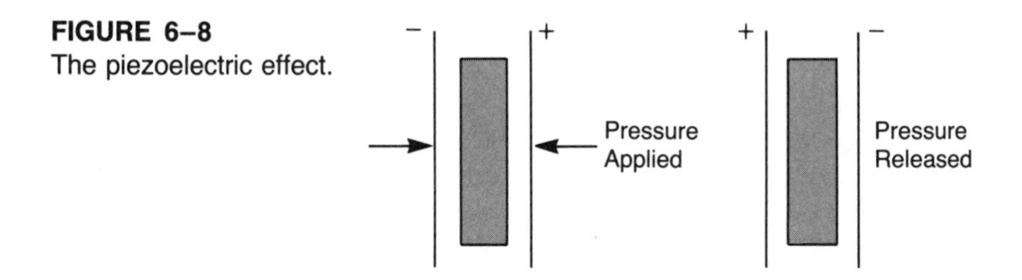

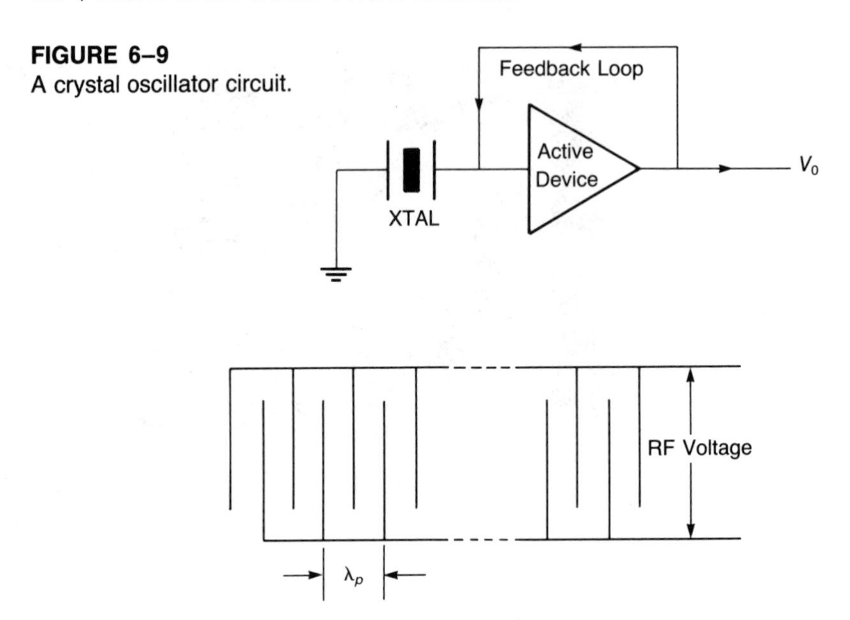

FIGURE 6–9
A crystal oscillator circuit.

FIGURE 6–10
The SAW device.

Conversely, if the crystal is shock excited by a momentary voltage pulse, the crystal will begin to oscillate mechanically at its natural resonant frequency, producing an electrical oscillation of the same frequency. These oscillations will be sustained if a properly phased feedback voltage is applied to the crystal. The result is a very stable oscillator, as shown in figure 6–9.

Above about 50 MHz, quartz crystals must be cut too thin, and, therefore, break because of their fragile nature. However, it is possible to utilize the piezoelectric principle purely as a surface phenomenon without requiring the entire crystal to flex mechanically.

If one of the faces of a crystal of quartz or lithium niobate is coated with a metallic film and etched with "fingers" appropriately spaced (see figure 6–10), it is possible to stimulate the surface of the device into mechanical resonance by the application of an RF voltage. Surface mechanical waves, traveling in both directions, will produce a pattern of standing acoustical waves by reflections from the crystal ends. The standing acoustical waves thus produced will excite the crystal piezoelectrically into oscillation at the frequency for which the crystal was prepared. Frequencies differing substantially from this frequency will obviously not be as successful in sustaining oscillations as intense as the intended (design) resonant frequency and, therefore, the SAW device may function as a narrow-band bandpass filter.

The SAW filter has a high mechanical Q and is essentially a low-loss device, thus making it very suitable as a microwave-range filter.

From the formula for wavelength, $\lambda = v/f$, it is apparent that as v decreases, λ will also decrease. And since the speed with which the surface (acoustical) wave is propagated is about 3,000 m/S (speed of sound), the distance between the device fingers will be considerably smaller than for the free-space velocity c. At 5 GHz, for example, the distance λ in figure 6–10 is:

$$\lambda = \frac{v}{f} = \frac{3{,}000}{5 \times 10^9} = 0.6\ \mu\text{m (about 24 } \mu\text{in)}$$

Consequently, the SAW filter is an extremely small package compared to a resonant cavity of the same frequency. Currently, the upper frequency limit for commercial SAW devices seems to be not much more than 5 GHz. This upper limit is a function of the present photoetching technology which places practical constraints on finger width and spacing dimensions. The lower frequency limit (50 MHz) exists only because conventional quartz crystals may be employed in this range.

REVIEW OF SEMICONDUCTOR THEORY

Since semiconductor action occurs at the atomic level, it is appropriate that we begin our discussion at that point. A brief overview of doping concepts will be followed by an explanation of general bipolar transistor behavior from the energy level perspective, instead of a circuit viewpoint. This omission is made since semiconductor circuit theory is beyond the scope of this text. Furthermore, it is assumed that the student is already familiar with such concepts. However, if the student feels the need for a review, there are many fine texts[4] that will adequately fulfill the need.

Doping of Semiconductor Materials

It is the outermost shell of electrons in the atom that interests us and that is responsible for semiconductor behavior. Therefore, we will concentrate on the configuration of the valence shell. An electrically neutral silicon atom had four valence electrons as shown diagrammatically in figure 6–11.

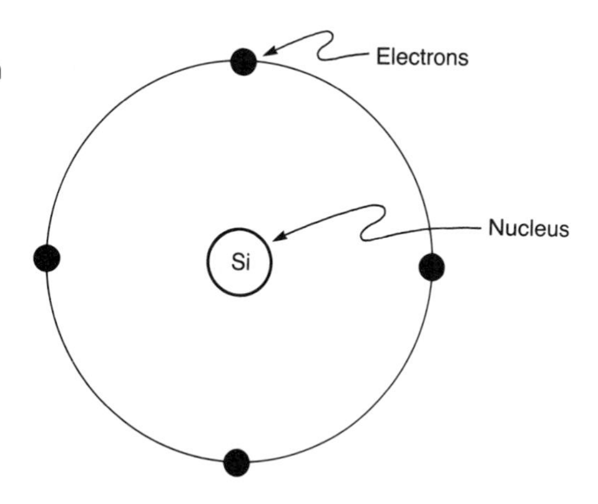

FIGURE 6–11
The valence shell of an electrically neutral silicon atom.

[4]One such text is Bogart, T. *Electronic Devices and Circuits,* Columbus, OH: Merrill Publishing Co., 1986.

In the crystal lattice, however, all the electrons of neighboring atoms are shared covalently, resulting in a complete outer valence shell of eight electrons, the maximum for this particular shell in any atom. Covalent bonding is diagrammed in figure 6–12 for five complete silicon atoms in the lattice structure. Note that there are eight (maximum) electrons surrounding each silicon atom in the valence shell.

Semiconductor doping results when impurity atoms having more or fewer valence electrons are introduced into the silicon lattice. If fewer valence electrons are present in the impurity, it will bond covalently with only seven, not eight, electrons from the surrounding silicon atoms. This incomplete sharing represents an electron deficit that acts like a positive charge, sometimes called a hole. Impurity atoms that create holes in the silicon lattice have only three electrons in their valence shell, and are called *trivalent* atoms. Examples of trivalent doping materials are aluminum and gallium. These impurities result in the formation of a type of semiconductor material called P-type, where the *P* stands for positive. Trivalent impurities bond with silicon atoms as shown in figure 6–13, where * represents a missing electron called a hole. Notice that there are only seven core atom electrons and one hole.

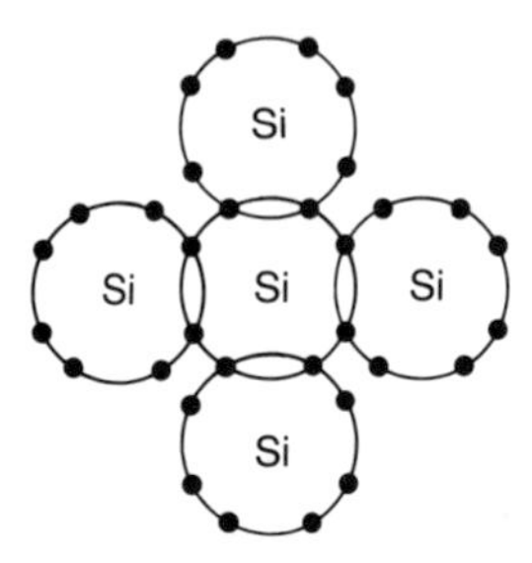

FIGURE 6–12
Covalent bonding of electrically neutral silicon atoms.

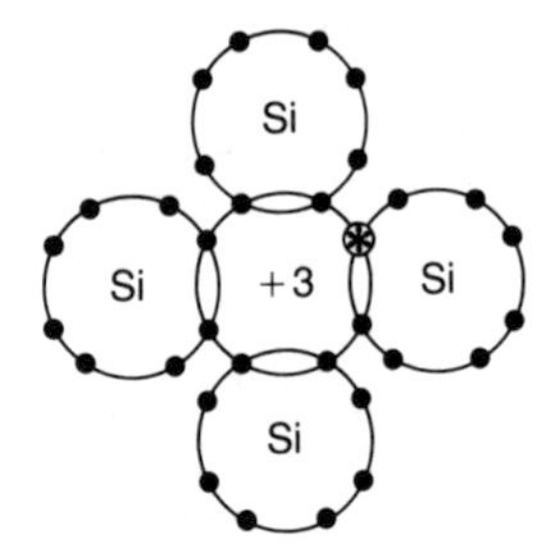

FIGURE 6–13
Formation of P-type material by trivalent doping.

In a similar manner, *pentavalent* impurities, which contain five valence electrons, will generate one extra electron in the bonding process. This leftover electron will reside in a higher-energy conduction band, producing what is called N-type semiconductor material. Typical N-type impurity atoms are phosphorus and arsenic. Pentavalent impurities will form covalent bonds with silicon atoms as shown in figure 6–14, where the * represents an extra free electron. Notice that there are nine core atom electrons, including the free electron.

Energy Levels of the P-N Junction

A bipolar transistor is formed whenever two distinct junctions are created by the union of three doped semiconductor materials alternating in type as shown in figure 6–15 for

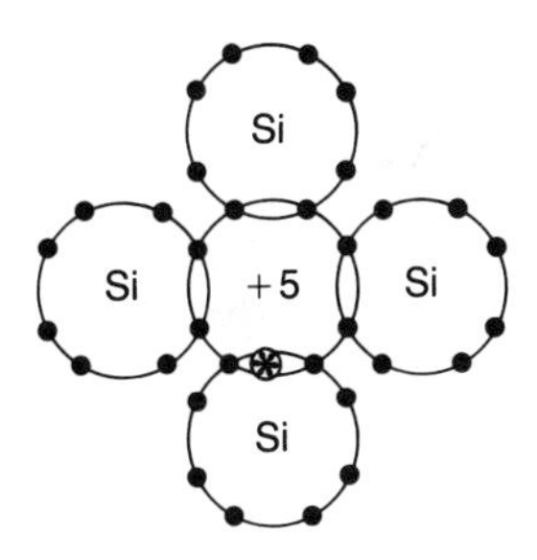

FIGURE 6–14
Formation of N-type material by pentavalent bonding.

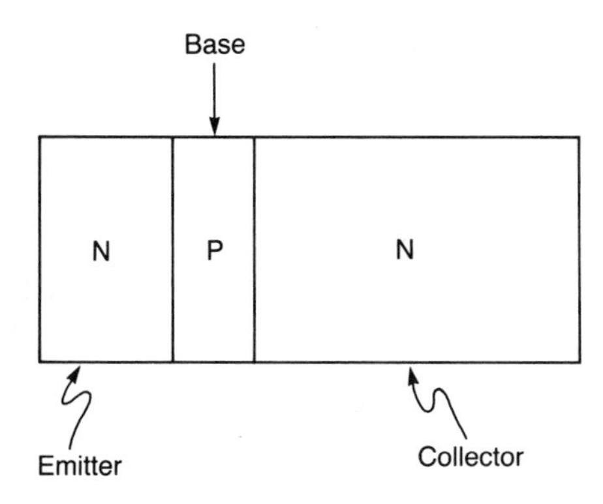

FIGURE 6–15
An NPN bipolar transistor.

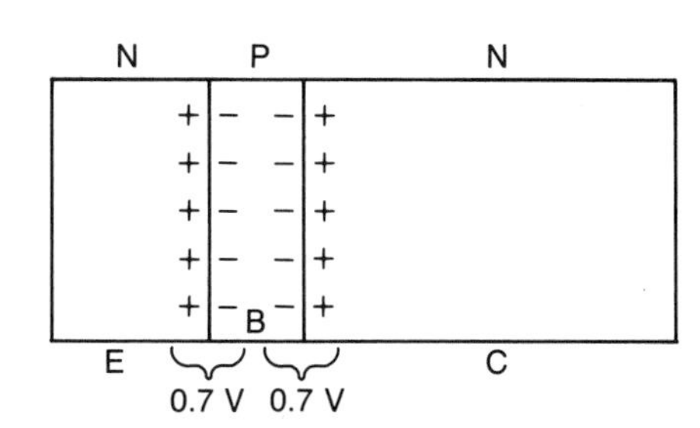

FIGURE 6–16
Barrier potential developed across junction depletion layer.

an NPN transistor. It is tempting to believe that after a short time, all the free electrons from the N-type material would migrate over to the holes in the P-type material and render the entire crystalline structure neutral. However, this sort of neutralizing migration will continue just long enough for negative ions to form in the P-type material sufficient to repel any further itinerant electrons away from the junction. For silicon-doped material, the ionic potential developed across the junction when all electron flow ceases is about 0.7 volts (at 20° C). The situation is illustrated in figure 6–16.

From the energy viewpoint, unbiased junctions appear as shown in figure 6–17 (p. 202). The P-type material has the higher energy level since, with only a +3 nucleus charge, the outermost electrons are only loosely held in orbit. The +5 of the N-type material, on the other hand, exercises a stronger hold on the outermost electrons.

Note that the conduction band free electrons in the emitter do not possess sufficient energy to climb the "energy hill" to enter the base and combine with the valence band holes. If the appropriate forward E–B and reverse B–C bias voltages are now

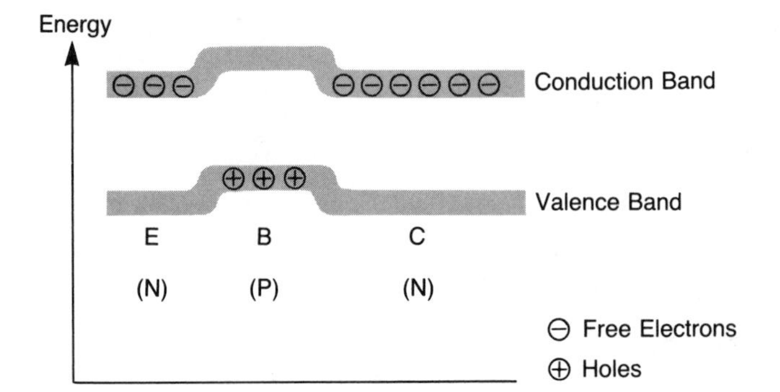

FIGURE 6–17
Energy levels of the unbiased bipolar transistor.

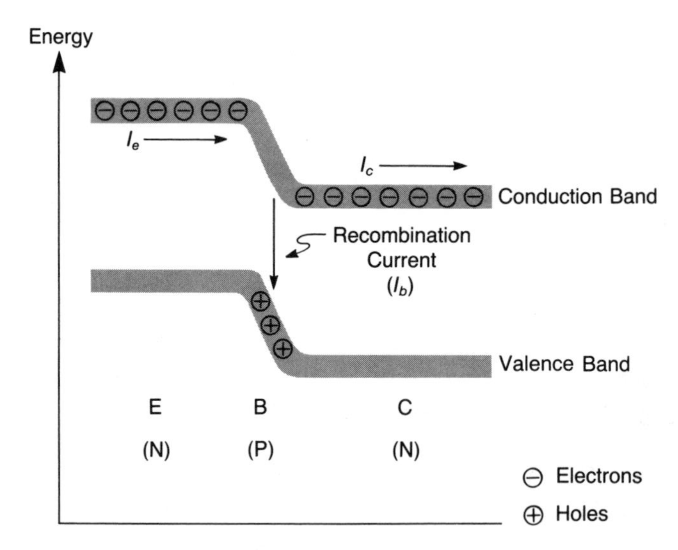

FIGURE 6–18
Energy conditions in a properly biased bipolar transistor.

FIGURE 6–19
Chip cross-section.

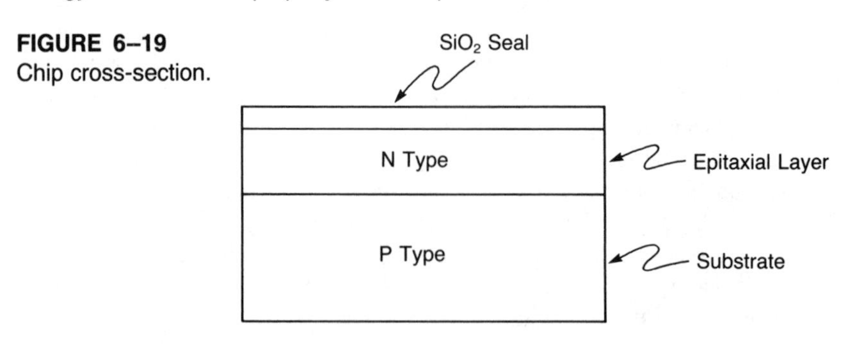

applied, the situation shown in figure 6–18 develops. Note that a small recombination current flows in the base, while most of the free electrons that enter the base region continue to flow through to the collector (i.e., the DC alpha is approximately unity). Only enough base current flows to insure that the transistor continues to conduct in step with base signal voltage variations.

Note, too, that the collector energy hill is very steep due to the reverse bias at that junction. Because of this, the electrons entering the collector descend through a high energy drop, which produces considerable heating of the collector as the electrons give up the kinetic energy of their fall. This is why the collector is always physically the largest region of the transistor.

The energy viewpoint is very useful in understanding the operation of most microwave semiconductor devices. We will have many occasions to use this energy model in subsequent sections. Therefore, the student should have a good grasp of this basic concept before going on.

FABRICATION OF THE PLANAR MICROWAVE TRANSISTOR

As we have seen again and again during the previous chapters, the function of any microwave device depends to a great extent upon how it is configured. That is, the role of the physical relationships between elemental parts is important to overall operation of the device. This was true with antennas, waveguides, resonant cavities, transmission lines, thermionic devices, circulators, and others. It is no less true with semiconductors, and an idea of fabrication techniques will prove helpful in understanding function as well as limitations of these devices. To this end, we will now briefly outline the steps in the fabrication of the planar microwave transistor, which is a diffused transistor in which the emitter, base, and collector regions are all brought out to the same plane surface.

The following steps give the general order followed in making the MIC (microwave integrated circuit) chip or individual transistor.

1. An *ingot* of P-type material is formed. Its dimensions may be on the order of several inches in diameter and 6 to 8 inches long.
2. The ingot is cut into *wafers,* which will become the chassis or substrate upon which the circuit elements will be formed. These wafers may be only 5 to 10 mils thick.
3. An *epitaxial* N-type layer is then formed by blowing a gas containing silicon atoms and pentavalent (+5) impurities across the surface of the wafer. This produces a 0.1 to 1.0 mil layer.
4. The epitaxial layer is then *passivated* by blowing pure oxygen across it, thus forming SiO_2, an inert *seal.*
5. The completed wafers are then cut into little square *chips* about 1/4 inch on a side. The resultant chip appears in cross-section as shown in figure 6–19.

1. Chip ready for etching.

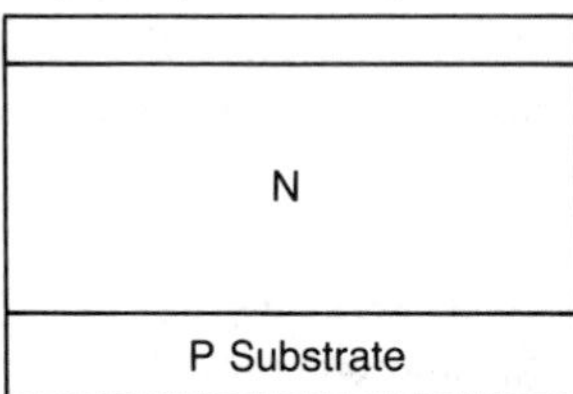

2. Part of SiO_2 layer is etched away.

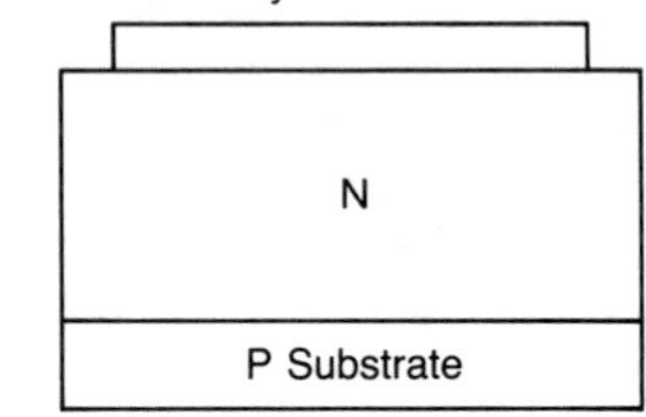

3. Trivalent atoms diffused into the epitaxial N layer changing it into P type.

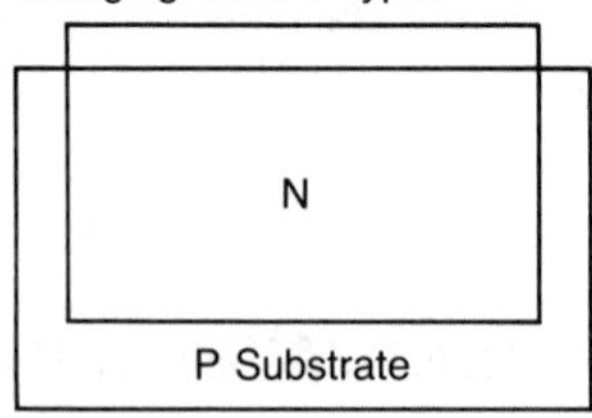

4. SiO_2 layer reformed.

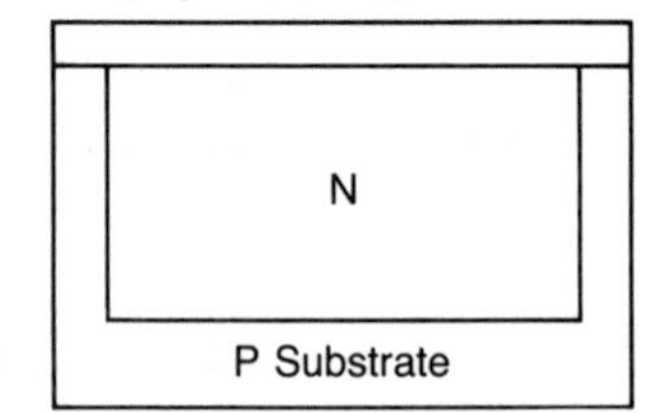

5. Window is etched into the SiO_2 layer. Now, diffuse Trivalent impurities into N layer, forming P layer.

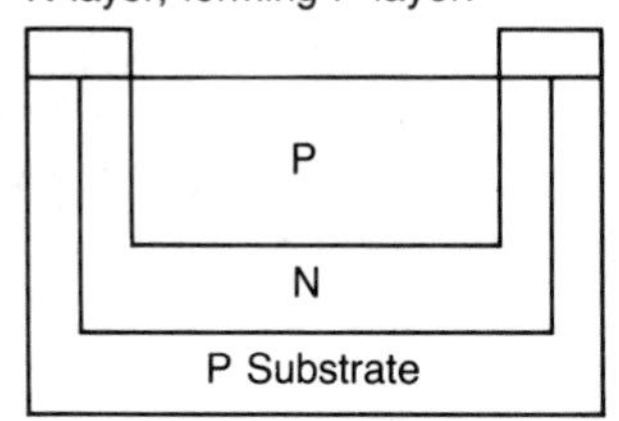

6. SiO_2 layer reformed.

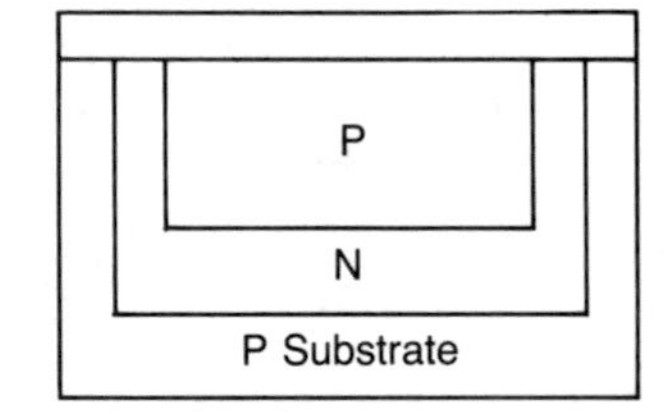

7. Window is formed, then diffuse Pentavalent impurities into P layer forming N type layer.

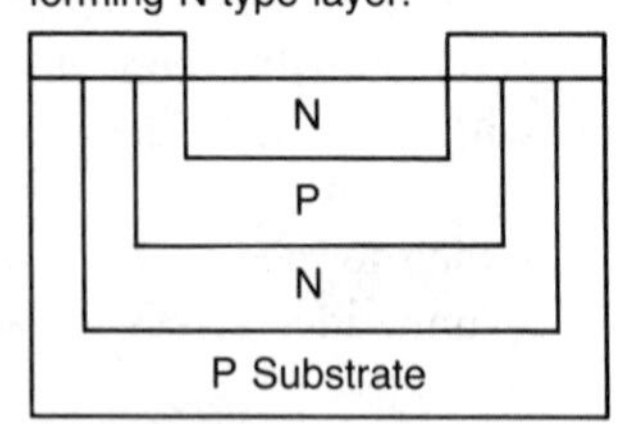

8. SiO_2 layer reformed.

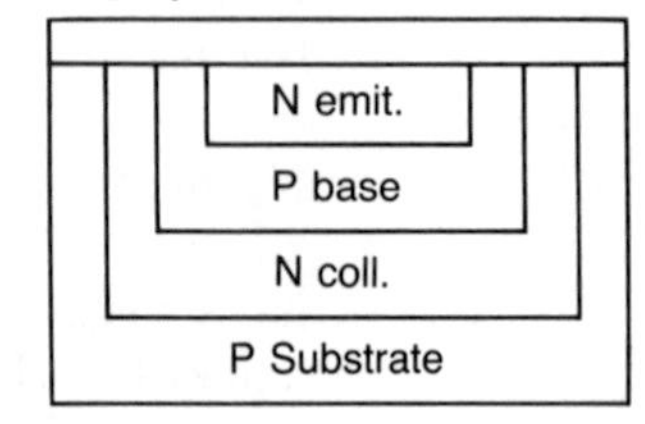

FIGURE 6–20
Steps in fabricating a planar transistor.

6. Figure 6–20 illustrates how an individual transistor would be fabricated on a single chip. It should be remembered that the foregoing operations as well as those of figure 6–20 are carried out at an elevated temperature that allows the chemical reactions to take place.

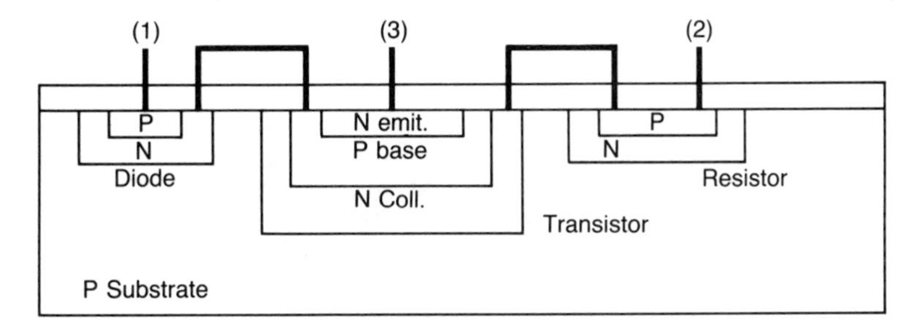

FIGURE 6–21
Fabrication of a simple transistor, diode, and resistor circuit.

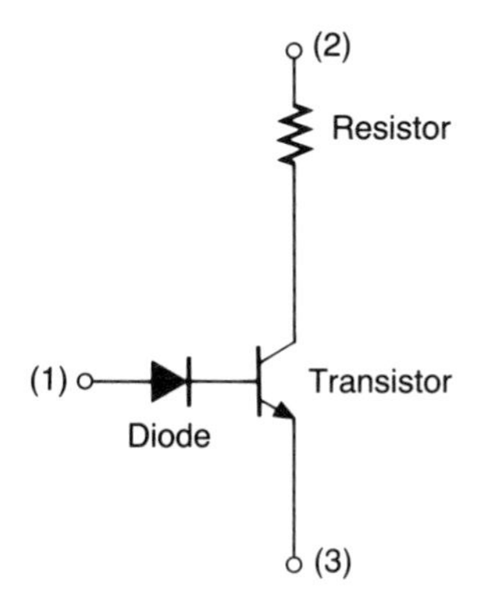

FIGURE 6–22
Transistor, diode, and resistor circuit schematic of monolithic MIC depicted in figure 6–21.

7. By adding metallic contacts at various stages of the fabrication process, we derive the three basic IC components. Figure 6–21 shows how the simple circuit of figure 6–22 might be fabricated.

THE MICROWAVE MESA FET

The invention of the bipolar transistor by Shockley and others of Bell Laboratories in 1948 revolutionized the electronics industry as a whole and extended the useful range of frequencies within the microwave region. Though vacuum tubes still dominate in high-power (both peak and average) applications, the development of the junction field-effect transistor (JFET)[5] and improved materials has enhanced both power and frequency performance of transistor devices. For example, the use of gallium arsenide (GaAs), with its higher ion mobility, has allowed the unipolar FET to outperform the

[5]The first field effect transistor (FET) was developed by Shockley and others in 1952.

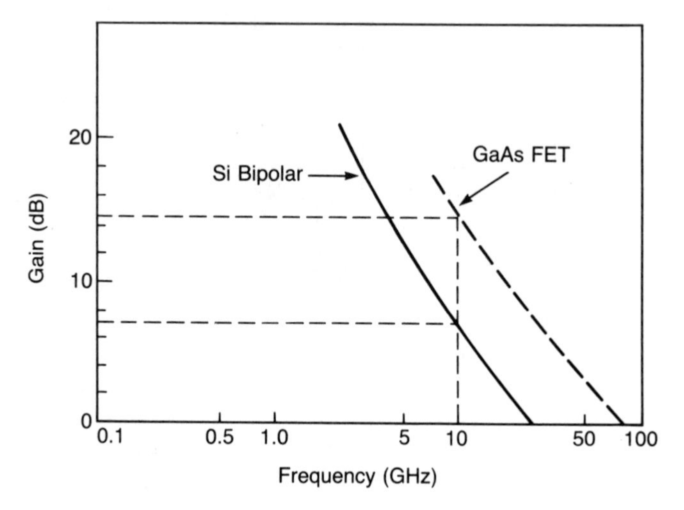

FIGURE 6–23
Comparison of gains of bipolar and FET devices.

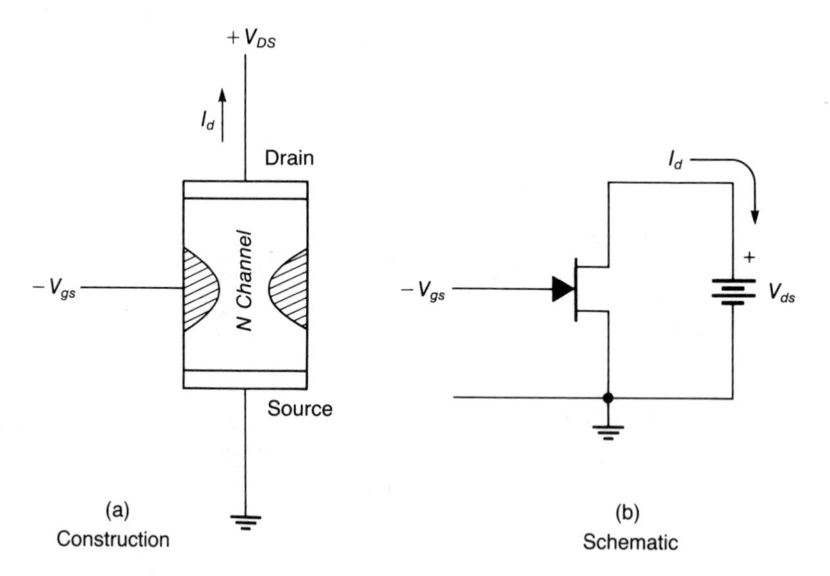

FIGURE 6–24
The N-channel JFET.

bipolar device at frequencies above 3 GHz, as shown in figure 6–23. Note the sharp decline of bipolar gains above 3 GHz. At 10 GHz, for example, the FET outperforms the bipolar by more than two to one (7 dB).

The JFET is a unipolar device which uses a transverse electric field to modulate the conductivity of a semiconductor channel through which current flows. The JFET

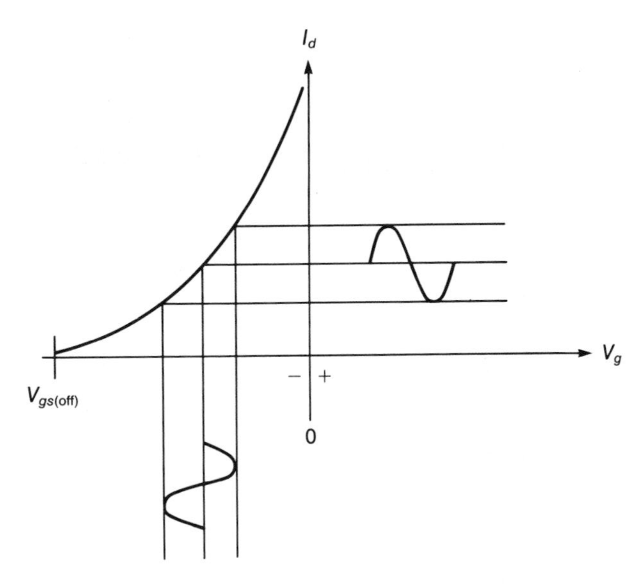

FIGURE 6–25
JFET characteristic curve.

symbol and usual circuit connection are shown in figure 6–24. In the figure, the gate-source junction is reverse biased so that gate current will not flow during the normal course of operation. Since no current flows, the device has an infinite input impedance, and is essentially a voltage-controlled device unlike the conventional bipolar which is a current-controlled semiconductor. The V_{gs}–I_d characteristic curve of a conventional JFET is shown in figure 6–25.

As the gate voltage is made more negative, the electric field traversing the N-channel between the gate halves becomes more intense, pinching the channel closed and reducing the flow of current between source and drain. Once again, it should be observed that gate current does not flow.

One of the biggest problems with semiconductor junctions at microwave frequencies concerns the idea of *charge storage*. For example, with a bipolar transistor, conduction-band charges exist for a finite interval before recombining with minority carriers in the valence band. As a result, a capacitive ''storage'' effect exists when an attempt is made to suddenly reverse bias the junction. In other words, it takes a certain amount of time to turn off a forward-biased junction, during which forward current continues to flow. At microwave frequencies, therefore, it is difficult or impossible for the device to respond to the rapid reversals of the signal voltage.

To get around the problem of charge-storage effect at microwave frequencies, the mesa field effect transistor (MESFET) using a Schottky barrier gate was introduced. In cross-section, the MESFET looks very much like a mesa (plateau), as shown in figure 6–26.

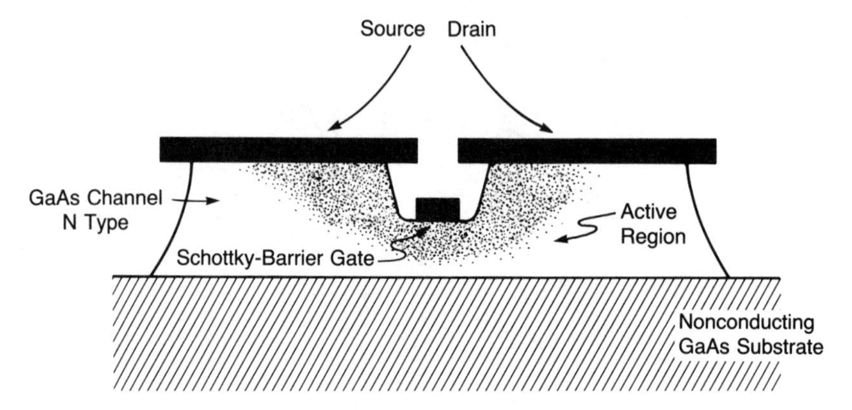

FIGURE 6–26
The microwave MESFET with Schottky barrier gate.

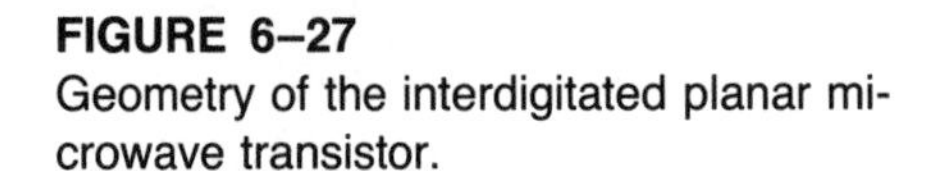

FIGURE 6–27
Geometry of the interdigitated planar microwave transistor.

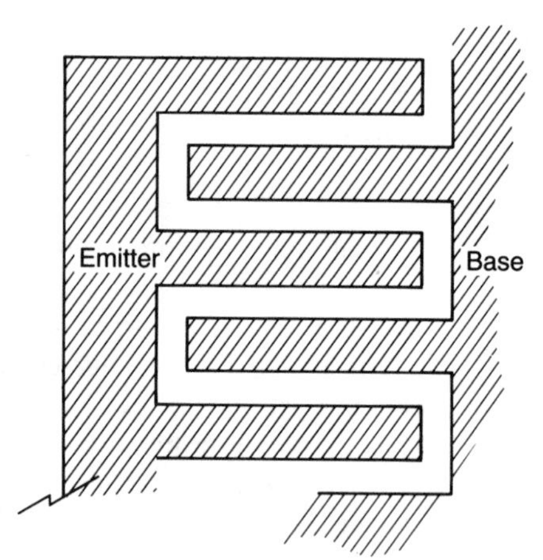

Because of the Schottky barrier gate, there are no barrier-type depletion layer and no stored charges. This is due to the fact that electrons are the majority carriers on *both* sides of the junction. Consequently, the on-off time is decreased many times over that of a conventional N-P bipolar junction. Hence, the upper-frequency limit of the device is extended considerably.

Before leaving this section, we should discuss one other interesting bipolar geometry which has led to both increased power handling as well as extended frequency operation of the microwave planar transistor. As shown in figure 6–27, the *interdigitated* geometry of the base-emitter region leads to an increased edge perimeter, which translates into higher current flow and more power. Moreover, the reduced emitter area means faster transit time through the device and, therefore, higher frequency of operation. The collector is located beneath the E-B fingers shown in figure 6–27.

FIGURE 6–28
Hybrid GaAs FET MIC power amplifier. (EPSCO, 31355 Agoura Rd., Westlake Village, CA 91361)

MICROWAVE INTEGRATED CIRCUITS

A *hybrid MIC* is one in which the passive components (e.g., resistors) are etched into a metallic film deposited onto a substrate. The active devices (transistors and diodes) are then added as discrete components, soldered or bonded into place on the chip. Figure 6–28 shows an EPSCO miniature GaAs FET hybrid MIC power amplifier with the top cover removed.

As an example of performance specifications, EPSCO's X-band hybrid MIC is rated at 30 dB with outputs as high as 8 watts. We might ask ourselves the following questions about this device:

1. What is the minimum amount of input power required to obtain the 8-watt maximum output?
2. What is the dBm rating of this device?
3. What output power would we expect if the input power was 1 mW?

To answer the first question, we proceed as follows

$$G' = 10 \log \frac{P_o}{P_i}$$

$$30 \text{ dB} = 10 \log \frac{8}{P_i}$$

$$P_i = 0.008 \text{ watt (8 mW)}$$

The answer to the second question is given by

$$G' = 10 \log \frac{P_o}{0.001}$$

$$= 10 \log \frac{8}{0.001}$$

$$= 39 \text{ dBm}$$

Finally, if 1 mW is the input power, the output power is

$$G' = 10 \log \frac{P_o}{0.001}$$

$$30 \text{ dB} = 10 \log \frac{P_o}{0.001}$$

$$P_o = 1 \text{ watt}$$

A microwave voltage-controlled oscillator (VCO) using hybrid GaAs FET technology is shown in figure 6–29.

Microwave circuits are also available as *monolithic* devices. Such a monolithic MIC (MMIC) has all devices fabricated directly on the substrate. Discrete devices are not used. The word *monolithic* itself literally means "one stone," and that definition is particularly descriptive of these devices. An example of a monolithic GaAs amplifier chip is shown in figure 6–30.

Gallium arsenide (GaAs) has become the material of choice over silicon because of its higher ion mobility, and FETs have the advantage of lower noise at microwave frequencies over bipolar devices of comparable gain. Currently, other materials are being explored in an effort to expand both power and frequency limits. One such material is gallium indium arsenide (GaInAs).

MICROWAVE DIODES

There is a variety of semiconductor materials whose bulk and junction properties have created a large and growing number of important microwave devices. Most of these devices are two terminal affairs and are, therefore, called diodes. This nomenclature is somewhat unfortunate, however, since these devices have little to do with ordinary diode operation. We will take a brief look at some of these "diode" properties and their applications to microwaves. We begin with an investigation of a reverse-biased P-N junction.

The Varactor

Figure 6–31 shows the energy level diagram of an unbiased P-N junction. Point b represents the physical junction of the N and P materials. The distance c–a is the width

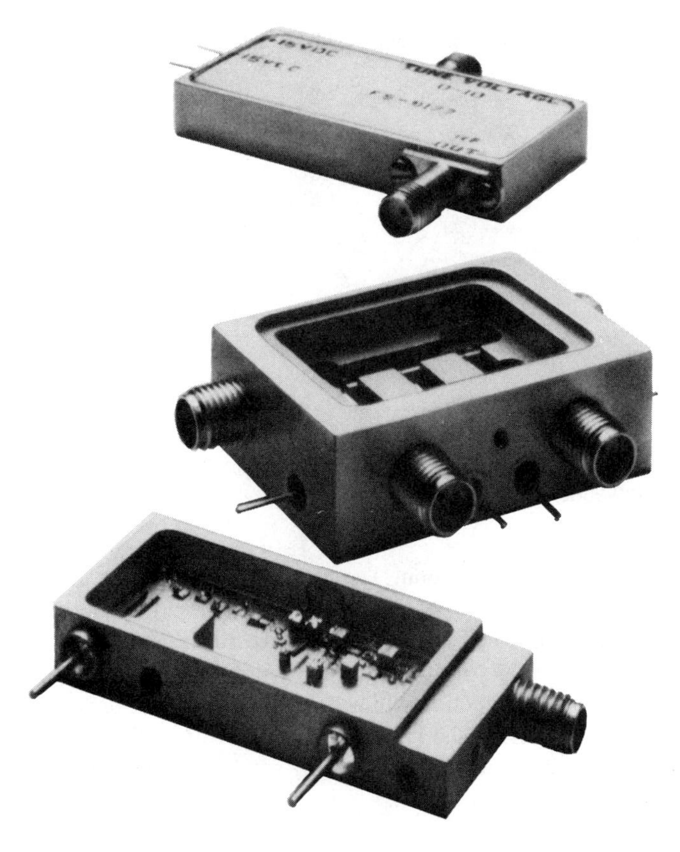

FIGURE 6–29
An MIC hybrid VCO. (Frequency Sources/subsidiary of Loral Corporation, 16 Maple Rd., Chelmsford, MA 01824)

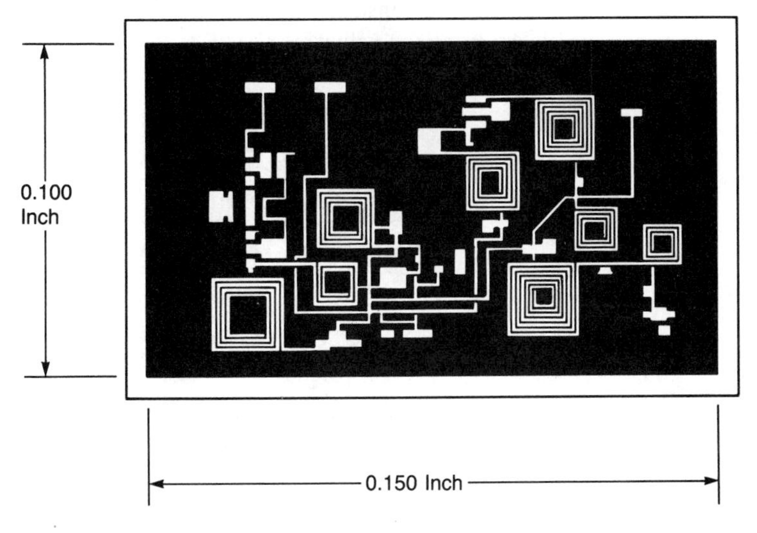

FIGURE 6–30
MMIC GaAs low-noise amplifier chip.

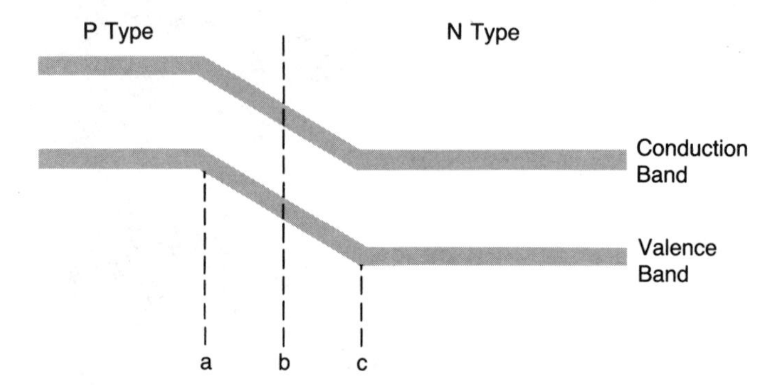

FIGURE 6–31
Unbiased P-N junction.

of the unbiased junction. The distances b–a and c–b are the distances into the P and N materials, respectively, where the electric field of the barrier potential permeates. The existence of the barrier potential is an artifact of ion formation within the junction region and has a value of about 700 mV. If an external voltage is applied to aid this barrier potential, the junction width c–a will increase. Conversely, if the applied voltage opposes the barrier, the junction width will diminish.

With the foregoing in mind, recall that the capacitance of a simple parallel-plate capacitor is given by

$$C = \epsilon_r\left(\frac{A}{D}\right) \tag{6–1}$$

where ϵ_r is the dielectric constant, A is the area of the plates, and D is the distance between the plates. For fixed values of ϵ_r and A, the capacitance becomes a function of D. Table 6–1 shows the dielectric constants for selected semiconductor materials.

TABLE 6–1
Dielectric constants of semiconductor materials.

Material	Dielectric Constant (ϵ_r)
Ge	16.0
Si	12.0
GaAs	13.5
InP	10.6
*InSb	16.5

*Indium antimonide.

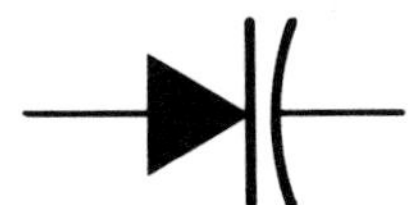

FIGURE 6–32
Schematic symbol of the varactor diode.

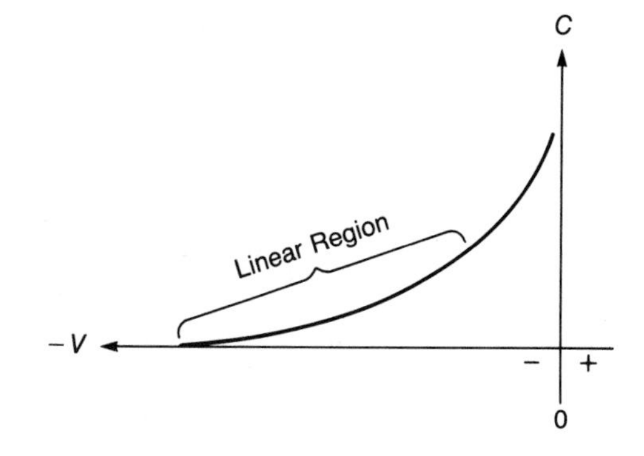

FIGURE 6–33
C versus V_r for varactor diode.

From the previous discussion concerning junction width and applied potentials, it should be obvious that a reverse-biased P-N junction may function as a voltage-controlled capacitance. Such a capacitive device is called a *varactor*. Its schematic symbol is shown in figure 6–32.

In practice, selective doping of the junction material optimizes the capacitance effect and creates a more linear response in the reverse-biased region. A typical capacitance versus reverse voltage characteristic curve of a varactor is shown in figure 6–33.

Varactors are frequently used as circuit elements in parametric amplifiers (PARAMPS), which are low-power, low-noise, narrow-band devices. They are low noise because resistors are never employed as circuit elements. The word *parametric* is used since amplifier operation depends on the change in some circuit parameter to achieve gain. In the following discussion, capacitance is the circuit parameter.

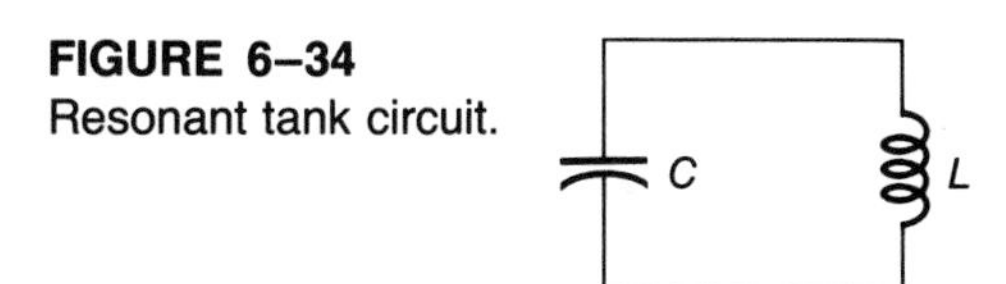

FIGURE 6–34
Resonant tank circuit.

A brief description of the principles of PARAMP operation will now be given. Consider the resonant circuit, shown in figure 6–34, which is oscillating at its natural resonant frequency. If the plates of capacitor C are suddenly pulled apart physically at the instant the charge Q is a maximum, then the voltage must increase since the value of Q has not changed, but the value of C has decreased. The truth of this assertion may be seen from the following simple relationship. Since the charge of a capacitor is given as

$$Q = CV \tag{6–2}$$

then, upon solving for V, we obtain

$$V = \frac{Q}{C} \tag{6–3}$$

Note that if the numerator (Q) remains fixed, the value of the fraction (V) must increase as the denominator (C) decreases.

Of course, the capacitance is not varied mechanically, but rather electronically using a "pump" circuit to modulate the varactor. In order to amplify without phase distortion of the input signal, the varactor capacitance must be varied at twice the resonant frequency. Consequently, PARAMPS are phase sensitive. The relationship between varactor capacitance, charge, and output voltage is shown in figure 6–35. Note that the pumping frequency (a), which is the same as the capacitance variation frequency, is exactly twice the signal voltage frequency (c).

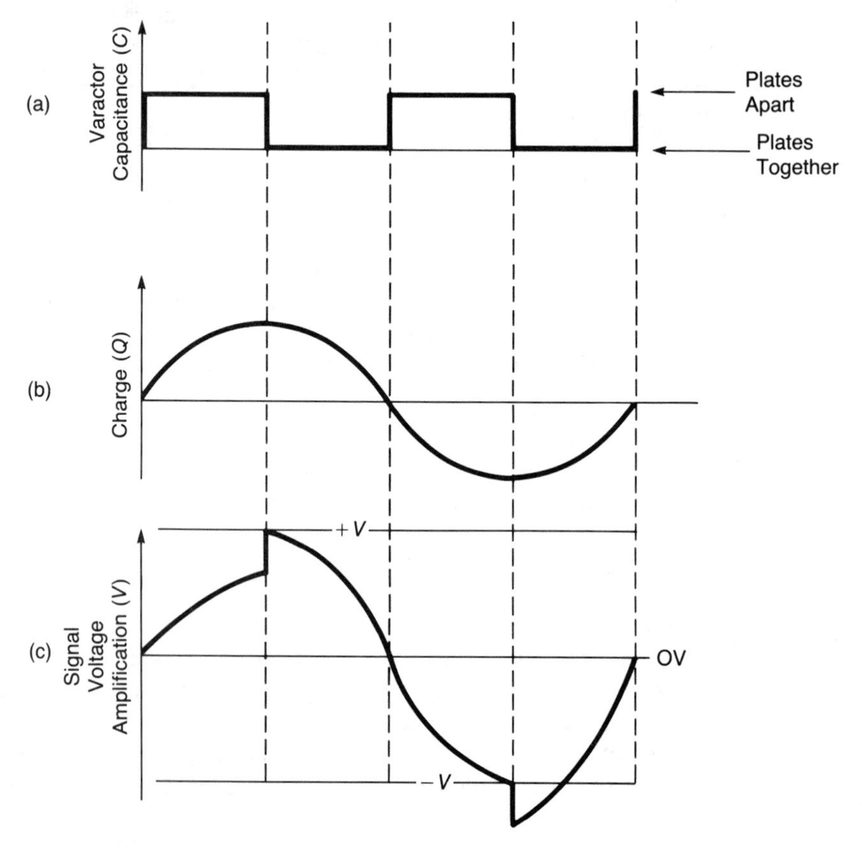

FIGURE 6–35
Pumping frequency diagram.

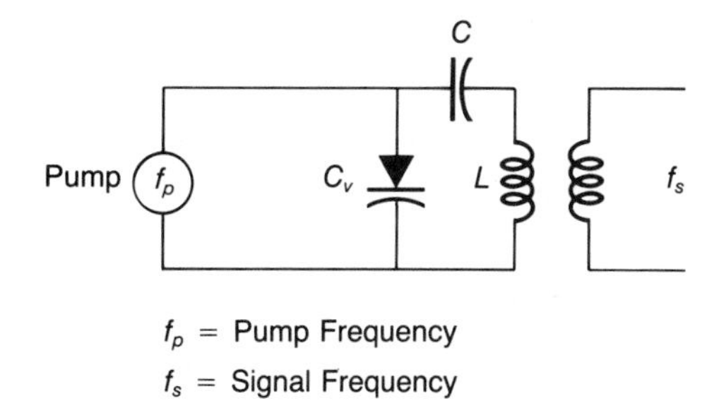

FIGURE 6–36
Basic circuit of degenerate-mode parametric amplifier.

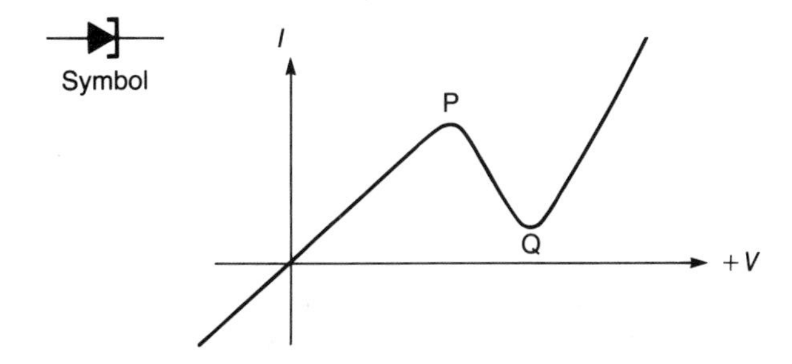

FIGURE 6–37
Tunnel diode characteristic curve and schematic symbol.

In this *degenerate* mode, the signal builds up with each step. This buildup would increase infinitely were it not for the fact that more energy is required to increase the voltage across the capacitor with each charge increment, and the pump energy is finite. The basic degenerate-mode parametric amplifier is shown in figure 6–36.

The Tunnel Diode

Before getting into a discussion of the operating principles of the tunnel diode, let us take a look at its somewhat unusual characteristic curve shown in figure 6–37. The first thing we notice about the curve is that for voltages below point P corresponding to less than about 50 mV (for germanium), the diode behaves linearly just as an ordinary resistor. From P up to about 300 mV at Q, however, the diode exhibits a negative resistance region wherein current actually decreases as forward voltage across the junction increases. Beyond point Q, the device behaves like an ordinary diode with the curve rising sharply. It is the negative resistance region lying between P and Q that is of interest to us and that makes the tunnel diode a useful device capable of gain. Before investigating how gain is possible with this device, we will attempt to explain the semiconductor properties responsible for such unique behavior.

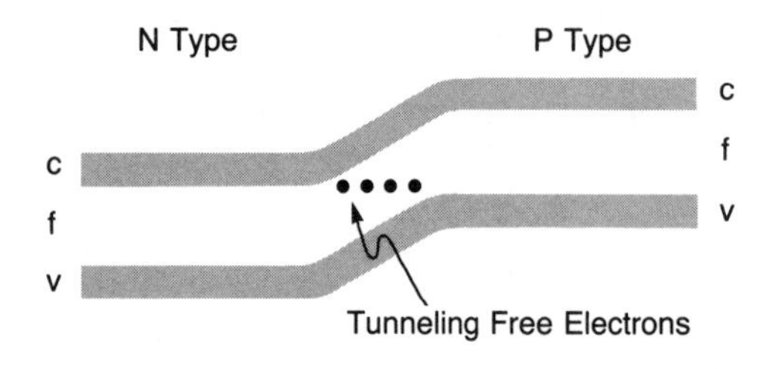

FIGURE 6–38
Energy levels at $V = 0$. Conduction band is labeled *c*, valence band *v*, and forbidden region *f*.

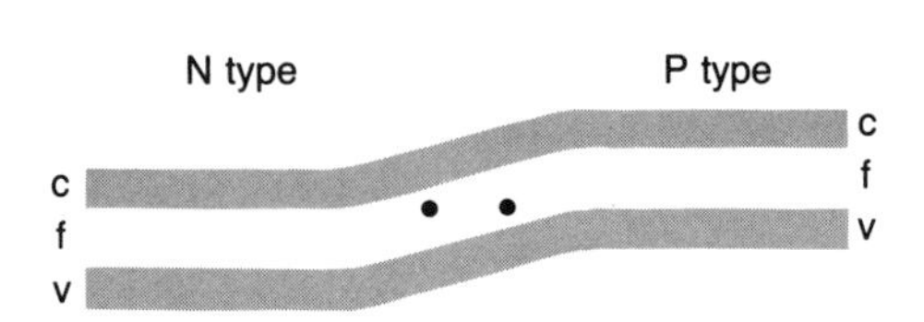

FIGURE 6–39
Energy level situation in negative-resistance region.

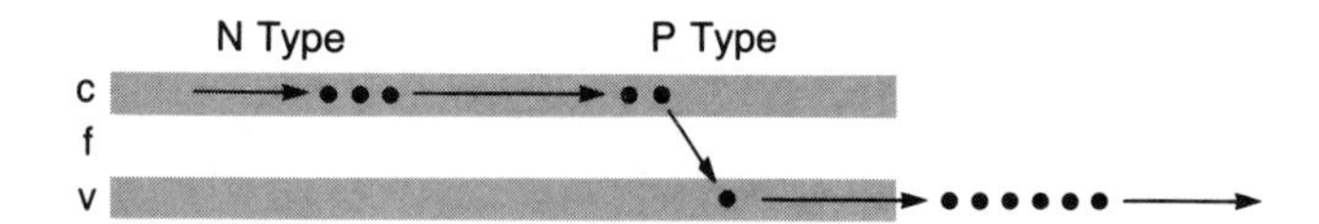

FIGURE 6–40
Energy level diagram for normal diode action.

To begin with, the tunnel diode is heavily doped so that free electrons may "tunnel" through the junction from the N-type conduction band to the P-type valence band on the other side in the absence of any forward voltage. However, since as many electrons move one way as the other, the net current flow is zero, as shown in the curve in figure 6–37. This condition of zero-bias zero current is shown in the energy level diagram of figure 6–38.

As the forward voltage is increased up to point P in figure 6–37, the N-type conduction band begins to line up with the forbidden region, and current flow declines as shown in figure 6–39. Fewer free electrons are able to tunnel through the junction. As the voltage is increased still further, the current continues to fall until the conduction band coincides with the forbidden region and current flow all but ceases at point Q of figure 6–37.

Finally, as the voltage is increased beyond Q, the conduction and valence energy levels coincide and normal diode action resumes as shown in figure 6–40. Electrons enter the conduction band, fall into holes in the valence band, and exit the diode.

With the foregoing understanding of the negative resistance effect in the tunnel diode, we will now provide the mathematical proof that gain is possible with this device. The parallel circuit shown in figure 6–41 will be used for convenience in the mathematical analysis of gain through negative resistance.

With the negative-resistance device removed from the circuit, the maximum power available from the source is obtained when $g_s = g_L$—that is, when the total conductance (G_t) equals $2g_s$. Also, the maximum power available to the load is one half the maximum power available from the generator. In other words,

$$P_{\max} = \frac{1}{2}\left(\frac{I_s^2}{2g_s}\right) = \frac{I_s^2}{4g_s} \tag{6–4}$$

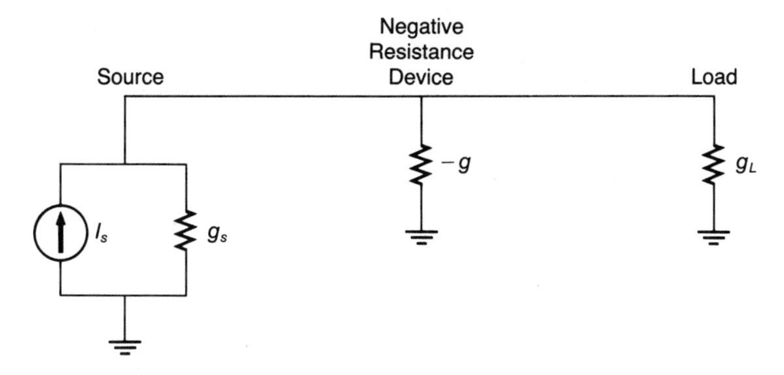

FIGURE 6–41
Conceptual negative-resistance amplifier circuit.

Putting the negative-resistance device back into the circuit will give a voltage drop across the load as determined by

$$v_L = \frac{I_s}{(g_s - g + g_L)} \tag{6–5}$$

Therefore, the power delivered to the load may be computed as

$$P_L = v_L^2 g_L = \left(\frac{I_s}{g_s - g + g_L}\right)^2 g_L \tag{6–6}$$

Now, if power gain (G) is to be a reality of the device, then by definition

$$\begin{aligned} G &= \frac{P_o}{P_i} = \frac{P_L}{P_{\max}} \\ &= \frac{I_s^2 g_L}{(g_s - g + g_L)^2} \times \frac{4g_s}{I_s^2} \\ &= \frac{4g_s g_L}{(g_s - g + g_L)^2} \end{aligned} \tag{6–7}$$

Therefore, in order to show that gain does, in fact, occur, it is only necessary to show that $G > 1$.

So, for maximum power transfer, $g_L = g_s$ as before. Therefore,

$$\begin{aligned} G &= \frac{4g_L^2}{(2g_L - g)^2} \\ &= \frac{4g_L^2}{4g_L^2 - 4g_L g + g^2} \\ &= \frac{4g_L^2}{4g_L^2 + g(g - 4g_L)} \end{aligned} \tag{6–8}$$

If the second term of the denominator of equation (6–8) is zero, $G = 1.0$. However, $G > 1$ if this term is negative—that is, if $g < 4g_L$. It is always possible in practice to arrange this ($R_L < 4R_{neg}$), so gain is always possible with a negative-resistance device. This concludes the first proof.

Note, too, that if the denominator of equation (6–8) is zero, the gain is infinite, and oscillations will occur. This happens when $g = 2g_L$. Substituting this into equation (6–8) gives

$$G = \frac{4g_L^2}{4g_L^2 - 4g_L(2g_L) + (2g_L)^2} = \frac{4g_L^2}{0} \tag{6–9}$$

Thus, we have just shown that any device that exhibits the property of negative resistance may be used actively as either an amplifier or an oscillator. Note that for amplification, $R_L < 4R_{neg}$, but for oscillation, $R_L = 2R_{neg}$. The implication here is that care must be taken in selecting R_L in order to avoid oscillation if amplification is desired.

The Gunn Diode

There is a group of microwave diodes whose interesting properties depend on bulk characteristics of the materials used rather than upon the action of any junction phenomenon. Accordingly, such devices are often called bulk-effect devices, and they make use of a variety of semiconductor materials including gallium arsenide (GaAs), indium phosphide (InP), cadmium telluride (CdTe), or indium arsenide (InAs). Because the properties of such materials depend primarily on what happens to the electrons as they are transferred to a higher-energy conduction band, these devices are also referred to as transferred-electron devices (TEDs). One such widely-used TED is the Gunn diode, which is used primarily as a low-power microwave oscillator. The Gunn diode is named after John Gunn of IBM who discovered the transferred-electron property in N-type GaAs in 1963.

The operating principle of the Gunn diode is as follows. If a voltage gradient in excess of 3.3 kV/cm is established across a thin section of N-type GaAs, the electrons will be transferred to a higher energy conduction where, because of their now higher effective electron mass, their motion slows down. The result of slower electron velocity under increased applied voltage is a negative-resistance effect. The characteristic curve of the Gunn diode is, therefore, very much like that of the tunnel diode, although for much different reasons.

Note that the voltage of 3.3 kV/cm mentioned above is not an *applied* voltage, but rather a voltage *gradient*. Therefore, for example, if a 6 μm thick Gunn slice is used, the applied voltage across the section would need to be only about 2 volts in order to maintain the required gradient.

It should be apparent that the use of such high-energy electrons in the TED materials generates a considerable amount of heat. In fact, such devices are said to operate with "hot" electrons, as opposed to conventional transistor and semiconductor devices,

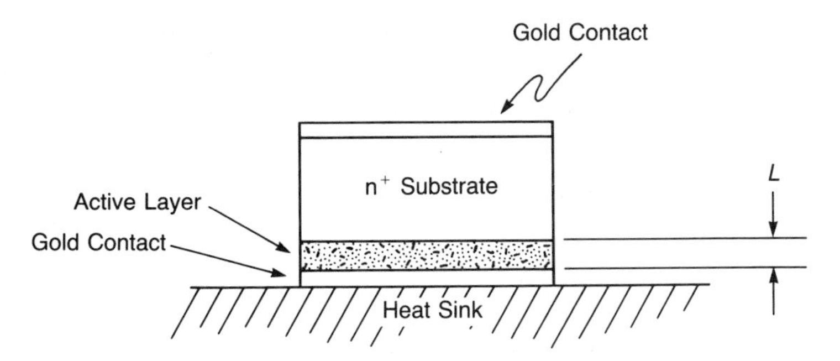

FIGURE 6–42
Typical GaAs epitaxial Gunn element.

FIGURE 6–43
Growth of the Gunn domain along the device.

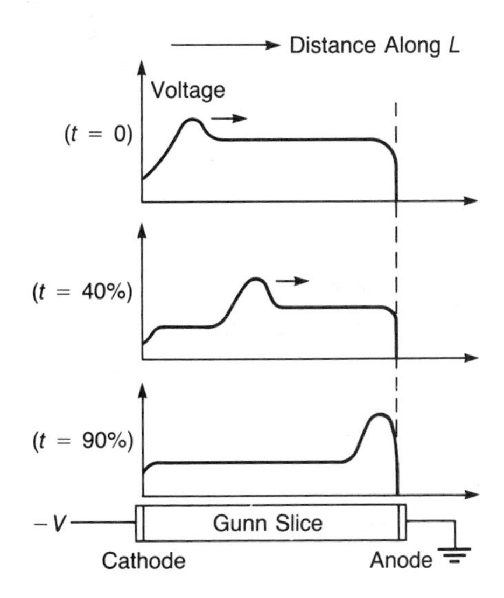

which operate closer to room temperatures. Accordingly, all TED applications necessitate the use of relatively massive heat sinks. Figure 6–42 shows a typical Gunn diode package. Note the relatively massive heat sink and the epitaxially grown N-type active layer.

The negative-resistance effect of the Gunn diode is not the only phenomenon at work in the TED. Indeed, the very cause of this effect is at the heart of the so-called Gunn mode. As mentioned earlier, the velocity of the high-energy electrons actually slows down as they are transferred between conduction bands. However, this effect is neither uniform nor simultaneous in all parts of the N-type material. As a result, an electron bunch, called a domain, is initiated at the cathode end and migrates rapidly toward the anode end of the device, as shown in figure 6–43.

The domain-progression process repeats itself once the anode is reached. It may be seen, then, that the Gunn frequency is a function of the total distance traversed by the electrons through the device. The Gunn oscillator frequency (f_g) is given by

$$f_g = \frac{V_s}{L} \tag{6–10}$$

where V_s is the saturation velocity of the Gunn material and L is the path length of the material. The average domain velocity is about 1×10^7 cm/S and L may vary between 1 μm and 40 μm at the lowest frequency end of the operating spectrum.

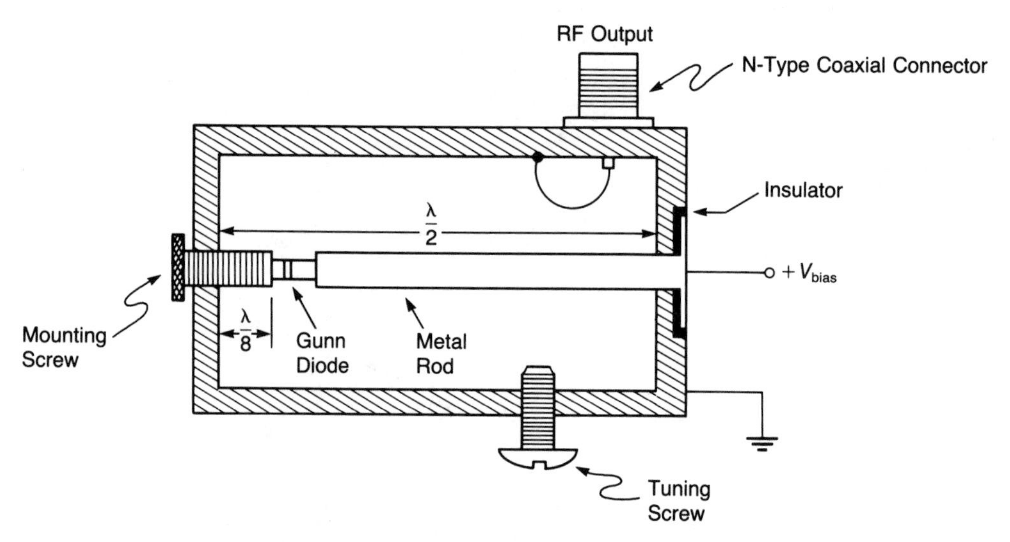

FIGURE 6–44
Cross-section of coaxial cavity with Gunn diode in place.

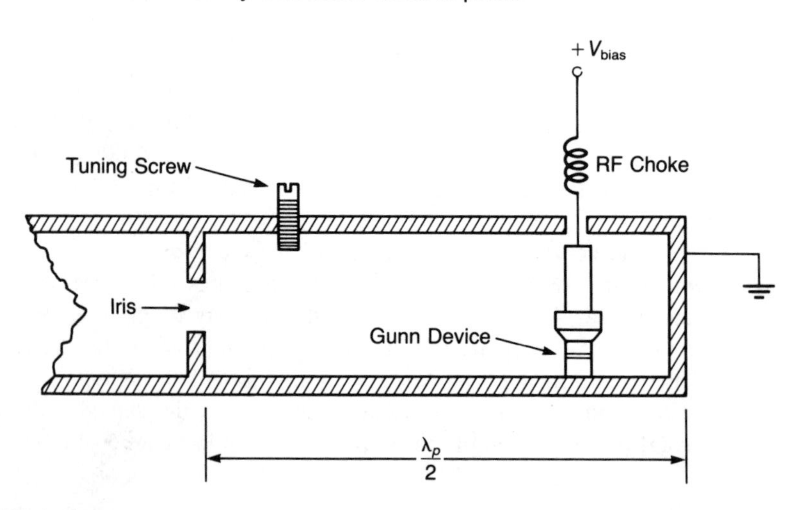

FIGURE 6–45
Gunn diode in a rectangular waveguide cavity.

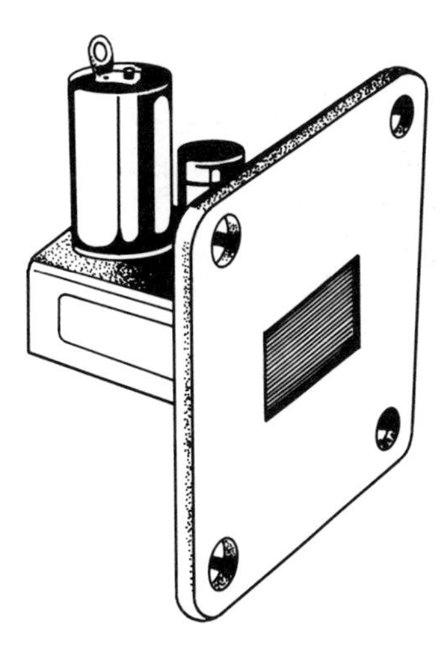

FIGURE 6–46
X-band Gunn oscillator. (Courtesy of M/A-COM GaAs Products, Inc.)

The arrival of the domain at the anode end shock excites any associated tuned circuit into oscillation. Note that the Gunn diode does not require an external resonant circuit to operate, but its frequency is less stable than if placed in a coaxial cavity as shown in figure 6–44 or rectangular waveguide cavity as shown in figure 6–45.

The RF choke is in series with the DC bias voltage and prevents the microwave energy from coupling back to the bias source. The iris forms one end of the tuned cavity and couples the microwave energy to the load which may consist of a horn antenna as used in hand-held police radar units. An X-band Gunn oscillator is shown in figure 6–46.

The IMPATT Diode

Although a practical IMPATT device did not become a reality until 1965,[6] the credit for a proposed design is given to W. T. Read of Bell Laboratories. Read suggested that a negative-resistance effect could be realized by an appropriate combination of delay in generating a current avalanche and its transit time through the device. This combination would produce the required 180 degree phase difference between voltage and current necessary for the negative resistance phenomenon. The device was given the acronym *imp*act *a*valanche and *t*ransit *t*ime (IMPATT) diode.

While GaAs and InP have both proven useful in IMPATT devices, ordinary silicon is preferred since it is more economical in terms of fabrication processes and raw material cost. Theoretically, however, GaAs will yield the most power at the highest frequencies with the least noise problem. Commercial devices have a range of between 3 GHz and 200 GHz, with power outputs varying all the way from 50 mW to 20 watts and efficiencies on the order of 15% up to around 50 GHz. Beyond that, the efficiency will fall, becoming less than 1% at the frequency extreme.

IMPATT devices are not generally used as oscillators because the avalanche process is inherently very noisy. Most IMPATT applications are for microwave amplifiers. It was shown in the early part of this section that any device exhibiting the property of negative resistance may be used as an amplifier. But since the IMPATT diode is a one-

[6]In 1965, R. L. Johnson, also of Bell Laboratories, successfully employed the Read model in a device producing 80 mW at 12 GHz.

port device, it must be coupled to a circulator (see chapter 4) in order to isolate amplifier input and output.

The required voltage gradient is considerably more than that of the Gunn slice. In fact, the Read gradient is close to 0.4 MV/cm, and operating temperatures of commercial devices near 500° F (260° C) are not uncommon.

The IMPATT phenomenon is based on the fact that an avalanche is a time-dependent process involving essentially the same multiplicative effect as a snowball rolling down a slope. As shown in figure 6–47, the actual avalanche begins near the reverse-biased anode and proceeds toward the cathode. The IMPATT operation principle may be appreciated by consulting figure 6–48 in conjunction with the following explanation.

FIGURE 6–47
The IMPATT diode.

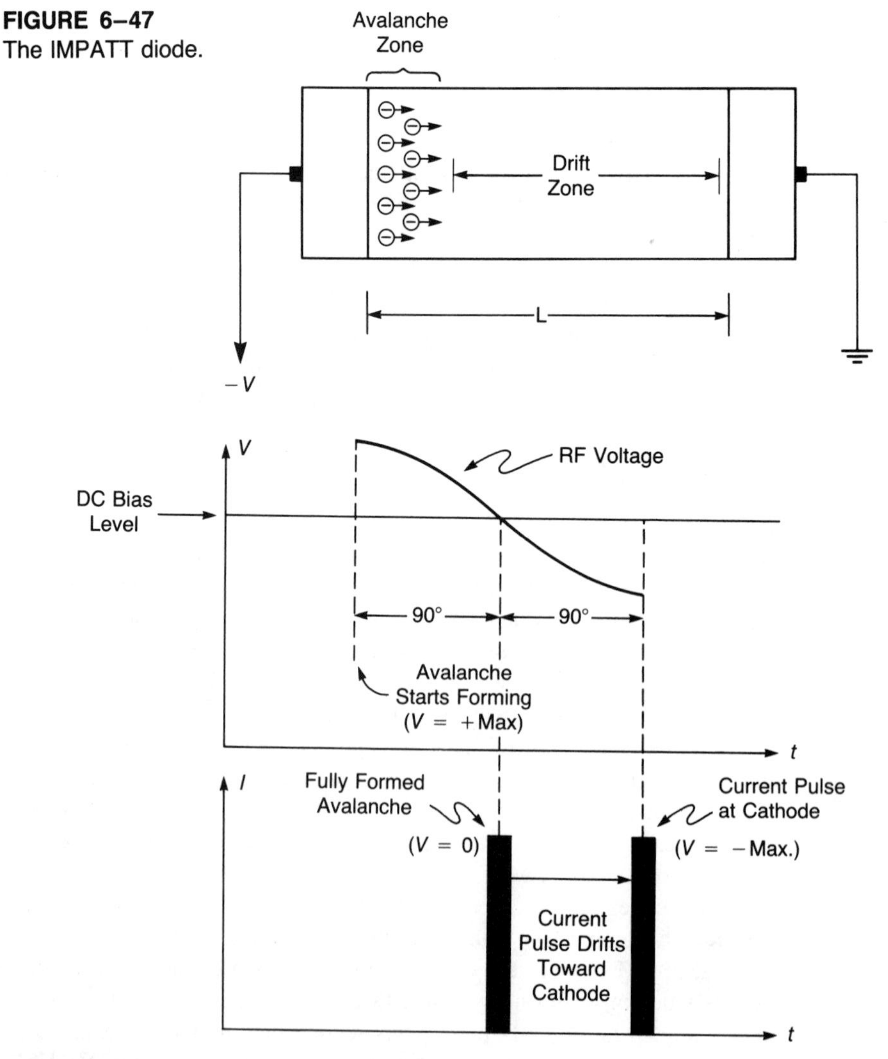

FIGURE 6–48
Operating principle of the IMPATT diode.

The DC bias voltage provides the avalanche threshold value, but the actual avalanche will not commence until the RF voltage is at $+V_{max}$. Of course, it is assumed that the device is already oscillating. The formation of the avalanche is site-specific and local, beginning at the anode junction and then drifting toward the cathode end of the device. As shown in figure 6–48, the fully-formed avalanche begins its journey toward the cathode end just as the RF voltage is zero and starting to go negative. One may already appreciate that a phase difference of 90 degrees has occurred. If, then, the length of the drift zone has been carefully selected, the avalanche will just be arriving at the cathode as the RF voltage is at its maximum negative value, thus providing an additional 90 degree delay. The sum of the two delays described above results in a total phase difference between the voltage and current of 180 degrees. Therefore, the required conditions for negative resistance have been met.

Note that since the arrival time of the avalanche is dependent on the length of the drift region, the IMPATT is basically a narrowband device. This is in contrast to the Gunn diode whose frequency of operation depends somewhat on the applied voltage (see equation 6–10). On the other hand, the progress of the IMPATT avalanche is not influenced much by voltage effects once the domain has fully developed.

SUMMARY

This chapter opened with a brief look at the microwave printed circuit board (MPCB) and its usefulness in providing printed counterparts of coaxial line (stripline) and parallel-line (microstrip). The chapter explained how small values of both inductance and capacitance could be achieved using etched copper techniques. The surface acoustical wave (SAW) device was introduced as a low-pass microwave filter.

An overview of semiconductor doping procedures was presented along with an energy-level explanation of solid state operation. The planar microwave transistor fabrication process using epitaxial growth methods was discussed.

The mesa field effect transistor (MESFET) as a microwave device was introduced, and the Schottky barrier gate was discussed as a means of offsetting the charge-storage effect, which limited the high-frequency performance of microwave transistors. Interdigitation was presented as an innovative means of increasing power handling of the transistor as well as extending its useful frequency range. Gallium arsenide (GaAs) was shown to be preferable to silicon as a semiconductor material due to its higher ion mobility.

Two packaging techniques for MICs were introduced: hybrid, with its discrete active components and monolithic, with no discrete components.

A variety of microwave diode-like devices were presented. The varactor was shown to be useful as a voltage-controlled capacitance with application capabilities in parametric amplifiers (PARAMPS). The tunnel diode with its negative-resistance effect was shown to be capable of gain in a properly designed active circuit, and mathematical evidence for such operation was presented and developed. The Gunn diode, a transferred-electron device (TED), was explored briefly as a low-power microwave power source. Finally, the IMPATT diode was shown to be useful as a narrowband amplifier in the 3–200 GHz region.

PROBLEMS

1. A stripline transmission path is to be fabricated into a MPCB such that the following conditions exist:

$$\begin{aligned} \text{board thickness } (b) &= 0.056 \text{ inch} \\ \text{trace width } (W) &= 0.168 \text{ inch} \\ Z_0 &= 75 \text{ ohms} \\ \text{dielectric constant of board material } (\epsilon_r) &= 2.15 \end{aligned}$$

   What weight of copper-clad board must be used to process the desired stripline? [2 ounce]

2. A thin-film chip inductor has a self-resonant frequency of 2.2 GHz and a stray capacitance of 0.15 pF. What is the circuit inductance? [35 nH]
3. The upper frequency limit of commercial SAW devices was stated to be about 5 GHz. If the minimum spacing between adjacent etched lines is limited to $\lambda/4$, to what maximum line width is state-of-the-art technology limited? [5.9 microinches]
4. An ingot of P-type material is formed with dimensions 2 inches in diameter by 8 inches long. If the chips are to be 1/4 inch square by 10 mils thick, how many such chips are theoretically obtainable from the ingot? [40,208]
5. A certain X-band hybrid MIC amplifier using GaAs FETs boasts of a 56 dB gain with a maximum output power of 4.5 watts.
   (a) What input power is required to achieve the maximum output? [11 μW]
   (b) What is the dBm rating of this amplifier? [36.5 dBm]
   (c) What power output would be expected if 0.22 μW is the input power? [about 88 mW]
6. A microwave source has an output impedance of 300 ohms and is completely matched to its purely resistive load. If a negative-resistance device of 175 ohms is shunted across the system turning it into an amplifier, what is the gain in dBs? [17.3 dB]
7. A Gunn slice is 2 μm thick. What minimum voltage is required to initiate the Gunn mode? [0.66 V]

QUESTIONS

1. In general, why does the power rating of a semiconductor device diminish as its frequency rating increases?
2. Why is GaAs superior to silicon as a microwave semiconductor material?
3. Explain the terms *seal, epitaxial layer,* and *substrate* as used in semiconductor manufacturing.
4. Why are pentavalent materials used in making N-type semiconductor material? Why is trivalent material used for P-type?
5. Draw a sketch showing how you would fabricate a bridge rectifier using the planar technique and method of epitaxial growth.
6. Explain how a SAW device operates. What limits its upper-frequency response?
7. Compare and contrast stripline and microstrip. Is one superior to the other? Why or why not?
8. Compare and contrast hybrid and monolithic MICs. Find trade journal articles advertising both types, and evaluate each one on the basis of your present knowledge.
9. Explain the operation of a varactor. How might one be used in an ordinary AM superheterodyne receiver?
10. How can any manufacturer of semiconductor devices logically justify as cost effective any device operating at an efficiency less than 5%? [Remember, the ordinary automobile engine is only about 35% efficient.]
11. What is the principle of operation of a PARAMP?
12. Why is the tunnel diode important in microwave work?
13. Explain the principle of operation of the Gunn diode.
14. What is the fundamental difference between bulk and junction-effect semiconductor devices?
15. Explain how the IMPATT principle is different from the Gunn mode.

7

RADAR

Radar sends out a radio wave and then ''listens'' electronically for the echo of that same signal. Through proper hardware and signal processing, the echo can be analyzed for information concerning the bearing, distance, and speed of the reflecting surface called the ''target.'' The word *radar* itself is an acronym, coined by the United States Navy in 1942, which stands for *ra*dio *d*etection *a*nd *r*anging.

One of the first documented observations of the echo effect of radio waves was made in 1922 by A. H. Taylor of the U.S. Naval Research Laboratories.[1] Similar observations and experiments with radio echos in other countries led to the almost simultaneous and independent development during the late 1930s of practical radar systems in Great Britain, France, and the United States. Germany, too, had developed a sophisticated radar system of its own by World War II, and had used it effectively and decisively during a sea engagement when the German battleship Bismarck sank the British cruiser HMS Hood.

Early radar systems were seriously hampered by large, unwieldy antennas with slow rotational speeds and severe power limitations. Some primitive radars operated at wavelengths exceeding 16 feet (60 MHz) due to a lack of adequate transmitting tubes capable of SHF operation. The invention of the cavity magnetron, however, made possible higher frequencies at unheard of power outputs.[2]

[1] *Radar Electronic Fundamentals,* Bureau of Ships. Navy Dept., Washington, D.C.: U.S. Government Printing Office, June 1944. NAVSHIPS 900,016.

[2] The magnetron is discussed in chapter 5.

FUNDAMENTAL IDEAS

Basically, a radar set consists of four main sections: (1) the transmitter (Tx); (2) the receiver (Rx); (3) a duplexer, which is sometimes referred to as an electronic transmit-receive (TR) switch; and (4) an antenna. The basic radar set is shown in figure 7–1.

In operation, a burst of SHF energy is delivered to the antenna from the transmitter via the TR switch while the receiver is disconnected. After a short time (usually a few microseconds), the TR switch disconnects the transmitter and connects the receiver. Any reflected pulse may then be detected by the receiver and subsequently displayed by some suitable means.

Since the outgoing and reflected pulses travel at the speed of light (300×10^6 m/S), an interesting time-distance relationship may be developed. First, it should be mentioned that all radar calculations are based on the nautical mile (1 nmi = 6,075 ft), not the statute mile (1 mi = 5,280 ft). The reason for this is that one nautical mile is exactly 1/21,600th of the earth's circumference as measured along any great circle.[3] In other words, one degree (60 minutes of arc) is equivalent to 60 nmi, or $1' = 1$ nmi.

We may now inquire, "How long will it take a pulse of SHF energy from a radar antenna to make a round-trip journey of 1 nmi?" From the method of analysis of units,[4] we may write,

$$\frac{\text{S}}{300 \times 10^6 \text{ m}} \times \frac{1{,}852 \text{ m}}{\text{nmi}} = 6.17\ \mu\text{S for a one-way journey}$$

Therefore, the round-trip time is 12.3 μS. So, for example, if the time between outgoing and return pulses is 615 μS, the target is 615/12.3 = 50 nmi away.

A similar question we might ask is, "How far will a pulse travel in 1 μS?" Again, from analysis of units, we write,

$$\frac{300 \times 10^6 \text{ m}}{\text{S}} \times \frac{1 \times 10^{-6} \text{ S}}{1} = 300 \text{ m}$$

If a pulse is received 1 μS after it was sent, the target is 150 meters distant.

NAVIGATIONAL CONCEPTS

Since one of the principal uses of radar is in navigation, it is prudent for us to digress somewhat at this point in order to develop a few fundamental ideas about navigation and identify appropriate terms and definitions.

Distance is not the only information available from the radar receiver display. Indeed, since the position of the antenna relative to the ship's heading is always known, it is possible to obtain directional information as well. Two terms are basic in navigation: heading and bearing. *Heading* refers to the direction in which the boat's bow is

---

[3]A great circle is a plane circle whose center passes through the center of the earth. For all practical purposes, every great circle is regarded as having the same circumference.

[4]Analysis of units is a computational strategy that treats the units of measurement like algebraic terms, and hence, they may be cancelled, squared, etc.

FIGURE 7–1
Basic elements of a radar set.

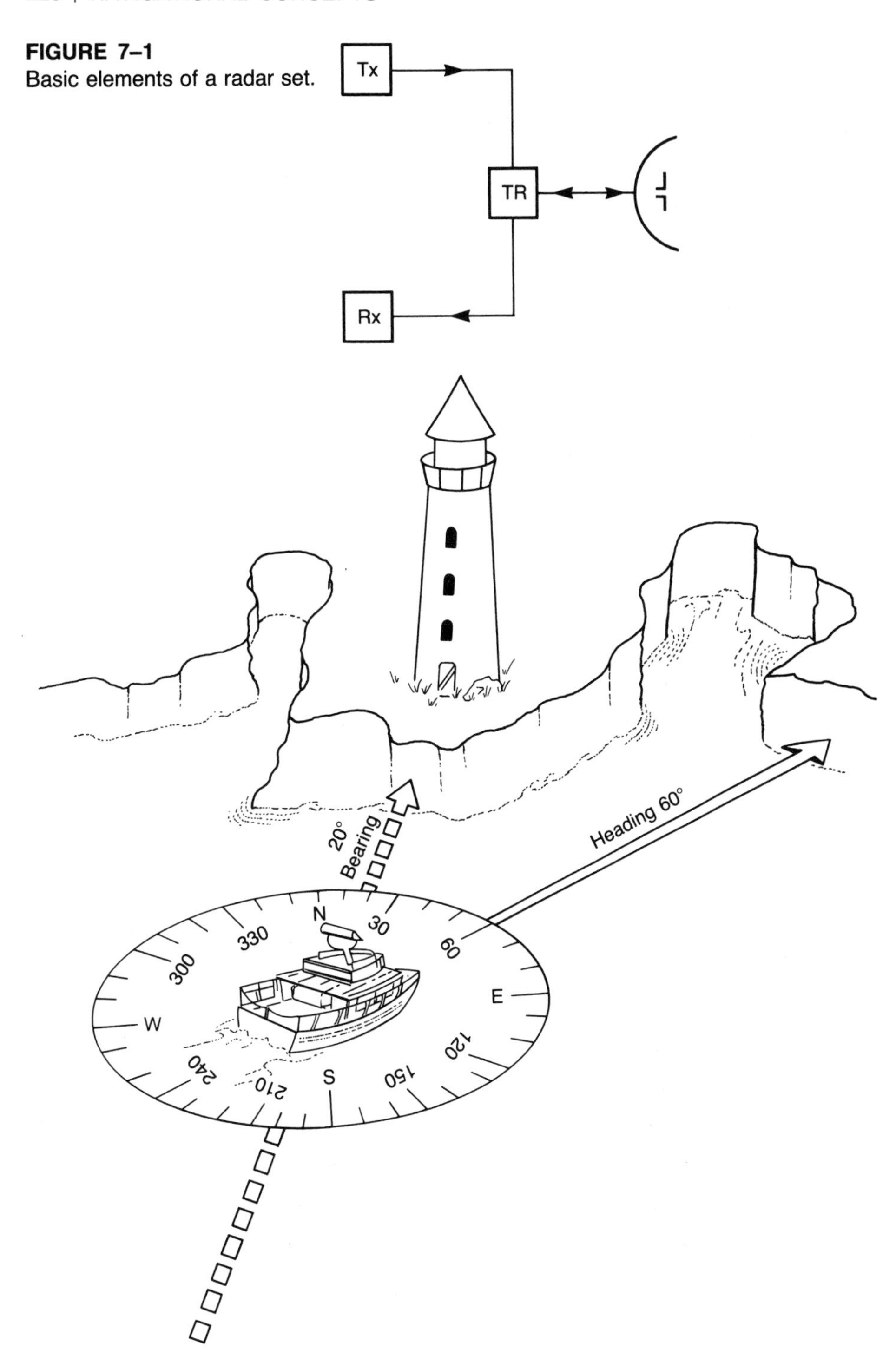

FIGURE 7–2
Heading and bearing.

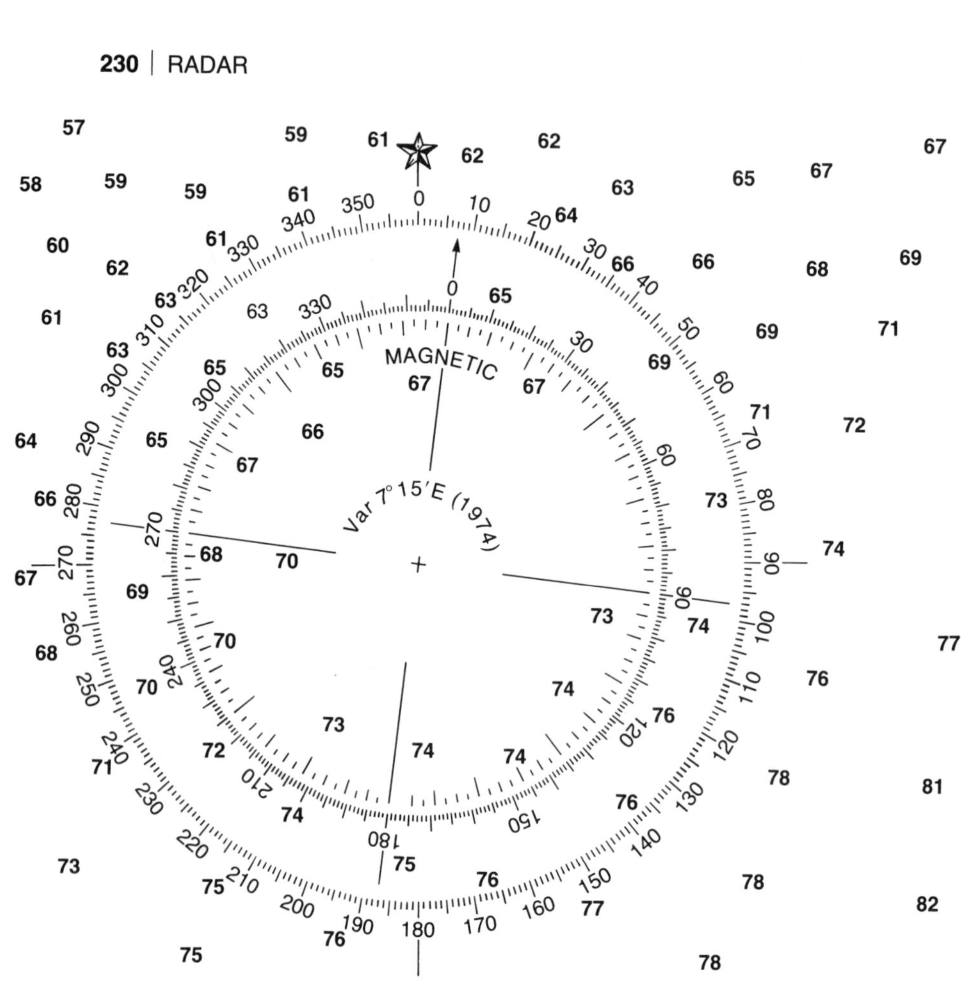

FIGURE 7–3
The compass rose.

pointed. *Bearing* is the direction of one object relative to another. Figure 7–2 shows the difference between the two terms. The boat is shown on a compass (magnetic) heading of 60°, and the lighthouse has a magnetic bearing of 20°. One might also have expressed the location of the lighthouse as 40° off the port bow. Note that the heading shown in the example was given in terms of degrees clockwise from magnetic north. Headings relative to true (geographical) north are referred to as *azimuth* and are given in degrees clockwise from true north.

Because of the erratic nature of the earth's magnetic field, the compass needle seldom aligns itself with true north. The difference between true and magnetic north is referred to as *variation*.[5] The amount of variation changes not only from one place to

[5]Variation must not be confused with *deviation,* which is the inherent compass error itself due to the magnetic peculiarities of the boat. Deviation is described as being easterly or westerly according to whether the compass needle is deflected east or west of magnetic north.

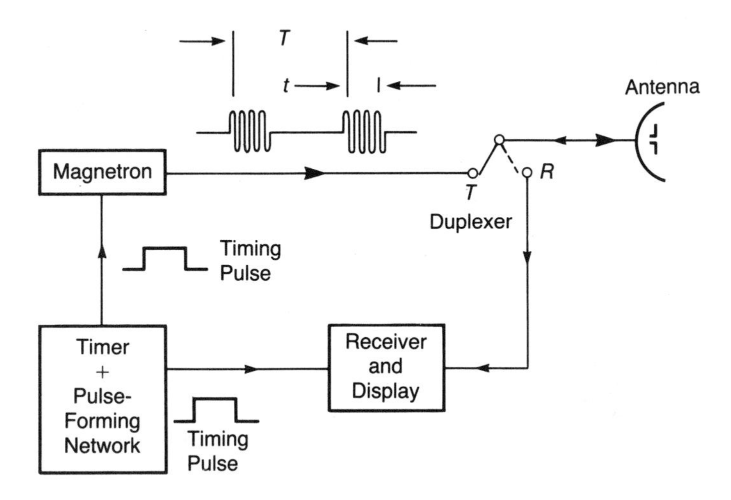

FIGURE 7–4
A basic pulsed radar set.

another on the earth's surface, but also from month to month in response to the shifting core of the earth. Therefore, the amount of such variation is included in navigational charts on the *compass rose,* as shown in figure 7–3. Note that the outer circle is graduated in degrees *true,* while the inner circle in degrees *magnetic*. In this particular example, the magnetic variation is indicated as 7° 15′ east. So, for example, if the boat's compass were indicating a heading of due west, the helmsman (person steering the boat) would know that his true course was 277° 15′. If the helmsman actually wanted to head due west, he would steer a compass heading of 270° − 7° 15′ = 262° 45′.

Note that the date on the compass rose is 1974. Obviously, then, navigational charts change and must, therefore, be updated or replaced periodically. Various governmental agencies have the responsibility of maintaining records on such variation as well as changes in coastal contours due to the natural attrition of the land by geological processes.

BASIC RADAR PRINCIPLES

A block diagram of a basic pulsed radar set is shown in figure 7–4. The timer produces a spiked pulse which is shaped into a square timing pulse of proper form and duration by the pulse-forming network. It is this pulse that actually fires the magnetron, causing it to supply a burst of SHF energy of proper duration to the antenna.

It is essential that the pulse be precisely shaped so that return echos will not give ambiguous target information. For example, consider the pulses shown in figure 7–5. It

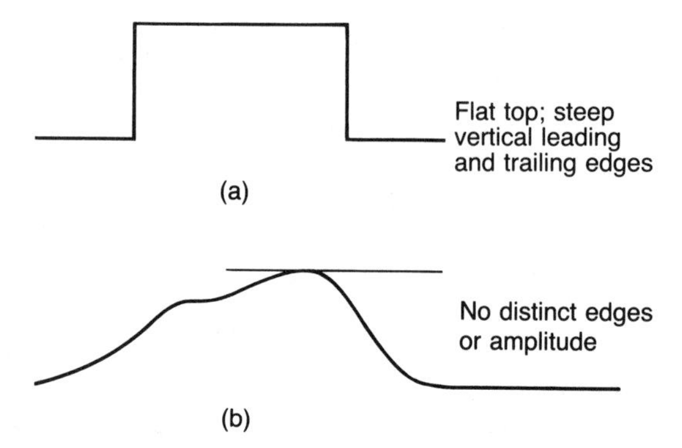

FIGURE 7–5
(a) Properly shaped pulse. (b) Improper pulse shape.

is obvious that the TR switch network will not be able to accurately switch the receiver over to the antenna at the proper moment since no distinct trailing edge exists for the pulse shown in part (b) of the figure. The TR switch will have no accurate indication of when the timing pulse has ended. Moreover, the duration of the magnetron burst will not be precisely controlled and neither will its initiation point. In addition, if the pulse does not have a flat top, frequency pushing of the magnetron will result since the frequency of the magnetron is voltage sensitive.

The timer also provides a simultaneous trigger pulse to the display section of the receiver. This pulse starts the electron beam of the CRT moving outward from screen center (your position). Any return echo (''blip'') will cause the trace spot to intensify at whatever position it happens to be. The distance to the target, then, is represented by how far the blip is from screen center.

In one form of display, the distance is determined by counting range rings out to the target blip. Each range ring represents a certain distance. One type of radar display showing the range rings is illustrated in figure 7–6.

In addition to providing range information, the display console also gives information concerning target bearing. Each time the antenna aligns itself with the boat's lubber line,[6] an electrical pulse (from either a snap switch or magnetic pickup) is sent to the display which then starts the electron beam of the CRT sweeping clockwise, and thus orients the ship's heading relative to the target. The pulse also produces a ship's heading marker on the display screen. As shown in figure 7–7, a target is on a bearing of 100° relative and a range of 3.4 nmi.

Note that the displays shown in figures 7–6 and 7–7 are polar displays with the radar antenna at the origin. This is the most common type of target representation since bearing and range are made simultaneously.

In connection with range and bearing, one must consider the ability of the radar system to discriminate between closely spaced targets. *Range resolution* refers to the

---

[6]A lubber line is actually a line or indication on the inside of the compass bowl that defines the fore-to-aft longitudinal axis (centerline) of the boat.

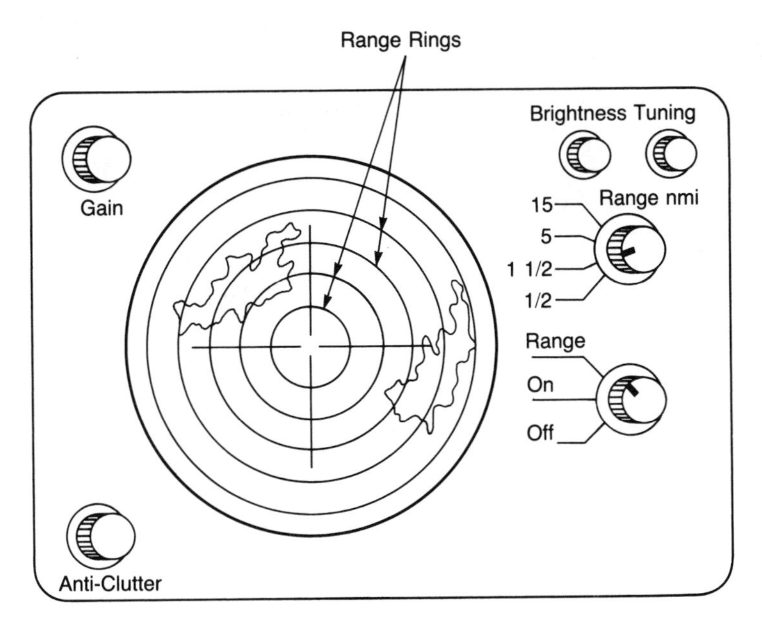

FIGURE 7–6
Shipboard radar display showing range rings.

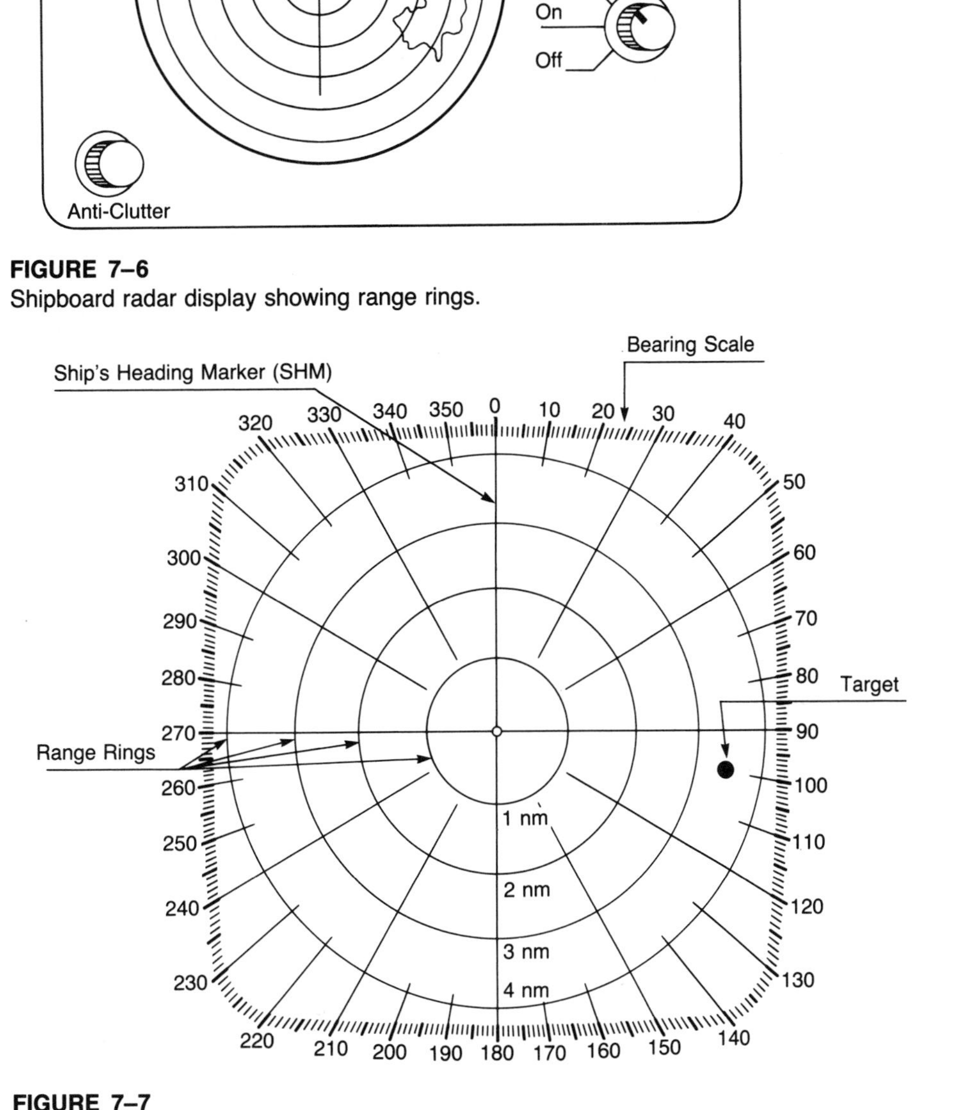

FIGURE 7–7
Display screen showing range rings, ship's heading marker, and target bearing.

ability of a radar system to resolve two or more targets on the same bearing. For example, figure 7–8 shows two targets, T_1 and T_2, on the same bearing, but separated by a distance. The Raytheon Mariners Pathfinder radar model 1200, for example, has a range resolution of better than 22 meters (about 72 feet).

Bearing resolution refers to the ability of the radar system to discriminate between two or more closely spaced targets at the same range. Figure 7–9 shows two targets at the same distance from the radar antenna, but close together. The Raytheon unit has a bearing resolution better than 1°.

Closely associated with radar resolution capability is antenna beamwidth and shape of the radiation pattern. Figure 7–10 shows a variety of antenna styles commonly used with radar systems. In order to resolve targets close together in bearing (bearing resolution), the beamwidth of the antenna must be very narrow. The cut paraboloid and the "pillbox" antennas will both provide the required radiation pattern characteristics. Typical horizontal beamwidths are on the order of 2–5°, and vertical width is greater than 20°. The narrow horizontal beamwidth is necessary to separate targets close together, and the broad vertical pattern allows the ship to pitch without losing the target. The wide vertical pattern also lessens the possibility of an airborne target avoiding detection by flying close to the water under the radar beam.

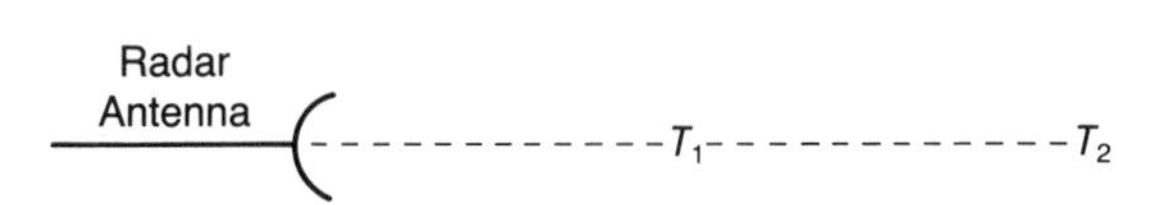

FIGURE 7–8
Two targets on the same bearing, but at different ranges.

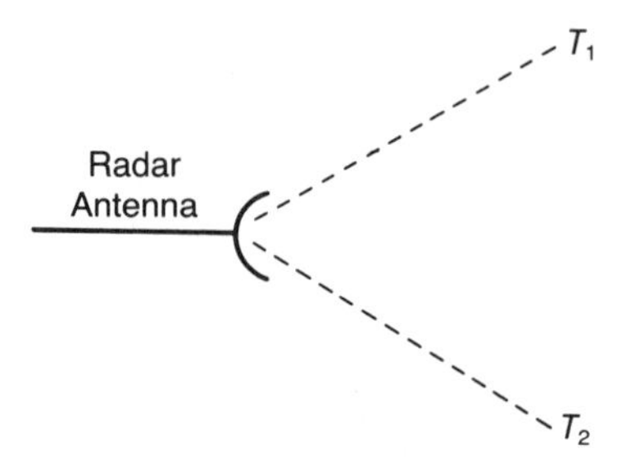

FIGURE 7–9
Two targets at the same range, but on different bearings.

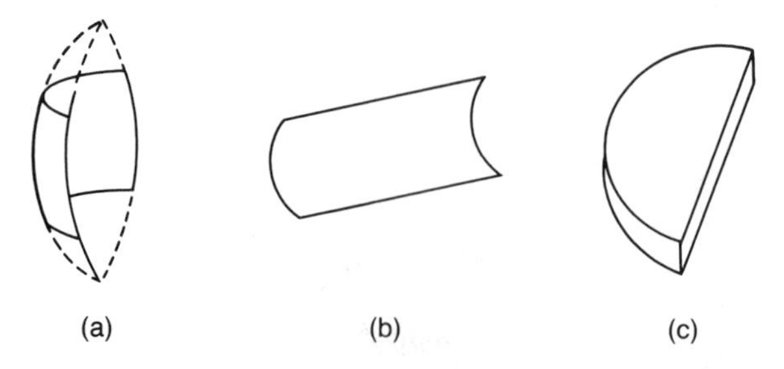

FIGURE 7–10
Paraboloid reflector types: (a) cut paraboloid; (b) parabolic cylinder; (c) "pillbox."

A common commercial marine radar antenna is the slotted waveguide antenna shown in figure 7–11. The slotted waveguide antenna consists of a section of waveguide terminated at one end with a microwave absorber which prevents reflections. Slots are precisely machined into one side of the guide, and a very narrow beam results having much the same characteristics as the pillbox antenna. Often, the slotted waveguide antenna is enclosed in a weatherproof housing as shown in figure 7–12. The entire antenna rotates atop the motor housing.

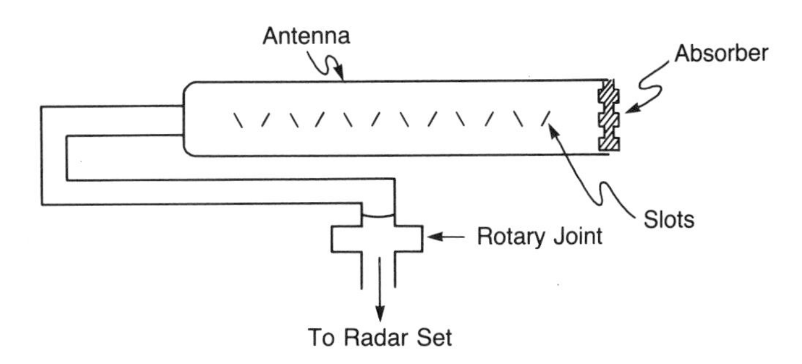

FIGURE 7–11
A slotted waveguide antenna.

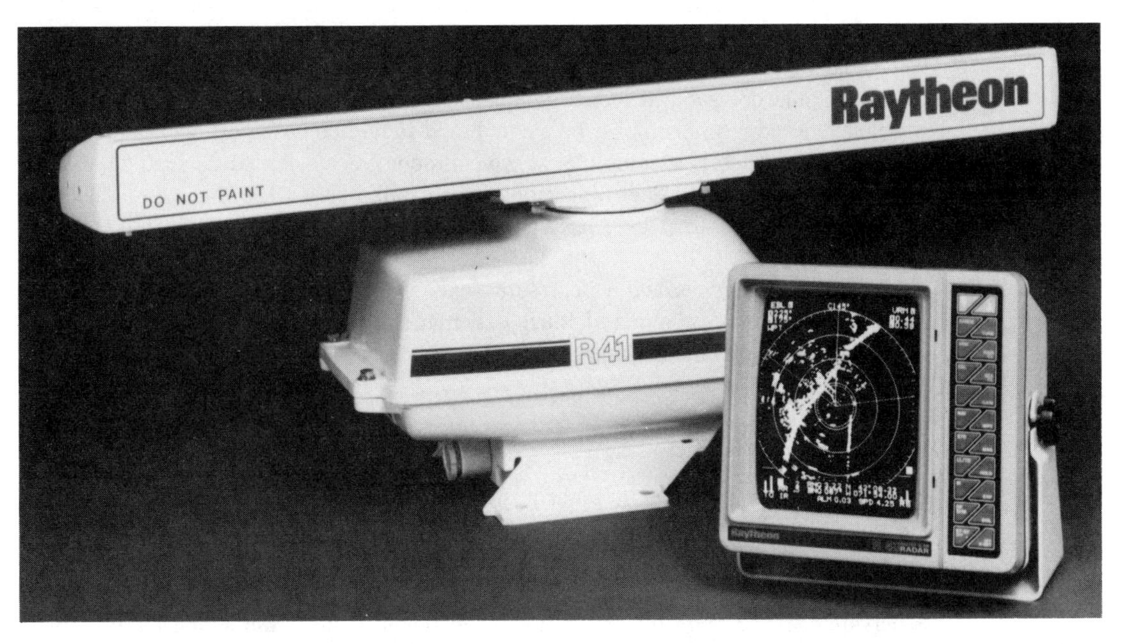

FIGURE 7–12
Slotted antenna enclosed in housing. (Courtesy of Raytheon Marine Co.)

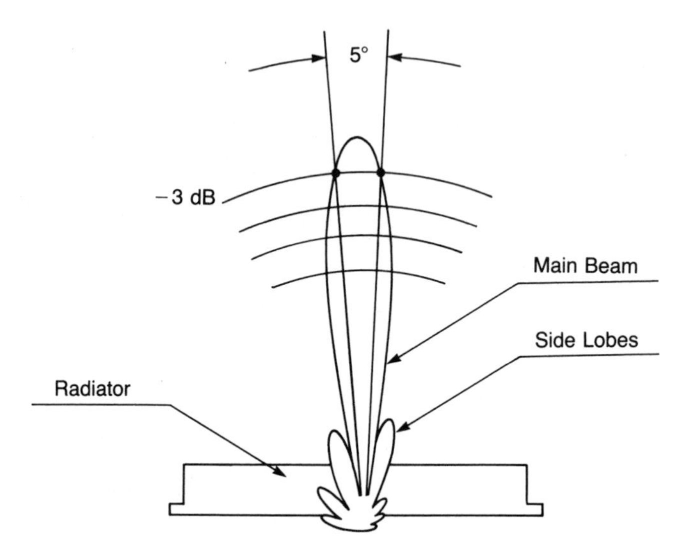

FIGURE 7–13
Radiation pattern of slotted waveguide antenna.

A typical slotted waveguide antenna radiation pattern in the horizontal plane is shown in figure 7–13. The horizontal beamwidth is 5° and the side lobes are more than 21 dB down from the main lobe. The vertical pattern's beamwidth (not shown) is 25°.

Radar antennas are often enclosed in *radomes,* which are plastic housings with smooth contours designed to reduce wind loading as well as protect the antenna from the weather. Radomes must never be painted, and should be hosed off occasionally with water as dust and debris accumulate. A typical radome for a slotted waveguide antenna is shown in figure 7–14. The dimensions are in millimeters. The diameter is about 2 feet and the height a little over 13 inches. The radome weighs slightly more than 4 pounds.

For larger antennas and where snow and wind loading is a crucial factor, the geodesic radome may be required. Such a dome, shown in figure 7–15, combines the lightest materials with the strongest structural shape.

RADAR RANGE EQUATIONS

While the maximum range of target acquisition is a function of many variables including peak output power, antenna directivity, weather conditions, and many others, in the final analysis the actual range is determined by the faintest detectable echo to which the receiver can respond. We will now attempt to develop a mathematical model to give us the maximum range under ideal conditions. In practice, then, the actual range will be much less than that indicated by the model.

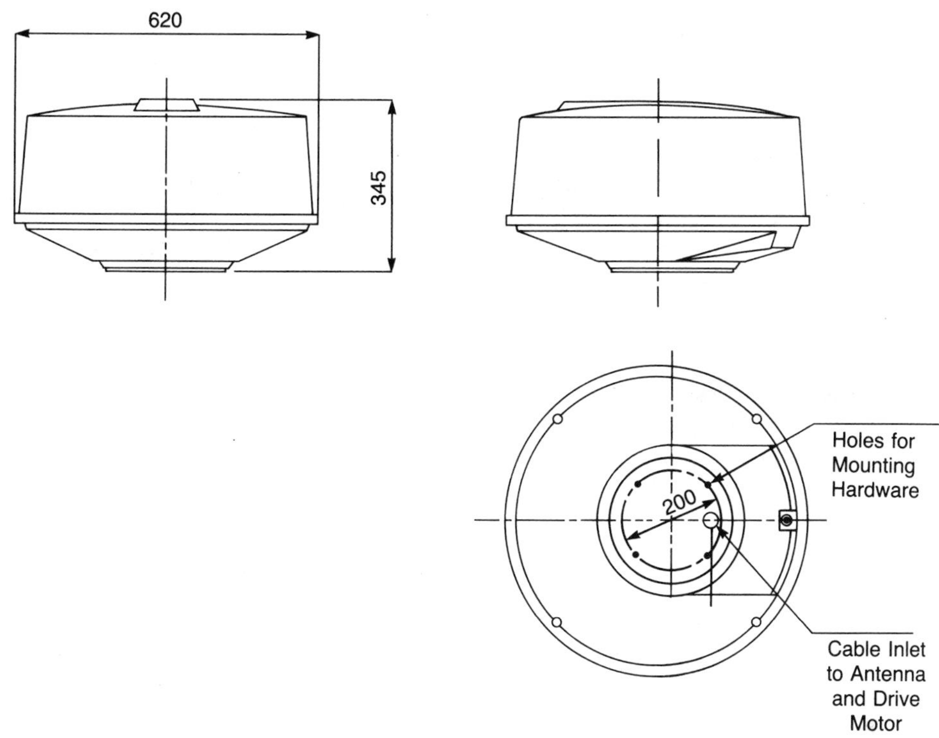

FIGURE 7–14
Radome for slotted waveguide antenna.

FIGURE 7–15
The geodesic radome.

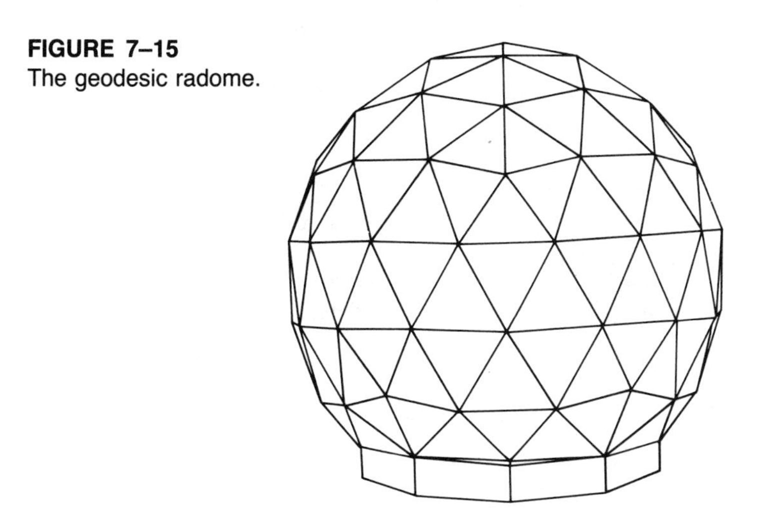

From chapter 2, we know that the power density is given by

$$P'_t = \frac{P_t}{4\pi r^2} A_p \tag{7–1}$$

where P_t is the peak pulsed power of the transmitter, P'_t is the power density at the target, and A_p is the power gain of the transmitting antenna.

Since the target has an effective radar cross-section, it will receive a power given by

$$P_{tgt} = P'_t A_{tgt} \text{ watts} \tag{7–2}$$

where A_{tgt} is the radar cross-section of the target.

The target will now reflect this power isotropically, since the target is regarded as having no appreciable directivity, and the power density received back at the receiving antenna of the radar set is

$$P'_r = \frac{P_{tgt}}{4\pi r^2} \tag{7–3}$$

The power that impinges upon the receiving antenna will be

$$P_{rcv} = P'_r A_r \text{ watts} \tag{7–4}$$

where A_r is the effective or capture area of the receiving antenna. Equation (7–4) may be rewritten using the appropriate substitutions as

$$P_{rcv} = \frac{P_t A_p A_{tgt} A_r}{(4\pi r^2)^2} \tag{7–5}$$

From chapter 3 we know that the gain of an ideal paraboloid with uniform illumination and no losses is given as

$$A_p = \left(\frac{\pi D}{\lambda}\right)^2 \tag{7–6}$$

where D is the dish diameter and $D^2 = 4A_r/\pi$ (see footnote 9 in Chapter 3). Upon substituting for D^2 in equation (7–6) we obtain

$$A_p = \frac{4\pi A_r}{\lambda^2} \tag{7–7}$$

where A_r is the aperture area. We may now substitute for A_p in equation (7–5) and write

$$P_{rcv} = \frac{P_t A_{tgt} A_r^2}{4\pi r^4 \lambda^2} = P_{min} \tag{7–8}$$

Solving equation (7–8) for the maximum range gives

$$r_{max} = \left[\frac{P_t A_{tgt} A_r^2}{4\pi\lambda^2 P_{min}}\right]^{1/4} \quad \textbf{(7–9)}$$

We could have also solved equation (7–7) for A_r and substituted this into equation (7–9), giving

$$r_{max} = \left[\frac{P_t A_{tgt} A_p^2 \lambda^2}{(4\pi)^3 P_{min}}\right]^{1/4} \quad \textbf{(7–10)}$$

A closer look at equation (7–10) reveals some interesting facts. First, because the maximum range is proportional to the fourth root of P_t, in order to double the range, the power must be increased sixteenfold, all other factors remaining fixed. Another interesting fact is that since the value of P_{min} appears in the denominator, a more sensitive receiver will lower this value and increase the magnitude of r_{max}.

Equation (7–9) shows that the range is also proportional to the square root of the capture area of the receiving antenna provided the wavelength (hence, frequency) remains constant. Alternately, the wavelength may be decreased (frequency increased) to extend the range. From equation (3–20), we see that as the *L/D* ratio decreases, beamwidth will also decrease, making for better bearing resolution.

DOPPLER RADAR

Targets moving toward the radar signal source may have their echos obscured if there is a sufficiently large, stationary background behind the target. For example, low-flying aircraft with mountains or hills in the background may not be detected. Any background reflections that obscure the target are called clutter.

By using the Doppler effect, however, it is possible to make the radar blind to stationary clutter and respond only to moving targets. The Doppler effect is the apparent change in frequency due to the relative motion between a fixed observer and a moving frequency source. Nearly everyone is familiar with the apparent change in pitch of an auto's horn as it approaches and recedes from the observer.

As it specifically applies to radar, the Doppler effect is as follows. If a moving target whose velocity is V_t approaches a stationary signal source of frequency F_s, the moving target "sees" a new frequency given by

$$F_t = F_s + F_s\left(\frac{V_t}{c}\right) \quad \textbf{(7–11)}$$

where F_t is the apparent frequency and c is the velocity of light. The whole point of equation (7–11) is that only *moving* targets undergo a Doppler process; stationary targets do not. Consequently, radar detection circuits that are phase sensitive will discern phase differences between recurrent scannings of the target area. Stationary targets that do not exhibit phase changes will be rejected, and only the moving targets will be visibly displayed on the screen. Note, however, that targets moving across the field of view, neither approaching nor receding, will not undergo a Doppler effect in the direction of the signal source and, hence, will not be visible.

THE PULSE-FORMING NETWORK

Now that we have a basic understanding of radar principles, let us take a somewhat more detailed look at a few of the subsystems involved. It was mentioned in an earlier section that the SHF pulse produced by the magnetron was initiated by the timer unit. The oscillator that initiates this timing pulse does not produce a square wave, and cannot accurately regulate the duration or amplitude of the SHF pulse which, as stated earlier, must be precise if unambiguous range indications are to be obtained. However, the oscillator's output can (1) establish the pulse repetition rate (PRR) which determines output power, among other things, and (2) trigger a pulse forming (shaping) circuit which can provide precise on-off energy bursts from the magnetron. A conceptual block diagram of the timer unit is shown in figure 7–16.

The pulses from the pulse-forming network are of very precise magnitude and duration and are used to trigger the firing of the magnetron through the modulator, which is basically a high-speed electronic switch. The heart of the timer unit is the pulse-forming circuit, which will now be examined.

The pulse-forming network is actually an artificial transmission line in the sense that (1) it causes a delay in the transfer of energy from input to output in a much shorter distance and (2) its characteristic impedance is due to lumped (discrete) components rather than distributed constants. The artificial line is shown in figure 7–17.

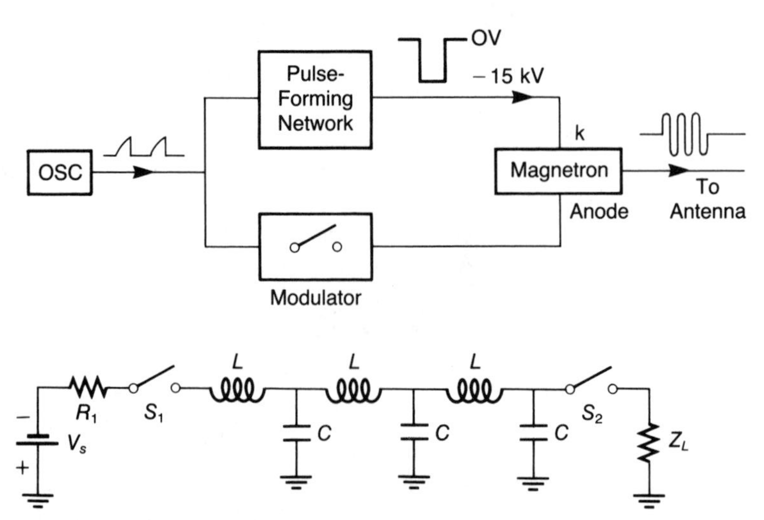

FIGURE 7–16
Timer unit block diagram.

FIGURE 7–17
An artificial transmission line.

Before beginning our explanation of the pulse-forming network itself, we should review the fact that energy-transfer delays are, in fact, possible with ordinary transmission lines. For example, if a pulse were sent along an air-dielectric line 300 m in length, the pulse would not be detected at the output until 1 μS had elapsed. This delay property was implied by equation (1–7)

$$v_p = \frac{c}{\sqrt{\epsilon_r}}$$

where $c = 300 \times 10^6$ m/S and ϵ_r is the dielectric constant. Obviously, as values of ϵ_r increase, so will the amount of delay. If the transmission line was terminated in an open circuit, all the energy reaching the end of the line would be reflected, producing an additional 1 μS delay, or a total lag of 2 μS.

Note the impracticality of using an actual length of transmission line to obtain delays. The line described above would have to be rolled up into a coil nearly 1,000 feet long! Consequently, we make use of artificial delay lines using discrete components to achieve the same effect.

Referring to figure 7–17, when switch S_1 is closed (S_2 is open), the capacitors in the line will charge through R_1 to the battery voltage V_s. If S_1 is then opened and S_2 is closed, the line begins to discharge through Z_L, whose value has been set equal to Z_0 of the line. However, two interesting things are now observed. First, the voltage across Z_L is *not* V_s as you might have expected. Secondly, the voltage does *not* discharge according to the predicted exponential curve so often associated with R-C circuits. Instead, Z_L appears in series with the line's Z_0, forming a voltage divider, as shown in figure 7–18.

As shown in figure 7–18, V_s divided equally, with $V_s/2$ appearing across Z_L. Moreover, this condition exists as long as any energy remains on the line. As a result, the pulse of energy formed across Z_L exists with constant amplitude for a precise duration as determined by the values of L and C. The pulse across Z_L is shown in figure 7–19.

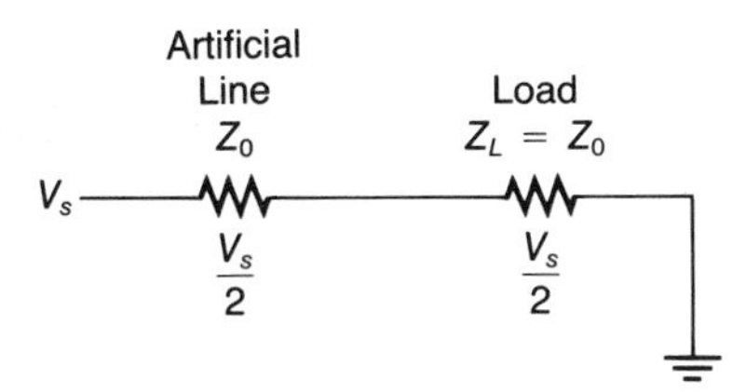

FIGURE 7–18
Equivalent circuit.

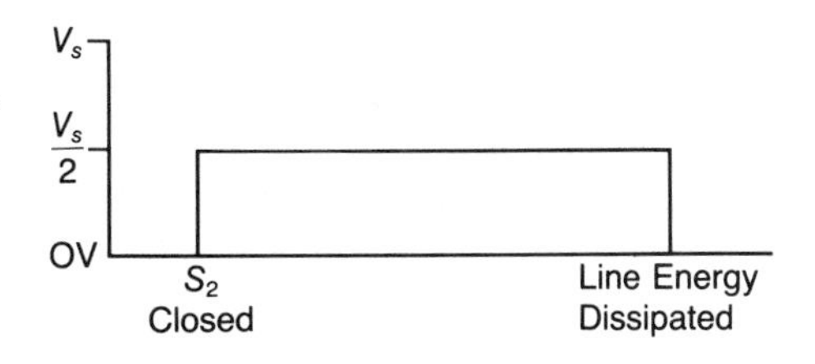

FIGURE 7–19
Pulse shape across Z_L.

The foregoing explains the constant *amplitude* and flat-top *shape* of the pulse output of the artificial line. The following will make clear the reason for the time *delay*. Refer to figure 7–20 throughout the explanation.

Part (a) of figure 7–20 shows the voltage conditions on the line at total charge with S_1 closed and S_2 open. This is the static, pre-pulse condition. In part (b) of the figure, two voltage waves (W_1 and W_2) start moving with the same velocity in opposite directions at the instant S_2 is closed (S_1 previously opened). Since $Z_L = Z_0$, the line appears infinite to the right. Therefore, wave W_2 continues to move into Z_L. However, wave W_1 moving to the left will eventually encounter an open circuit (S_1 open), and will be reflected back toward the right. Since the line to the right appears infinitely long, no further reflections will occur, and the pulse continues as long as there is energy in the line.

The artificial line, therefore, is capable of delivering the well-defined pulses of precise amplitude and duration required for accurate firing of the magnetron. The effect of the line is the same as very rapid on-off switching of a DC source having a voltage of $V_s/2$. An artificial line thus described will produce a delay time (t) given by

$$t = \sqrt{LC} \text{ seconds} \tag{7–12}$$

PULSE REPETITION RATE CONSIDERATIONS

The pulse repetition rate (PRR) determines how closely spaced the SHF pulses from the magnetron will be. For a fixed rate, the pulse repetition interval (T) is as shown in figure 7–21. The repetition interval (T) is given by

$$T = \frac{1}{\text{PRR}} \tag{7–13}$$

The relationship between pulse repetition interval (T) and pulse width (t) is illustrated in figure 7–22. As may be seen from the figure, the only mathematical requirement is that $t < T$. There are, however, several practical requirements, which will be discussed later.

The selection of the PRR and pulse duration (t) is governed by several concerns. As discussed in chapter 5, the magnetron is capable of delivering very high-power pulses for only brief intervals without overheating. This sort of performance is tied up with the concept of duty ratio as given by

$$\text{duty ratio} = \frac{t}{T} \tag{7–14}$$

Duty ratio is often expressed as a percent by multiplying equation (7–14) by 100. In this case, it is called *duty cycle*.

The relationship between average power and peak pulsed power is expressed as

$$\text{peak power} = \frac{\text{average power}}{\text{duty ratio}} \tag{7–15}$$

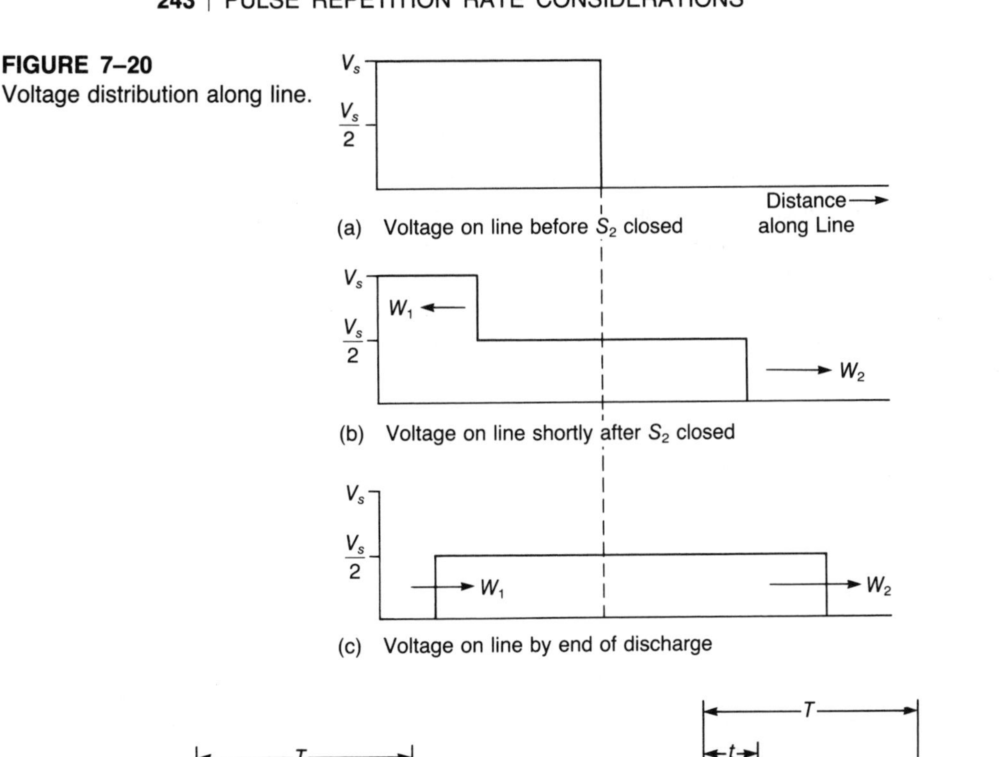

FIGURE 7–20
Voltage distribution along line.

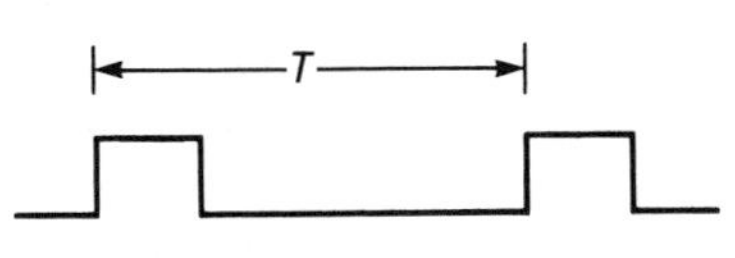

FIGURE 7–21
Relationship between PRR and pulse repetition interval (T).

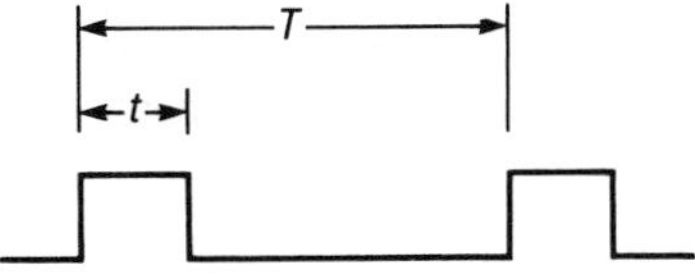

FIGURE 7–22
Relationship between pulse repetition interval (T) and pulse duration (t).

For example, suppose an average power of 5 watts is delivered at a PRR of 800 pulses per second and the pulse width (t) is 0.5 μS. The peak pulsed power given by equation (7–15) is

$$P_{\text{pk}} = \frac{5}{\text{PRR} \times t} = 12.5 \text{ kW}$$

where PRR $\times\ t$ = duty ratio = t/T.

Suppose the pulse duration in the above example was increased to 1 μS. The peak power would drop to 6,250 watts. Similarly, if the pulse duration was 2 μS, the peak power would be 3,125 watts. It may be noticed that as the pulse duration interval (t) increases, the peak power will decrease. This makes sense, since increasing (t) while keeping the pulse rate ($1/T$) constant increases the duty ratio. Consequently, less peak power must be delivered in order to keep the magnetron from overheating.

We saw in equation (7–10) that a sixteenfold increase in the power must occur before the range is doubled. Therefore, the reduction in peak power is not a primary

FIGURE 7–23
Time between pulses or pulse interval.

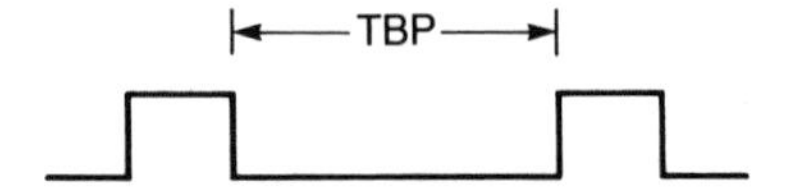

concern. What is more of a problem, however, is the range-limiting effect of the time between pulses (TBP) or pulse interval, as shown in figure 7–23.

As was determined earlier, a burst of SHF energy requires 12.3 μS to complete the round-trip distance to a target 1 nmi away. Therefore, if the radar set is required to have a maximum range of, say, 30 nmi, the minimum TBP is $30 \times 12.3 = 369$ μS. If the pulse width (t) is set at 2 μS to insure adequate peak pulse power, then the pulse repetition interval (T) will be set at 371 μS which, in turn, establishes the PRR at $1/T = 2{,}695$ pulses/sec.

In practice, commonly used PRR values are on the order of 200–10,000. These values yield pulse repetition intervals (1/PRR) between 5,000 μS and 100 μS, which correspond to ranges (T/12.3) of 406 nmi to 8 nmi.

THE DUPLEXER

The duplexer[7] is essentially a high-speed transmit-receive (TR) switch which allows a single radar antenna to serve the dual function of both receiver and transmitter antenna. There are several variations of the duplexer. For low-powered radar sets, a circulator may be used. Recall from chapter 4 that a circulator has the property that only rotationally-adjacent ports are connected. In figure 7–24, the circulator shown in the diagram will allow the high-energy SHF burst from the magnetron to transfer from port 2 to port 3, which is connected to the antenna. A returning echo will enter port 3, which couples only to port 1. In this way, the outgoing high-energy pulse is kept out of the sensitive receiver circuitry where it would cause damage.

Another type of duplexer used in high-power radar sets is shown schematically in figure 7–25. When the magnetron fires its burst of SHF energy, both spark gaps A and D are operated, effectively producing a short circuit at these two locations, which are $\lambda/4$ away from the main waveguide arm. As a result, an open circuit (infinite impedance) is reflected into the waveguide at B and C.[8] Therefore, the full energy of the SHF pulse proceeds to the antenna, and none enters the receiver.

During the "listening" period, both gaps are opened, and the short at S, $\lambda/2$ away from the main arm, is reflected to B as a short. In turn this short is reflected $\lambda/4$ away to C as an open circuit (infinite impedance load), which effectively diverts the weak energy of the echo to the receiver.

---

[7]The *diplexer,* on the other hand, allows two signals to be fed simultaneously to the same antenna without intermixing. Such a device is commonly used with TV broadcasting antennas where the AM video and FM audio are sent out over the same antenna. See chapter 3 and, in particular, figures 3–36 and 3–37.

[8]Recall that a short at the end of a waveguide section $\lambda/4$ long is reflected as an open circuit.

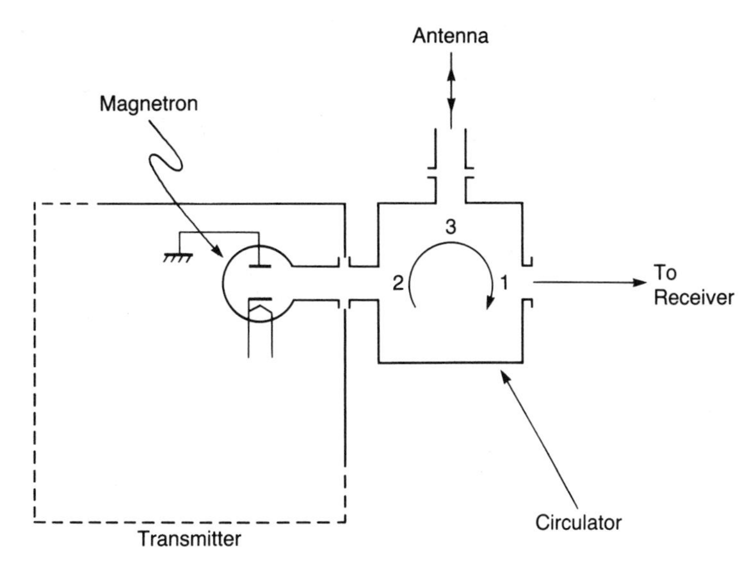

FIGURE 7–24
Circulator used to isolate transmitter from receiver.

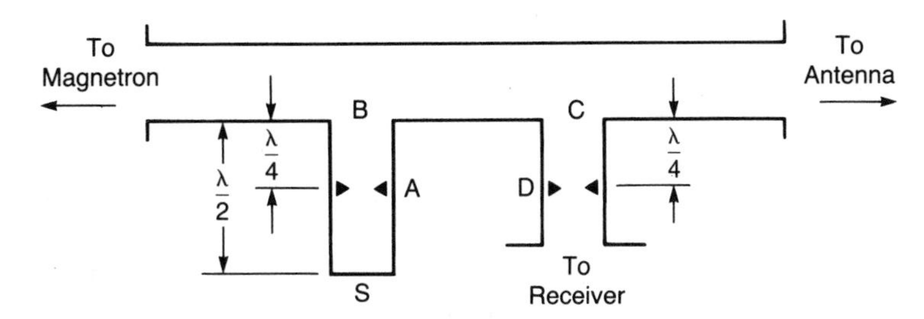

FIGURE 7–25
A waveguide duplexer using spark gaps.

The spark gap at D (called a TR cavity) is a gas-filled resonant cavity biased to the threshold of conduction by a "keep-alive" voltage, which makes the device extremely sensitive. Only the slightest signal energy from the magnetron is required to trigger this cavity. The spark gap at A, called an anti-TR (ATR) box, requires no biasing voltage since it has no circuits to protect.

BASIC RADAR RECEIVER

The block diagram of a simplified radar receiver is shown in figure 7–26. The radar receiver is a superheterodyne with an intermediate frequency (IF) on the order of 30

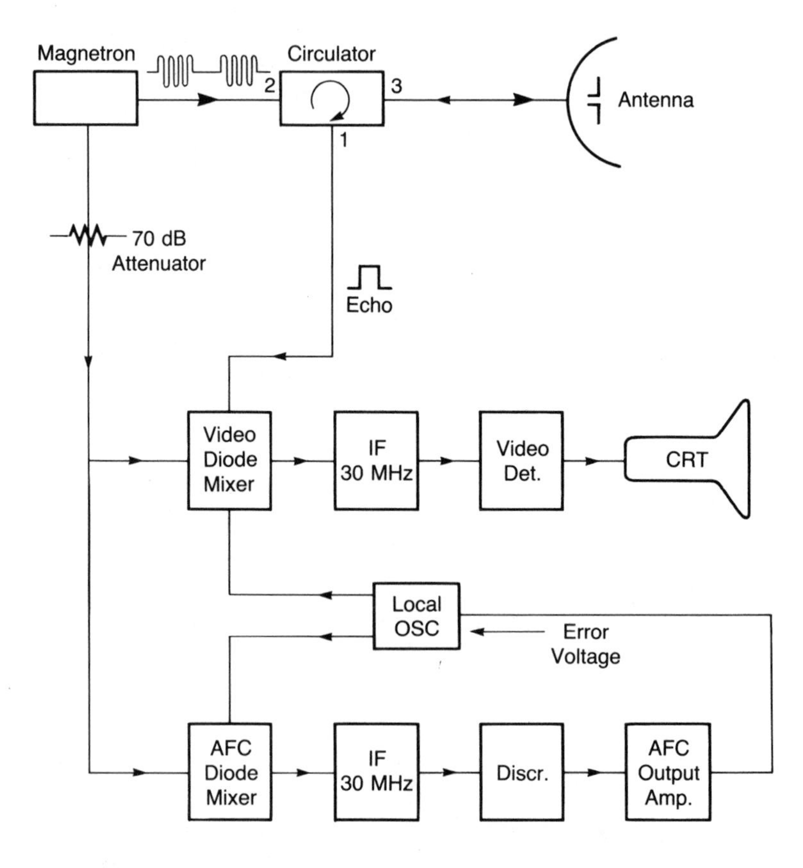

FIGURE 7–26
Basic radar receiver.

MHz and an IF bandwidth in the range of 5–10 MHz.[9] The narrower the pulses, the wider the bandwidth requirements. The operating frequency of the magnetron oscillator falls within precisely prescribed limits for commercial marine radar as defined by the FCC. The commercial bands for ship radar operation are shown in table 7–1.

TABLE 7–1
FCC marine radar bands.

Band	Frequency
S band (10 cm)	2.90-3.10 GHz
C band (5 cm)	5.46-5.65 GHz
X band (3 cm)	9.30-9.50 GHz

Kaufman, M. *Radio Operator's License Q & A Manual,* 10th ed., New Jersey: Hayden, 1985.

[9]The Raytheon model 1200 Mariner Pathfinder mentioned earlier has an IF center frequency of 38 MHz. The required bandwidth is 6 MHz, and the pulse width is between 0.12–0.5 μS. The magnetron frequency is X-band, 9.445 GHz ±30 MHz.

A greatly attenuated (−70 dB) sample signal voltage from the magnetron is heterodyned with the local oscillator (Gunn diode or reflex klystron) in the video diode mixer circuit to produce the required 30 MHz IF for the video display. It is essential to maintain accurate control over the intermediate frequency value at all times in order to provide maximum response of the display. This control is exercised by an automatic frequency control (AFC) circuit since the frequency of the magnetron oscillator tends to drift. The AFC diode mixer produces the same 30 MHz IF by mixing the sampled magnetron voltage with the local oscillator. If both the magnetron and local oscillator are operating at the proper frequency, the output of the discriminator will be zero volts. However, if either the magnetron or the local oscillator frequency drifts, an error voltage will be generated by the discriminator and applied to the local oscillator frequency-control circuit, bringing the IF back to its center frequency.

The Display Section

As mentioned earlier in connection with figures 7–6 and 7–7, most radar displays are represented using a polar coordinate scheme that allows for the simultaneous display of target bearing and range information. Such an arrangement is often called a plan-position indicator (PPI) and uses intensity modulation of the CRT's electron beam to display a ''map'' of the surrounding targets including stationary land contours. The bearing marks are usually printed directly on the scope graticule, while range rings and the ship's heading marker (SHM) are displayed using the rotating trace of the electron beam, which moves in synchronization with the rotating antenna.

The CRT uses a high-persistance phosphor which holds residual images for up to 10 seconds as compared with the 1/100th second phosphors encountered in ordinary television receiver tubes. Deflection is by means of an electromagnetic yoke surrounding the neck of the CRT, and is very similar to those found in TV receivers. In older radar systems, the yoke was required to turn in order to sweep the electron beam around the CRT. This method was a mechanical nightmare, requiring a slip-ring arrangement to connect the deflection signals to the spinning display tube. More modern techniques use either a stationary deflection yoke whose magnetic fields are rotated electronically or an on-board computer which simulates the rotating beam by appropriate digital signal processing, thus eliminating all moving parts.

In order to prevent strong returning echos of nearby waves (called ''sea return'') from blinding the radar set, a sensitivity time-control (STC) circuit is utilized. The circuit temporarily disables the IF circuits for a few microseconds immediately after the transmitted SHF pulse leaves the antenna. Therefore, large blurred, bright areas near screen center, which would mask a real target, are eliminated. A user-operated ''sea clutter'' control is also used to reduce gain level for short range only. In addition, a ''rain clutter'' control may be adjusted to mitigate the returns of snow or rain, thus allowing weaker targets to be displayed.

In radar sets using computers to digitally process target data, a ''freeze'' pushbutton is often used to store the last trace on the display so that the operator has time to make bearing or range estimates. The display is often presented in green or amber which not only helps eliminate eye fatigue, but also makes target ''blips'' more visible under conditions of high ambient illumination as might be encountered in bright daylight.

THE RADAR HORIZON

In chapter 2, equation (2–20) was given as an empirical expression for calculating the curvilinear distance between a transmitting antenna and a receiving antenna. It should be noted that the results obtained from this expression are approximations which, while providing good estimates, are subject to wide variations due to changes in the refractive index between the antennas. Such changes might be due to temperature differences between the two sites; rain or other inclement weather; or destructive wave interference. Note, too, that unless the antennas are very high in relation to the intervening space, the curvilinear distance is essentially the same as the rectilinear distance. In determining the radar horizon, we use the conventional formula, but make use of the conversion factor 1 nmi = 1.852 km. Therefore, we obtain the situation shown in figure 7–27 and give the revised distance formula as

$$D_t = D_1 + D_2 = 2.16\sqrt{h_1} + 2.16\sqrt{h_2} \qquad \textbf{(7–16)}$$

where D_1 and D_2 are in nautical miles and h_1 and h_2 are in meters.

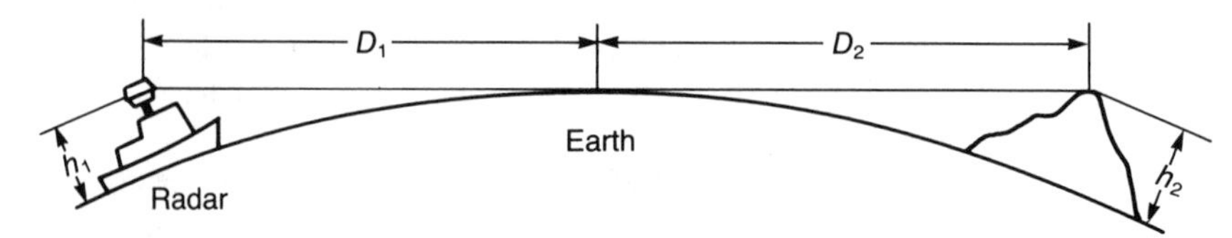

FIGURE 7–27
The radar horizon.

As an example, if the radar antenna is 12 meters above the water line, a 12-meter-high target should just be coming into view at a distance of $D_t = 4.32\sqrt{h} = 4.32\sqrt{12} = 15$ nmi. As another example, suppose we were required to find the height of a radar antenna with a radio horizon of 8.5 nmi. We would make use of equation (7–16) by solving $D = 2.16\sqrt{h}$ for the value of h. On solving for h, we obtain $h = (D/2.16)^2 = 15.5$ meters. Careful comparison between this example and the previous one might lead you to question the results. For while the heights of the antennas differ by nearly 11 1/2 feet, the difference between the horizons is only one nautical mile. The disappointingly small difference in the distance may be accounted for by noticing that the range varies as the square root of the antenna height. Looking at it another way, in order to double the distance to the horizon, the antenna height must be increased by a factor of 4.

PHASED ARRAYS

It was mentioned in the opening of this chapter that early radar installations were severely handicapped due to the slow rotational speeds of their massive antennas. Indeed,

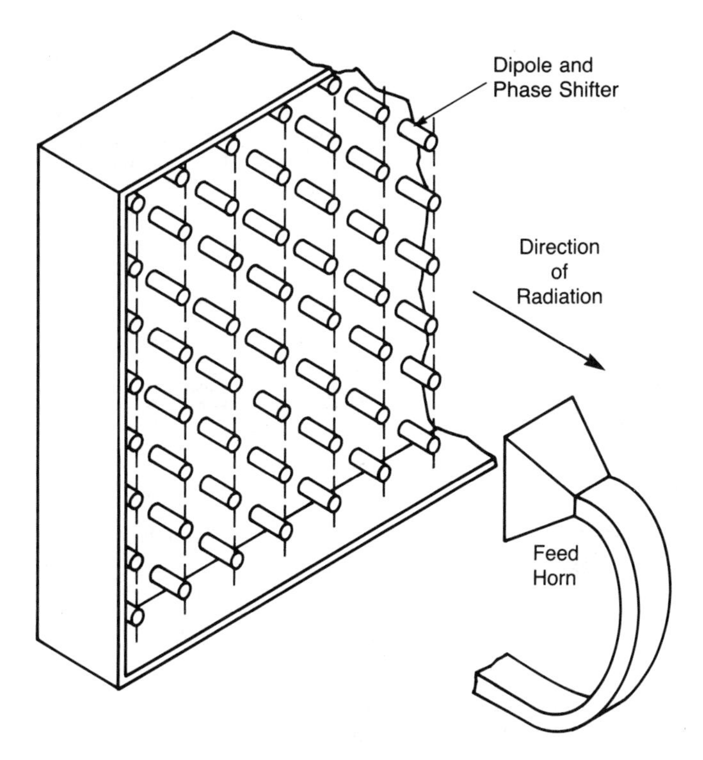

FIGURE 7–28
A phased array showing dipoles and associated phase shifters.

even though antenna sizes have declined considerably and lighter materials have been employed in their construction, rotational speed limits imposed by inertial constraints have nonetheless restricted scanning rates. Fortunately, however, due mostly to advances in ferrite devices, ways have been found to steer an antenna's zero inertia beam while allowing the physical antenna to remain motionless. This is made possible by nesting many small dipoles (often thousands) in a single physical array, then feeding each elemental dipole with a differently phased signal such that the combined effect is a narrow main lobe which changes its position in response to the total radiation. Antenna assemblies thus designed are often referred to as electronically-steered array (ESA) systems. The ferrite phase shifters, using essentially the same principles outlined in chapter 4, are small enough to be mounted directly on the array along with the individual dipoles.

A section of an ESA radar antenna is illustrated in figure 7–28. The main lobe of the beam is in a direction normal to the surface of the array. Two important beam configurations and their associated antenna geometries are the *linear* and *planar* arrays as shown in figure 7–29. The linear array provides a fan-shaped radiation pattern, while the planar array results in a pencil beam.

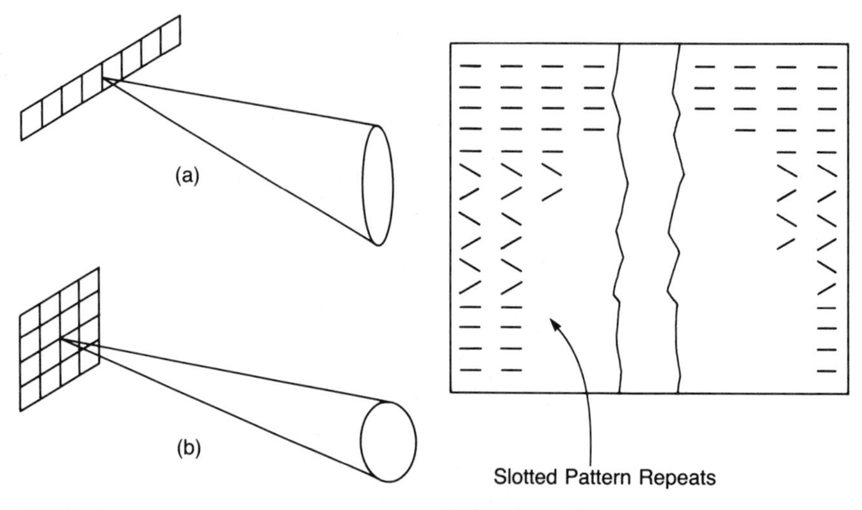

FIGURE 7–29
(a) Linear radiation pattern. (b) Planar ESA pattern.

FIGURE 7–30
A slotted waveguide phased-array antenna.

Dipole antennas are not the only possible elemental radiators used with ESA systems. Figure 7–30 illustrates a radiator using slotted antennas with the angle of each slot positioned so that the resultant radiation pattern is a narrow azimuth beam for scanning in the horizontal plane. Again, the beam is swept across the horizon using ferrite phase shifters. The aperture size of the antenna illustrated is roughly 4 feet square.

SUMMARY

Radar was defined as a means of determining the range and bearing of a target by means of emitting a short SHF pulse, then electronically evaluating the echo data. The four main systems of the radar set were the transmitter, the receiver, the duplexer, and the antenna.

Important navigational terms were defined. Azimuth and relative bearing were distinguished from each other, and heading was discussed in terms of true and magnetic north geographic references. The compass rose was described, and magnetic variation was compared to magnetic deviation due to boat-property anomalies.

Bearing and range resolution were described in terms of the radar's ability to distinguish between targets in close proximity to each other in either angular separation (bearing) or distance (range). The pillbox or the cut paraboloid antennas were shown to provide the narrow beam required for high bearing resolution. The radar range equation was developed, and maximum range was shown to depend primarily on (1) transmitted power, (2) receiver sensitivity, and (3) capture area of the receiving antenna.

The Doppler radar equation was also presented. It was explained that the Doppler property of moving targets could be used in removing clutter through phase-sensitive receivers.

The pulse-forming network was shown to be at the heart of the radar transmitter, and was implemented by an artificial transmission line capable of forming precisely-shaped pulses having the appropriate duration. Factors affecting the pulse repetition rate (PRR) were discussed in relation to duty cycle and peak power of the SHF energy burst.

Two versions of the duplexer were discussed briefly. The circulator was a ferrite device capable of isolating the transmitter and receiver signals. The spark gap duplexer was described as a higher-power version of signal isolator which employed waveguide structures and sensitive spark gaps.

The radar receiver was discussed as a system, and the plan-position indicator (PPI) was presented as a polar plot of target bearing and range.

A method of calculating the estimated radar horizon was given. It was shown that estimates could be thrown off considerably by signal-path anomalies affecting the refractive index.

Finally, the phased-array was presented as an effective method of sweeping the radar beam electronically. Such antenna systems were referred to as electronically-steered arrays (ESAs), and both linear and planar patterns were depicted.

PROBLEMS

1. A target echo is processed by the radar display computer once every 852 μS. How far away is the target? [about 69 nmi]
2. At a certain location, the magnetic variation is 11° 16′ west. The boat's compass has a deviation of 0.5° east. What must the compass heading be if the helmsman wishes to steer due east? [100° 46′]
3. A mid-S-band radar set has a peak pulse power of 2.5 MW and is required to have a range of 158 nmi for targets whose radar cross-section is 1.18 square meters. If the receiving antenna is regarded as ideal (i.e., 100% illumination and no spillover) and the minimum received power is 0.16 pW, what is the diameter of the smallest paraboloid reflector that can be used? [3.00 m]
4. If a practical antenna is assumed for problem 3, what is the paraboloid diameter? [3.85 m]
5. A low-power radar set has a peak pulse power of 3 kW and a pulse duration of 2 μS. If the PRR is 8 kHz, what must the average power be? [48 W]
6. What is the theoretical maximum range of the radar in problem number 5? [10 nmi]
7. If the peak pulse power of the magnetron is 20 kW, what power will arrive at the video or AFC mixers in figure 7–26? [2 mW]
8. A short-range radar set requires a radar horizon of 10 nmi. At what height must the radar antenna be erected? [21.4 m]

QUESTIONS

1. Why is time-domain reflectometry (TDR) sometimes referred to as ''closed-loop radar''?
2. What are the four basic systems in every radar set?
3. Why is the nautical mile used in navigation as opposed to the statute mile?
4. What is the difference between heading, bearing, and azimuth?
5. What is the difference between magnetic variation and deviation?
6. What is the purpose of the two concentric circles in the compass rose?
7. The timer section of a radar set sends a trigger pulse to two main subassemblies to initiate two simultaneous events. What are these two events, and why must they be synchronized?
8. Give two reasons why the pulse-forming network must produce a flat-top, steep-sided timing pulse.
9. What is the purpose of the range rings on a radar set display?
10. What is the purpose of the ship's heading marker (SHM)? What is the point of its origin?
11. What is meant by the polar display of a PPI?
12. What is the most important factor in determining bearing resolution?
13. Why is the vertical beamwidth of a ship's radar so many times greater than its horizontal pattern? Could you obtain this same pattern with a full paraboloid antenna? Why or why not?
14. Describe the radiation pattern obtained with the parabolic cylinder illustrated in figure 7–10(b).
15. Why must radomes never be painted?
16. Define and describe each of five factors influencing the maximum range of a radar set (see equation 7–10).
17. Why will Doppler radar *not* acquire a target moving normal to the scan field?
18. Explain in your own words the operation of the artificial delay line.
19. Draw a diagram illustrating the relationship between pulse duration (t), PRR, pulse repetition interval (T), time between pulses (TBP) or pulse interval. In what way does each factor affect peak pulse power and duty cycle?
20. Explain the operation of two important types of duplexer.
21. Explain the function of each functional block (stage) in the radar set diagram shown in figure 7–26.
22. Why is a high-persistance phosphor CRT required in radar displays?
23. Explain the terms *sea return* and *clutter*.
24. Explain the relationship between the SHM, the timer unit trigger pulse, and the CRT display.
25. What factors may introduce errors in determining the radar horizon using equation (7–16)?
26. Why does moving the radar antenna up or down a few feet have little effect on the radar horizon distance?
27. Briefly explain the operation of an ESA antenna system.
28. In what situations would planar and linear arrays be best suited?

INTRODUCTION TO AMPLITUDE MODULATION

BASIC IDEAS

Amplitude modulation (AM) is a method of propagating a lower frequency signal by superimposing it "piggyback" style on a higher frequency signal. To determine the reason this might be desirable, consider the height of an ideal Marconi antenna required to propagate a 1 kHz signal.

$$\frac{\lambda}{4} = \frac{c}{4f} = 75 \text{ kilometers, or } 46.5 \text{ miles}$$

On the other hand, if this frequency could be superimposed on a much higher frequency in the microwave range, say 10 GHz, a similar Marconi would only need to be a little over one inch long.

One way to accomplish such modulation is called series modulation. In this method, the modulator, power supply, and modulating voltage are all in series. Consider the simplified block diagram transmitter circuit shown in figure A–1.

The RF oscillator produces a radio-frequency voltage (f_c) which is fed to the modulator. The modulator is an active device such as a vacuum tube or VMOS. The current flowing in the primary of the coupling transformer would remain a constant value were it not for the modulating voltage (f_m), which adds algebraically to the output signal of the modulator. Therefore, the *amplitude* of the RF voltage is made to vary in step with f_m.

The result of all this is that the energy radiated from the antenna is an RF signal containing both *pitch* (frequency) information as well as *loudness* (amplitude) information which can be received at a distant point and "decoded" as intelligence. Basically,

FIGURE A–1
A simplified AM transmitter.

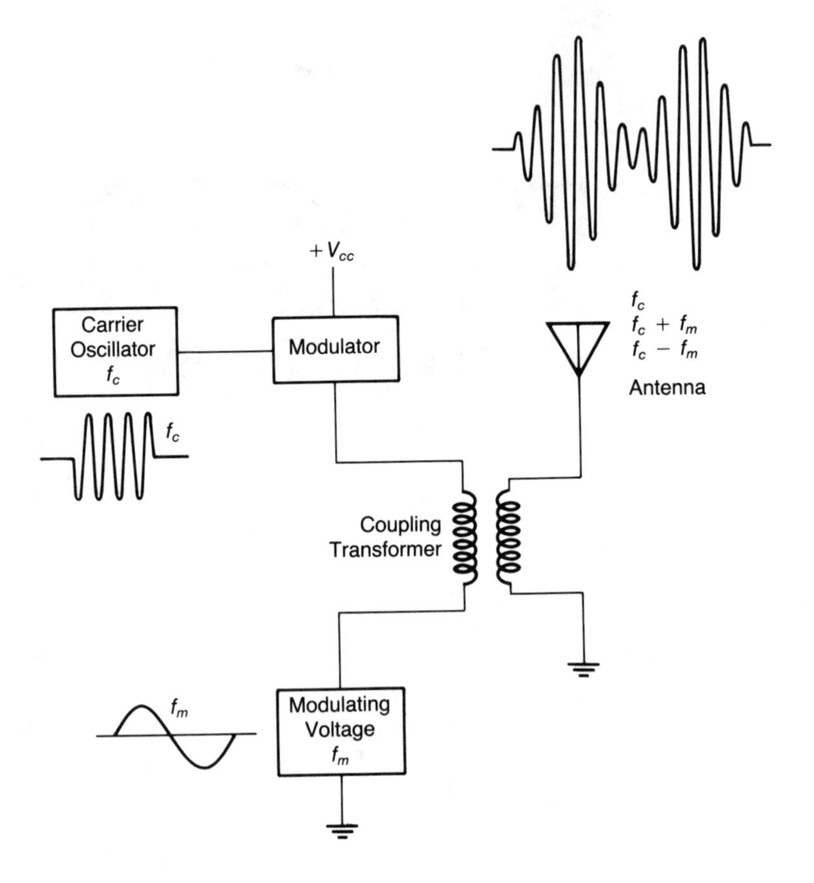

FIGURE A–2
The AM process.

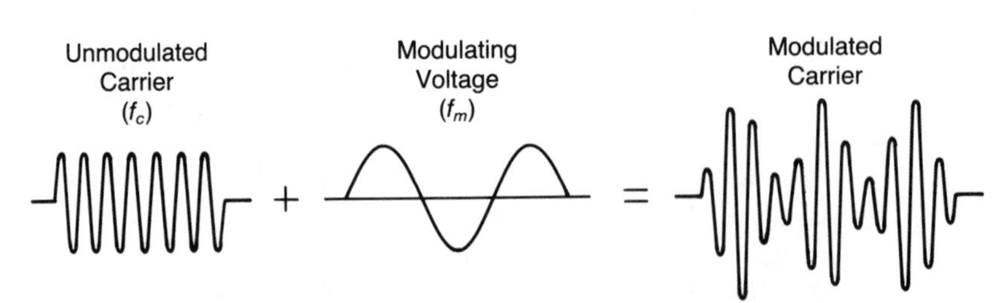

FIGURE A–3
The AM carrier.

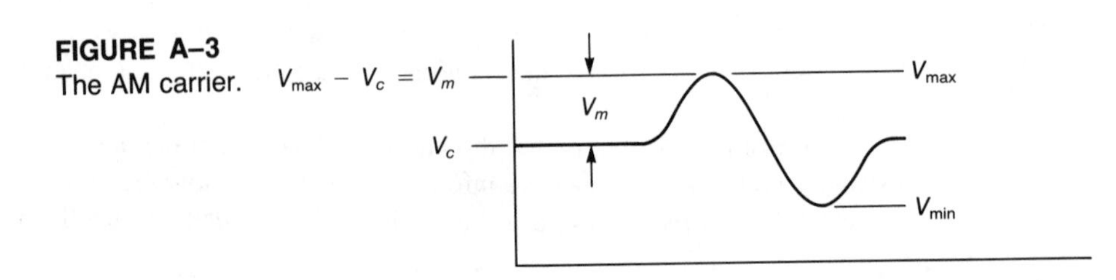

this is what is called radio communication. A carrier voltage (f_c) modulated by a lower-frequency voltage (f_m) would appear as shown in figure A–2.

THE MATHEMATICAL MODEL OF AM

Purpose: To show that both amplitude as well as frequency may be encoded as variations in amplitude of the carrier, and that two sideband frequencies are created as a natural by-product of the AM process.

1. The instantaneous RF carrier voltage, which is almost always a sine wave, is given by

$$v_c = V_c \sin \omega_c t \qquad \textbf{(A–1)}$$
$$\omega_c = 2\pi f_c$$

2. The instantaneous modulating voltage will be assumed to be sinusoidal for convenience of analysis, and is given by

$$v_m = V_m \sin \omega_m t \qquad \textbf{(A–2)}$$
$$\omega_m = 2\pi f_m$$

3. The amplitude-modulated carrier is illustrated in figure A–3. The ratio of V_m to V_c is defined as the percent of modulation (m), also called modulation index or depth of modulation

$$m = \frac{V_m}{V_c} \qquad 0 < m < 1 \qquad \textbf{(A–3)}$$

The extremes of modulation are shown in figure A–4.

4. The instantaneous value of the modulated carrier is given by

$$v = (V_c + v_m) \sin \omega_c t \qquad \textbf{(A–4)}$$

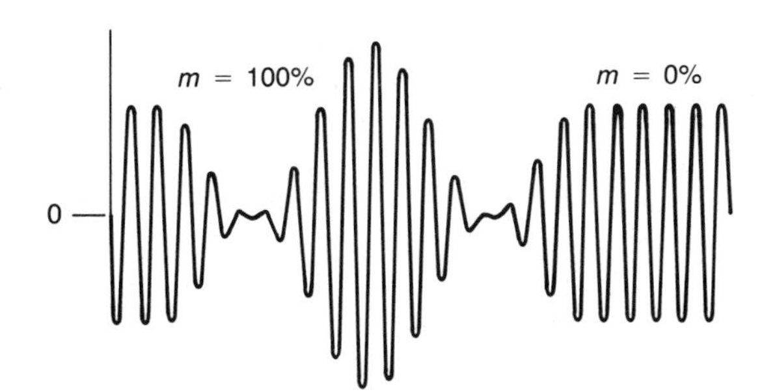

FIGURE A–4
Extremes of modulation ($0 < m < 1$).

Equation (A–4) shows that the amplitude of the *modulated* carrier will vary according to the instantaneous changes in f_m.

From equation (A–2) we may substitute for v_m to obtain

$$v = (V_c + V_m \sin \omega_m t) \sin \omega_c t \qquad \textbf{(A–5)}$$

We may now substitute $V_c \times m$ from equation (A–3) for V_m in equation (A–5) to get

$$v = (V_c + V_c m \sin \omega_m t) \sin \omega_c t \qquad \textbf{(A–6)}$$

On factoring out V_c, we have

$$v = V_c(1 + m \sin \omega_m t) \sin \omega_c t \qquad \textbf{(A–7)}$$

Equation (A–7) may be expanded to yield

$$\begin{aligned} v &= V_c \sin \omega_c t + V_c \sin \omega_c t\ (m \sin \omega_m t) \\ &= V_c \sin \omega_c t + V_c m\ (\sin \omega_c t)(\sin \omega_m t) \end{aligned} \qquad \textbf{(A–8)}$$

The second term may be rewritten using the trig identity

$$\sin x \sin y = \frac{1}{2}[\cos(x - y) - \cos(x + y)]$$

This gives us the equation

$$v = \underbrace{V_c \sin \omega_c t}_{\text{instantaneous value of unmodulated carrier}} + \underbrace{V_c m/2 \cos(\omega_c - \omega_m)t}_{\text{instantaneous value of lower sideband}} - \underbrace{V_c m/2 \cos(\omega_c + \omega_m)t}_{\text{instantaneous value of upper sideband}} \qquad \textbf{(A–9)}$$

Here, we may see that there are three different components in the AM wave, the last two of which have the same absolute value, but differ in frequency ($\omega_c \pm \omega_m$). Note that the first term is simply equation (A–1), which is the unmodulated carrier. Note, too, that the first term does not contain ω_m in the argument, but that the other terms do. This illustrates the fact that intelligence is carried *only* in the sidebands, not in the carrier. It should also be apparent that as the depth of modulation (m) increases, the magnitude of the sidebands also increases. Finally, note that since the low-frequency signal (f_m) does *not* appear as an individual component of the modulated signal, transmitting antennas may assume reasonable dimensions, as shown in the opening paragraph.

This concludes the proof of the assertion that the AM process may be used to encode both frequency and amplitude variations, and that the process yields two sidebands as a natural consequence of the modulation phenomenon.

AM IN THE FREQUENCY DOMAIN

A spectrum analyzer display of AM is shown in figure A–5. Note that bandwidth is

$$(f_c + f_m) - (f_c - f_m) = f_c + f_m - f_c + f_m = 2f_m$$

That is, the bandwidth is always twice the frequency of the modulating voltage

$$BW = 2f_m \tag{A–10}$$

For commercial AM, the FCC specifies bandwidth as 10 kHz because adjacent stations are allocated every 10 kHz. Consequently, if $f_m > 5$ kHz, interference with an adjacent station will occur.

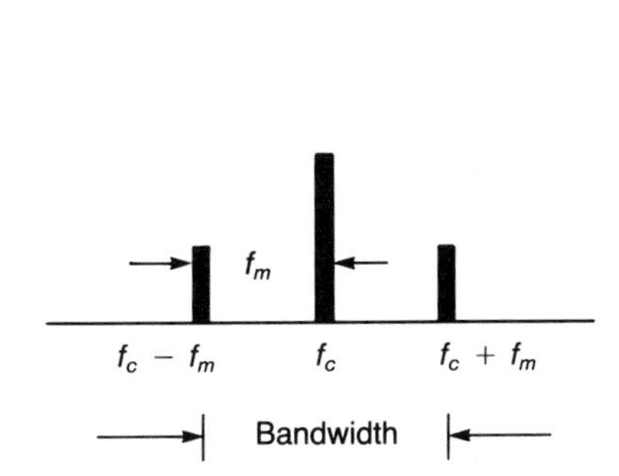

FIGURE A–5
Spectrum analyzer display of AM.

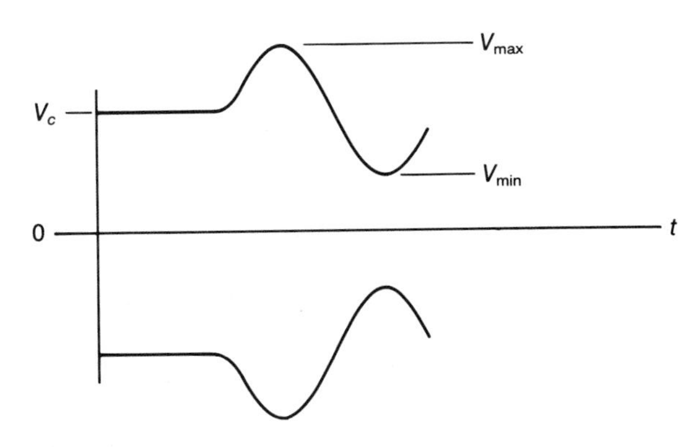

FIGURE A–6
Time domain display of AM.

DETERMINING m DIRECTLY FROM A CRO DISPLAY IN THE TIME DOMAIN

Consider the scope display shown in figure A–6. From the definition of % modulation, $m = V_m/V_c$.

$$V_m = \frac{(V_{max} - V_{min})}{2} \tag{A–11}$$

$$\begin{aligned} V_c &= V_{max} - V_m \\ &= V_{max} - \frac{(V_{max} - V_{min})}{2} \\ &= \frac{2V_{max} - V_{max} + V_{min}}{2} = \frac{(V_{max} + V_{min})}{2} \end{aligned} \tag{A–12}$$

Therefore, $m = V_m/V_c = (V_{max} - V_{min})/2 \times 2/(V_{max} + V_{min})$ which equals

$$m = \frac{V_{max} - V_{min}}{V_{max} + V_{min}} \tag{A–13}$$

For best intelligibility and loudness at the receiver, $m = 100\%$. We may calculate the modulation index directly from the display as

$$m = 2 \text{ antilog } \frac{V_{sb(dB)} - V_{c(dB)}}{20} \tag{A–14}$$

For 100% modulation, the spectrum analyzer would display sidebands that differed from the carrier by −6 dB. For example, figure A–7 shows the spectrum analyzer display for $m = 1.0$.

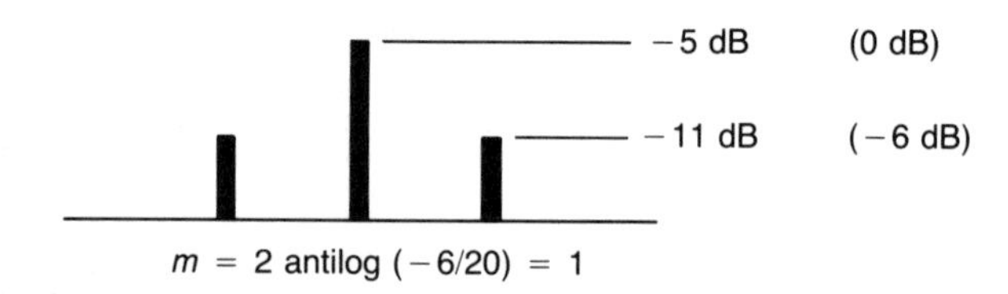

FIGURE A–7
Spectrum analyzer display for 100% modulation.

TOTAL POWER CONTAINED IN THE SIGNAL

$$P_t = P_c + P_{LSB} + P_{USB} \tag{A–15}$$

This equation follows from equation (A–9), which shows that the modulating signal *adds* to the carrier, and does not merely change its form. The RMS power in the carrier is

$$P_c = \frac{\left(\frac{V_c}{\sqrt{2}}\right)^2}{R_r} \tag{A–16}$$

where $V_c/\sqrt{2}$ = RMS value of V_c, and R_r is the antenna radiation resistance.

The RMS power in *either* sideband is

$$P_{SB} = \frac{\left(\frac{mV_c}{2}\Big/\sqrt{2}\right)^2}{R_r} \tag{A–17}$$

where $mV_c/2$ is the same peak *sideband voltage* given by equation (A–9).

Therefore,

$$P_c = \frac{(V_c^2)}{2R_r} \text{ and } P_{SB} = \frac{(m^2V_c^2)}{8R_r} \quad \textbf{(A–18)}$$

Finally,

$$P_t = \frac{V_c^2}{2R_r} + \frac{V_c^2}{2R_r}\frac{m^2}{4} + \left(\frac{V_c^2}{2R_r}\frac{m^2}{4}\right) \quad \textbf{(A–19)}$$

We have intentionally arranged equation (A–19) in this way since P_c is the coefficient of each term $P_c = V_c^2/2R_r$. Then, $P_t = P_c + P_c(m^2/2) = P_c[1 + (m^2)/2]$. And finally,

$$\frac{P_t}{P_c} = \frac{1 + (m^2)}{2} \quad \textbf{(A–20)}$$

This last result is important for 2 reasons:

1. It shows that m may be found when P_t and P_c are known, while V_{max} and V_{min} are *not* known. In microwave work, it is often a simpler matter to measure power rather than voltage.
2. It shows that for 100% modulation, $P_t = 1.5P_c$, which is the maximum power that various stages of amplification must be able to withstand without distortion.

AM RECEIVER FUNDAMENTALS

The simplest radio-signal receiver is shown in figure A–8. The incoming signal is rectified so that the pull on the earphone diaphragm is always in the *same* direction and has a magnitude that varies at the audio rate. If it were not for diode D_1, one would not be able to hear the audio.

The next most obvious step would be to amplify the AF signal so that a louder sound may be heard. One way to do this is with a power detector—that is, a detector

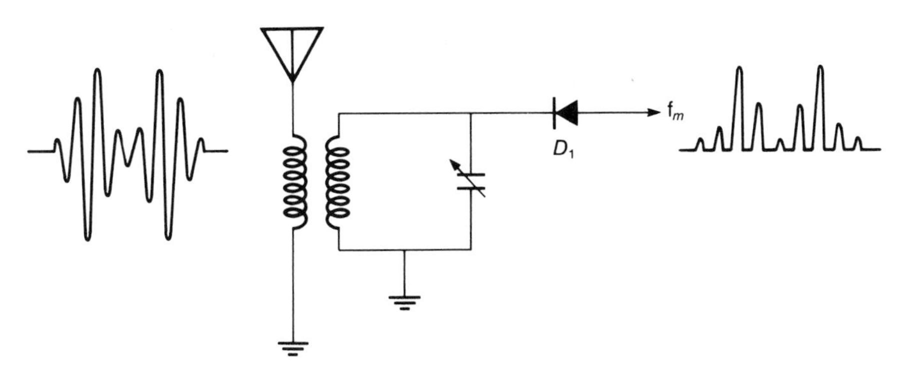

FIGURE A–8
Simplest AM demodulator.

and amplifier in one active device. Any active device biased just at cutoff (i.e., class B) will operate as a power detector (see figure A–9).

The Superheterodyne Receiver

The *mixer* is fed two signals, one (F_s) from the *RF amplifier* and the other (F_0) from the local oscillator. Whenever two such signals are combined electronically, 4 distinct signal frequencies emerge. They are F_s, F_0, and $F_0 \pm F_s$. It is the difference frequency (usually, but may also be the sum) that interests us here.

To tune from one station to another, the RF amplifier, mixer, and local oscillator are all tuned simultaneously (tracked) to produce the intermediate frequency (F_{IF}):

$$F_{IF} = F_0 - F_s \quad \textbf{(A–21)}$$

The *superhet* is the most practical tuneable receiver *primarily* because the IF amplifiers and subsequent stages do *not* require tracking. They are all adjusted to the IF and remain that way regardless of the frequency at the front end.

A block diagram of a simple AM superhet is shown in figure A–10. The example is that of a station signal of 1,070 kHz. The local OSC is tuned to 1,525 kHz. Therefore, from equation (A–21), the IF is 455 kHz.

Since the standard AM broadcast band is 535–1,605 kHz, the local oscillator must track between 990–2060 kHz to get the required IF.

Double Conversion

The superhet described in the previous section may experience image frequency problems. For example, the signal of 1,070 kHz is not the only signal capable of producing the IF when beat against the local oscillator frequency of 1,525 kHz. If a frequency of, say, 1,980 kHz should find its way into the mixer, the IF of 455 kHz will also be produced: 1,980 − 1,525 = 455 kHz. The mixer has no way of telling which signal is the desired one. Consequently, both frequencies will compete for dominance in the receiver's output with the result that neither signal achieves clarity.

The 1,980 kHz signal is called the *image frequency,* which is defined as any frequency which, when beat against the local oscillator, will produce the IF. It will be noted that any frequency 455 kHz above or below the local oscillator frequency will produce the IF. This statement leads to the expression

$$F_i = F_0 \pm F_{IF} \quad \textbf{(A–22)}$$

From equation (A–21), $F_0 = F_{IF} + F_s$. Then, substituting for F_0 in equation (A–21) we obtain

$$\begin{aligned} F_i &= (F_{IF} + F_s) \pm F_{IF} \\ &= 2F_{IF} + F_s \end{aligned} \quad \textbf{(A–23)}$$

As shown in figure A–11, the RF amplifier has a rather broadband response, and since the image frequency (F_i) is only $2F_{IF}$ (910 kHz) away from the signal frequency (F_s), it falls well within the band pass of the RF input stage.

FIGURE A–9
Demodulated output of power detector.

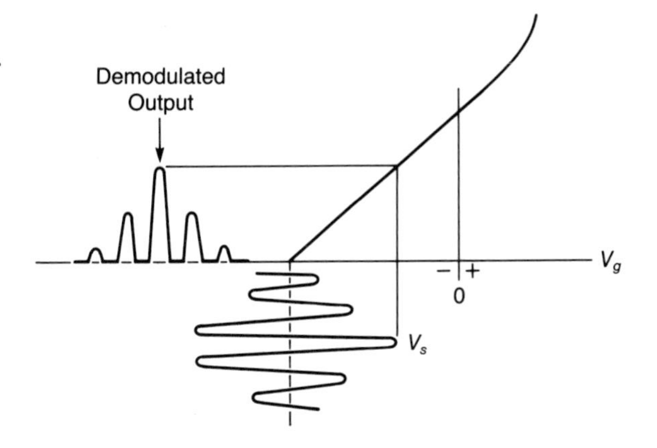

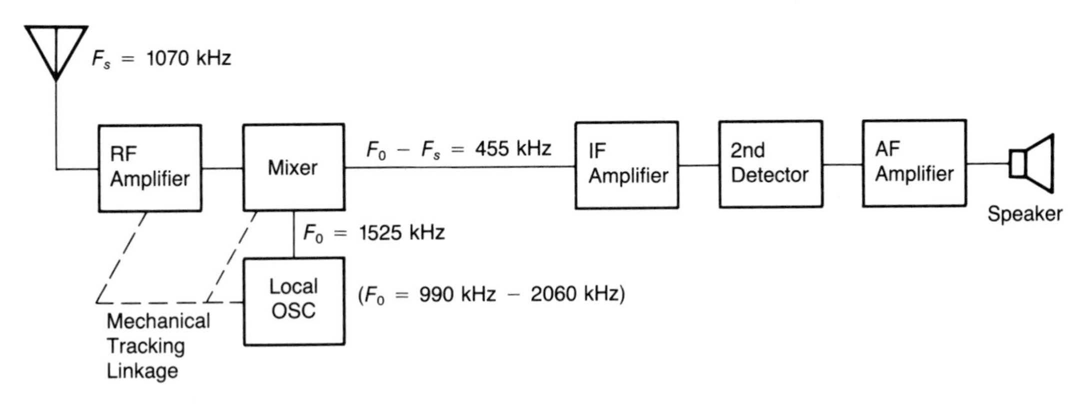

FIGURE A–10
Block diagram of superheterodyne receiver.

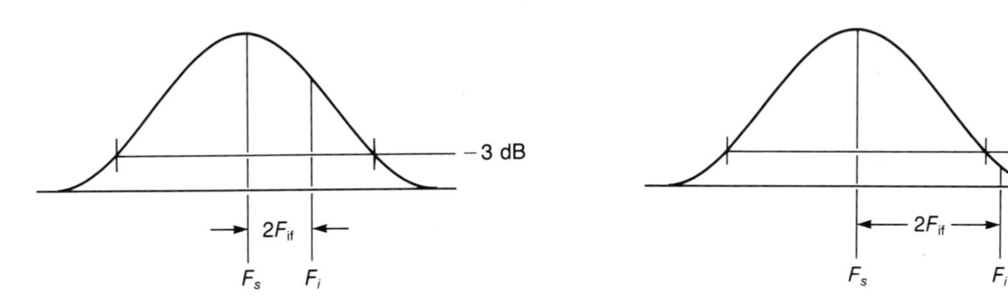

FIGURE A–11
RF amplifier bandwidth.

FIGURE A–12
IF outside of RF amplifier passband.

Since the receiver is usually required to tune over a wide band of frequencies, it is not always practical to eliminate the image-frequency problem by using high-Q circuits in the receiver front end. A more practical approach would be to use a higher value for $F_{IF,}$ thus placing the image well outside the bandwidth, as shown in figure A–12.

A second mixer may now be used to get back down to the 455 kHz IF where high-Q circuits are easier to realize. The image-frequency solution outlined above is called *double conversion,* and may require a first IF of several megahertz.

INTRODUCTION TO FREQUENCY MODULATION

BASIC IDEAS

Recall that in the process of amplitude modulation (AM), both amplitude and frequency information were encoded as amplitude variations. With FM, however, these qualities are encoded as *frequency* variations.

FM was first introduced commercially in 1936 in an attempt to decrease noise. Since most man-made and atmospheric noise have strong amplitude components, they have their greatest effect on other amplitude phenomena, to wit, AM. FM, on the other hand, is virtually immune to these noise sources.

Consider the simple FM transmitter circuit shown conceptually in figure B–1. The *amplitude* of the oscillator is always constant, but the *frequency* of the carrier is determined by the tank circuit whose center frequency is set to that of the carrier (F_c).

FIGURE B–1
Conceptual diagram of an FM transmitter.

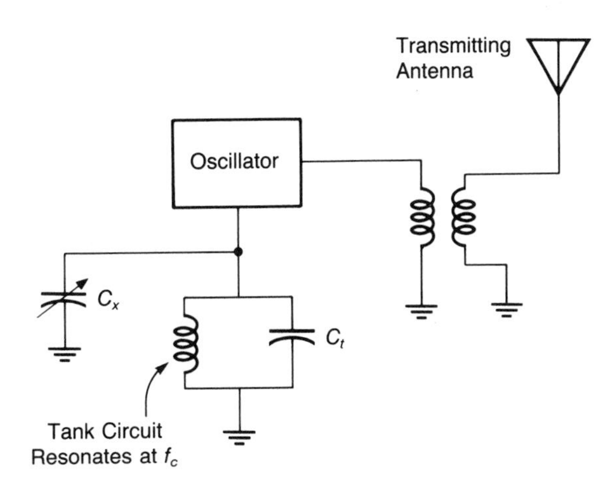

However, since C_x and C_t are in parallel, changing the value of C_x will obviously change the value of F_c.

The value of C_x, which may be a capacitor microphone or varactor, is a function of the *amplitude* of the modulating signal (F_m). As the amplitude of this signal changes, F_c will change. Moreover, the rate at which F_m changes determines the *frequency* at which F_c swings above and below the carrier frequency. Consequently, both amplitude and frequency may be encoded as frequency variations about a fixed center carrier frequency (F_c) without ever affecting the amplitude of the oscillator voltage. Figure B–2 illustrates how amplitude changes are encoded as changes in the carrier frequency.

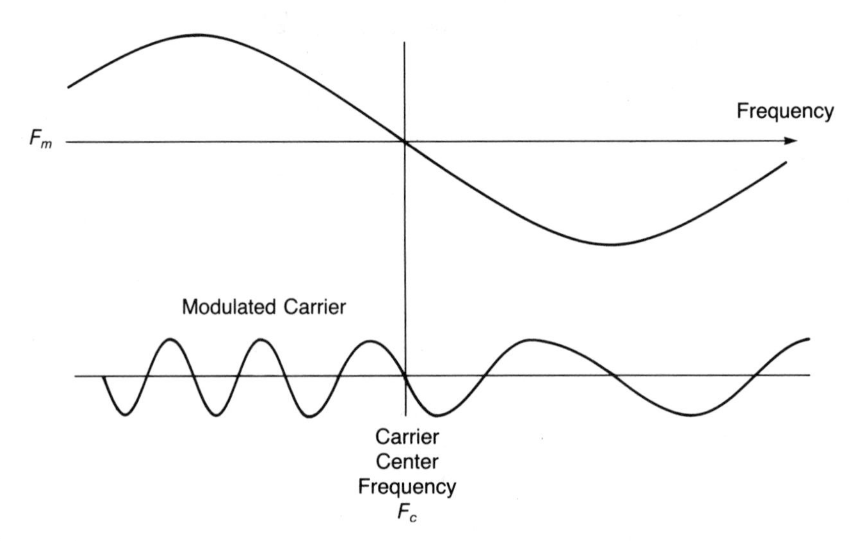

FIGURE B–2
Amplitude encoded as frequency variation in the FM process.

THE MODULATION INDEX

The modulation index (m_f) for FM is defined as

$$m_f = \frac{\delta}{f_m} \tag{B–1}$$

where δ = maximum amount by which carrier deviates from its unmodulated value due to amplitude variations
f_m = the modulating frequency

For commercial FM broadcast stations, the FCC has set δ = 75 kHz as 100% modulation. For wideband FM, the FCC has also set f_m as 15 kHz. Therefore, the modulation index is

$$m_f = \frac{\delta}{f_m} = \frac{75}{15} = 5$$

As will be seen later, this value of m_f will produce 8 *pairs* of significant[1] sidebands for a total bandwidth of $15 \times 16 = 240$ kHz. This bandwidth is a little wider than the maximum channel allocation of 200 kHz because of the 15 kHz limit set on f_m. However, this represents an extreme case, and since the fundamental of any musical instrument is less than 5 kHz, it is unlikely that much weaker harmonics would be strong enough to produce a bandwidth as great as 240 kHz.

The spectrum analyzer display for this 240 kHz wide band with modulation index of 5.0 would appear as shown in figure B–3.

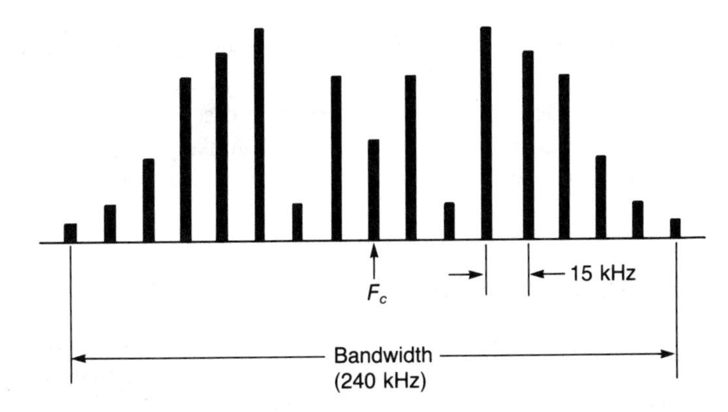

FIGURE B–3
Spectrum analyzer display for $\delta_{(max)}$ [75 kHz], $f_{m(max)}$ [15 kHz], and $m_f = 5.0$.

As was mentioned above, the FCC has *defined* 100% modulation to be $\delta = 75$ kHz maximum excursion of the carrier on either side of F_c. Therefore, for example, if $\delta = 50$ kHz, then

$$\%\text{-mod} = \frac{50}{75} = 67\%$$

The deviation required for 50% modulation is

$$\%\text{-mod} = \frac{\delta}{75}$$

Therefore, $\delta = (0.50)(75 \text{ kHz}) = 37.5$ kHz.

COMMERCIAL FM CHANNEL ALLOCATIONS

The FM commercial broadcast band (located in the middle of the TV broadcast band between channels 6 and 7) extends from 88 MHz to 108 MHz, giving a total frequency

---

[1]A significant sideband is one whose amplitude is $>$ 1% of the unmodulated carrier voltage V_c (i.e., $A' = 20 \log 0.01 = -40$ dB). Therefore, any sideband < -40 dB is not considered significant.

allocation of 20 MHz. As stated above, the FCC has set 150 kHz as the *total* maximum carrier deviation, plus a 25 kHz ''guard'' channel on *each* side, resulting in a total channel width of 200 kHz. Therefore,

$$\frac{20 \text{ MHz}}{200 \text{ kHz}} = 100 \text{ broadcast channels}$$

However, in an effort to reduce interference, adjacent channels are not assigned in the same locality. Therefore, 50 channels of commercial FM are available in a single location.

Earlier we saw that if $\delta = 75$ kHz and $f_m = 15$ kHz (both *maximums*), then the bandwidth is 240 kHz. This value obviously exceeds the channel bandwidth allocation of 200 kHz. However, this is an extreme case, and since the fundamental frequency of any musical instrument is < 5 kHz, it is unlikely that the weaker harmonics would ever be strong enough to produce a channel this wide as previously stated.

MATHEMATICAL ANALYSIS OF FM

In FM, the voltage of the modulating signal (v_m) may be used to change the instantaneous carrier frequency (f_c). For example, a varactor may be used to change the capacitance of the carrier frequency oscillator circuit. The instantaneous value of the carrier frequency then becomes

$$f_c = F_c + kv_m \tag{B–2}$$

where k is the *frequency deviation constant* and may be considered the amount by which F_c changes per volt of modulating signal.

If $v_m = V_m \cos \omega_m t$, (cosine chosen for mathematical simplicity), then equation (B–2) may be written as

$$f_c = F_c + kV_m \cos \omega_m t \tag{B–3}$$

Since there is some maximum amount by which f_c will change in response to v_m, then we define peak frequency deviation as:

$$\delta = kV_m \tag{B–4}$$

Equation (B–3) may now be written as

$$f_c = F_c + \delta \cos \omega_m t \tag{B–5}$$

Figure B–4 shows how the frequency of the carrier varies cosinusoidally with time.

The instantaneous amplitude of a frequency-modulated carrier is given in general form by

$$v = A \sin \theta \tag{B–6}$$

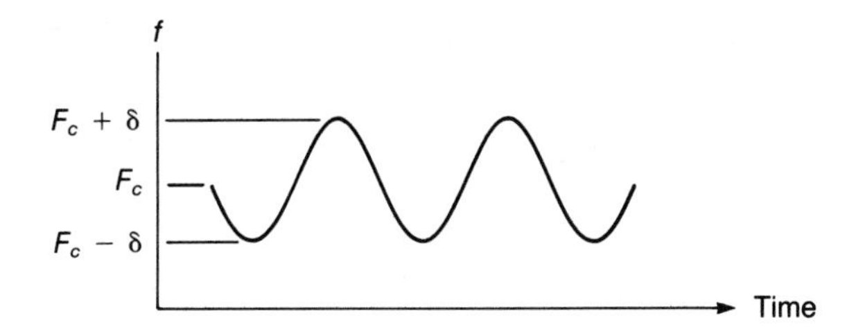

FIGURE B–4
Cosine variation of F_c with time.

Equation (B–6) requires an instantaneous angular velocity ω_i whenever θ is not a constant as is certainly the case with FM. In particular

$$\omega_i = 2\pi(F_c + \delta \cos \omega_m t) \qquad \textbf{(B–7)}$$

where the instantaneous frequency (f_c) was given by equation (B–5). In other words, ω_i equals the instantaneous rate of change of θ. Therefore

$$\omega_i = \frac{d}{dt}\theta \qquad \textbf{(B–8)}$$

Finally, we may equate equations (B–7) and (B–8) and integrate with respect to t:

$$\frac{d}{dt}\theta = 2\pi(F_c + \delta \cos \omega_m t)$$

Integrating, we get

$$\begin{aligned}\theta &= \omega_c t + \frac{2\pi\delta}{\omega_m} \sin \omega_m t + K \\ &= \omega_c t + \frac{2\pi\delta}{2\pi f_m} \sin \omega_m t + K \\ &= \omega_c t + \left(\frac{\delta}{f_m}\right) \sin \omega_m t + K \\ &= \omega_c t + m_f \sin \omega_m t + K \qquad \textbf{(B–9)}\end{aligned}$$

where K is a constant of integration which may be set equal to zero by an appropriate choice of the X-axis.

Equation (B–9) may now be substituted into (B–6) to obtain:

$$v = A \sin (\omega_c t + m_f \sin \omega_m t) \qquad \textbf{(B–10)}$$

An attempt to expand equation (B–10) shows that the instantaneous carrier amplitude depends on the sine of a sine. Therefore, we will find it convenient, from a computational perspective, to resort to Bessel functions[2] in arriving at a value for v. We now expand equation (B–10) to obtain:

$$\begin{aligned} v = {} & A\{J_0(m_f)\sin \omega_c t \\ & + J_1(m_f)[\sin(\omega_c + \omega_m)t - \sin(\omega_c - \omega_m)t] \\ & + \ldots J_n(m_f)[\sin(\omega_c + n\omega_m)t \pm \sin(\omega_c - n\omega_m)t]\} \end{aligned} \qquad \textbf{(B–11)}$$

where n is a positive integer denoting the order of the sideband pair ($n = 1$, 1st pair; $n = 2$, 2nd pair, etc.). The modulation index (m_f) is as defined in equation (B–1). The Bessel coefficient of the carrier is $J_0(m_f)$, and that of the nth sideband is $J_n(m_f)$. Bessel coefficients may be determined from convenient tables like the one shown in table B–1. Such a table gives the *relative* amplitudes of the carrier and sidebands only. Note the $\pm$ between the sines of the last bracketed term. The sign is $-$ for n odd and $+$ for n even.

FM RECEIVER BASICS

In order to detect the frequency variations in the carrier, we must have a circuit that is frequency sensitive rather than amplitude sensitive as was the case with AM. One such detector, called a discriminator in FM, is the phase-locked loop (PLL) discriminator shown in figure B–5.[3]

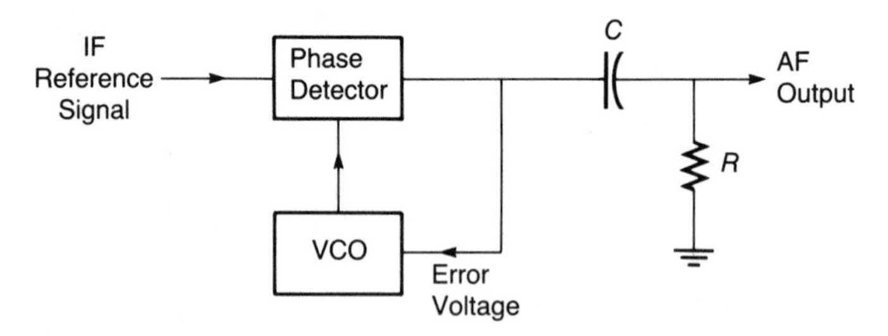

FIGURE B–5
The PLL discriminator.

The basic principle of the PLL is that two frequencies are simultaneously fed into a phase detector circuit. If both frequencies are exactly the same, there will be no change in the DC output voltage of the phase detector. Hence, no AC voltage will appear across R. However, if the reference frequency (the IF) changes, the phase detector will output an error voltage proportional to the difference between the two frequencies, which is then applied to a voltage-controlled oscillator (VCO) whose frequency

---

[2]Wylie, C. *Advanced Engineering Mathematics,* 3rd ed. New York: McGraw-Hill, 1966.

[3]Other detectors used are the balanced (stagger-tuned) and Foster-Seeley discriminators.

TABLE B–1
Bessel functions.

m_f	$J_0(m_f)$	$J_1(m_f)$	$J_2(m_f)$	$J_3(m_f)$	$J_4(m_f)$	$J_5(m_f)$	$J_6(m_f)$	$J_7(m_f)$	$J_8(m_f)$	$J_9(m_f)$	$J_{10}(m_f)$	$J_{11}(m_f)$	$J_{12}(m_f)$	$J_{13}(m_f)$	$J_{14}(m_f)$	$J_{15}(m_f)$	$J_{16}(m_f)$
0.00	1.00	—	—	—	—	—	—	—	—	—	—	—	—	—	—	—	—
0.25	0.98	0.12	—	—	—	—	—	—	—	—	—	—	—	—	—	—	—
0.5	0.94	0.24	0.03	—	—	—	—	—	—	—	—	—	—	—	—	—	—
1.0	0.77	0.44	0.11	0.02	—	—	—	—	—	—	—	—	—	—	—	—	—
1.5	0.51	0.56	0.23	0.06	0.01	—	—	—	—	—	—	—	—	—	—	—	—
2.0	0.22	0.58	0.35	0.13	0.03	—	—	—	—	—	—	—	—	—	—	—	—
2.5	−0.05	0.50	0.45	0.22	0.07	0.02	—	—	—	—	—	—	—	—	—	—	—
3.0	−0.26	0.34	0.49	0.31	0.13	0.04	0.01	—	—	—	—	—	—	—	—	—	—
4.0	−0.40	−0.07	0.36	0.43	0.28	0.13	0.05	0.02	—	—	—	—	—	—	—	—	—
5.0	−0.18	−0.33	0.05	0.36	0.39	0.26	0.13	0.05	0.02	—	—	—	—	—	—	—	—
6.0	0.15	−0.28	−0.24	0.11	0.36	0.36	0.25	0.13	0.06	0.02	—	—	—	—	—	—	—
7.0	0.30	0.00	−0.30	−0.17	0.16	0.35	0.34	0.23	0.13	0.06	0.02	—	—	—	—	—	—
8.0	0.17	0.23	−0.11	−0.29	−0.10	0.19	0.34	0.32	0.22	0.13	0.06	0.03	—	—	—	—	—
9.0	−0.09	0.24	0.14	−0.18	−0.27	−0.06	0.20	0.33	0.30	0.21	0.12	0.06	0.03	0.01	—	—	—
10.0	−0.25	0.04	0.25	0.06	−0.22	−0.23	−0.01	0.22	0.31	0.29	0.20	0.12	0.06	0.03	0.01	—	—
12.0	0.05	−0.22	−0.08	0.20	0.18	−0.07	−0.24	−0.17	0.05	0.23	0.30	0.27	0.20	0.12	0.07	0.03	0.01
15.0	−0.01	0.21	0.04	−0.19	−0.12	0.13	0.21	0.03	−0.17	−0.22	−0.09	0.10	0.24	0.28	0.25	0.18	0.12

changes by an amount necessary to reduce the error voltage to zero. Obviously, this error voltage will change at an audio rate, producing an output across the R-C voltage divider network. Thus, we have an effective method of decoding both amplitude as well as frequency variations.

Other than the discriminator circuit, the FM broadcast receiver is essentially the same as the AM receiver. An FM superheterodyne FM receiver is shown in figure B–6. Note that in the FM receiver, the intermediate frequency (IF) is given as $F_s - F_o$, whereas in AM, this frequency is $F_o - F_s$. The reason is that in AM, if the station being tuned to was 910 kHz, the local oscillator frequency would need to be 455 kHz, the IF itself. Therefore, the stronger local oscillator signal would completely block the weaker IF, and nothing would be heard on station 910. It has become standard practice to use a local oscillator frequency higher than the station frequency up to about 20 MHz.

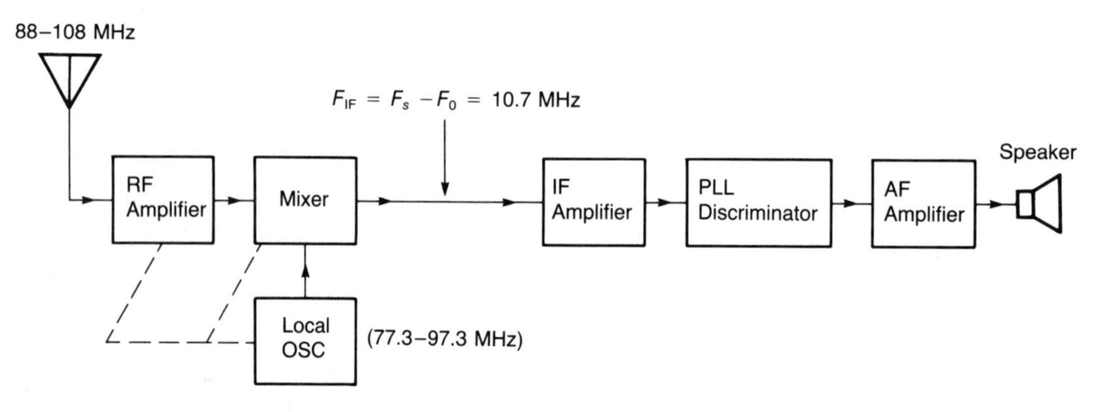

FIGURE B–6
Simplified FM receiver.

C

MICROWAVE SAFETY HAZARDS AND PRECAUTIONS

Microwave radiation, unlike X-rays or gamma rays, is not an ionizing radiation. That is, it does not dislodge electrons from tissue or alter DNA structure at the atomic level. Rather, it literally cooks it.[1] This cooking (heating) effect is most noticeable in the near-field region—that is, within one-half wavelength of the source. Two common pathologies associated with prolonged exposure to microwave radiation are cataracts and sterility. A cataract condition is usually treatable by surgical replacement of the opaque lens. Surgical treatment of sterility is not possible.

The *lowest* frequency producing harmful heating effects appears to be 50 MHz[2] (near field = 10 ft), and the *upper* safe exposure dosage has been set at 10 mW/cm^2 in the United States. However, recent experiments in the Soviet Union[3] have reduced the exposure level to around 50 $\mu W/cm^2$. The Soviet experiments have also shown evidence of behavioral anomalies in laboratory animals at 2.375–9.4 GHz.

The student should understand that the harmful effects of microwave radiation are a *combination* of exposure *time* as well as *magnitude*. For example, a few seconds of exposure within the near field of a 30 MW pulsed-radar antenna will probably result in death within a few hours or few days at most. On the other hand, it may take years of low-level exposure to produce the pathologies mentioned earlier. Moreover, as of this writing, there is no absolute certainty of the limits of either magnitude or duration of microwave exposure, and the data are inconclusive.

---

[1] Microwave ovens frequently operate at 2.45 GHz.

[2] For comparison, VHF television channel 2 begins at 54 MHz.

[3] *Behavioral Effects of Microwave Radiation Absorption*. U.S. Dept. of Health and Human Services. John Monahan, ed. U.S. Govt. Printing Office, August, 1985.

In light of these nebulous findings, there is some uncertainty as to the real limits of the danger from microwave exposure. However, the following are several precautions that should be taken when working with or around microwave devices.

1. **NEVER** make or break any connection carrying microwave energy when the system is under power.
2. **ALWAYS** terminate a microwave set-up in a suitable load.
3. If possible, work in an anechoic area where reflections from flat walls and other plane surfaces are reduced to a minimum. Not only is this a good safety practice, but it will eliminate contamination of experimental results.
4. **ALWAYS** place UHF/SHF generators on *standby* when not in use, when making or breaking connections, or when preparing an experimental set-up. An unterminated coaxial cable, for example, sprays microwave energy around like water from a garden hose.
5. Be certain to erect suitable physical barriers to prevent an unwary pedestrian from accidentally straying into your induction field of an antenna set-up. A physical barrier may consist of simply roping off the area with wide red or yellow tape.
6. **NEVER** walk or stand in front of or near any microwave antenna when it is radiating. This is especially true of high pulsed-power radar antennas.

D

HOW TO SPEAK dB-ESE

Perhaps no concept is quite so elusive to the burgeoning student technician as an understanding of the decibel. Used to specify a seemingly endless array of ideas from antenna gain to transmitter output power, the ubiquitous dB appears to defy virtually every tenet of human logic.

For example, while we cannot see current flowing in a circuit, we can certainly infer its existence through simple measurement with an ammeter. On the other hand, we can measure decibels, but they have no physical presence in the circuit itself. They do not "flow" through the circuit as electrons do. So how can one be expected to understand something that can be measured, yet doesn't exist?

Part of the problem lies in the fact that the student is usually introduced to the dB by what seems to be a cryptic mathematical formula without any appreciation of where the dB (or the formula, for that matter) comes from. The student is merely informed that the idea is in some way very important and should be learned. In other words, the dB is presented in a sort of historical vacuum. The second part of the problem rests in a complete lack of understanding and preparation on the part of many students concerning what should be very simple ideas about logarithms and exponents. Therefore, we are going to explore both a bit of history as well as some simple mathematics.

IN THE BEGINNING

Suppose you were some divine creator and had set for yourself the task of creating the first human being. Everything starts out very smoothly: hair, arms, nose, heart, muscle, teeth, liver, feet, elbows, armpits. All fit and function as intended. But now come the ears. How shall your creation hear? Let's try linear audition. Look at the graph in figure D–1.[1]

Notice that the change in intensity a–b of a very soft sound produces the same sense of loudness A–B as an identical change c–d of a very *loud* sound. Looking at it another way, if the intensity of a sound doubled, the ear would hear a sound twice as loud as before. Similarly, if the sound were 10 times as loud, the ear would hear a sound 10 times louder than before. While this seems to work out well for most medium-intensity sounds, consider what would happen if, say, a loud clap of thunder occurred which was 100,000 times louder than a clap of the hands. The result: a very painful experience and possible deafness for life. So if thunder happened to be the first sound your creation ever heard, it might well be its last. Linear audition is definitely out. So you decide to try some form of non-linear hearing.

In non-linear hearing, equal changes in absolute sound intensity do *not* produce corresponding changes in the perception of the sound.

Look at the graph in figure D–2. Notice that the change in intensity c–d of a very *loud* sound does *not* produce the same audible sensation as a like change a–b of a very soft sound. C–D is much smaller than A–B for the *same* change in intensity. In other

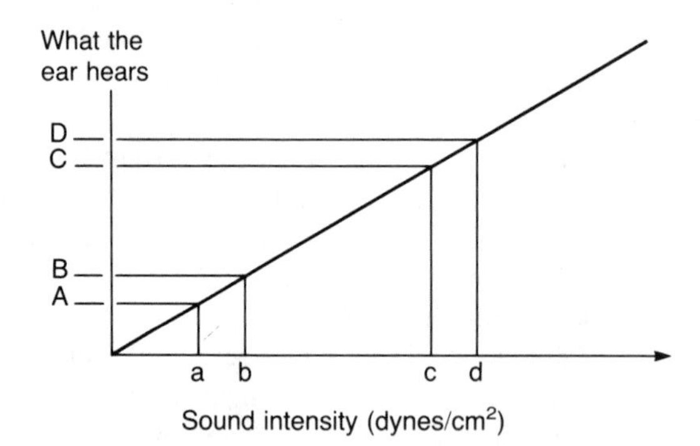

FIGURE D–1
Linear audition.

[1]The sensation of loudness is related to intensity, but is not the same. Loudness is the *psychological* perception of sound and should not be confused with the *physical* intensity of the sound pressure itself as measured on the SPL (sound-pressure level) scale in dB.

$$\text{dB(SPL)} = 20 \log \frac{\text{pressure \#1}}{\text{pressure \#2}}$$

where pressure #2 = 0.0002 dynes/cm², and represents the threshold of human audition.
From Morgan, C. and King, R. *Introduction to Psychology,* 5th ed. New York: McGraw-Hill, 1975.

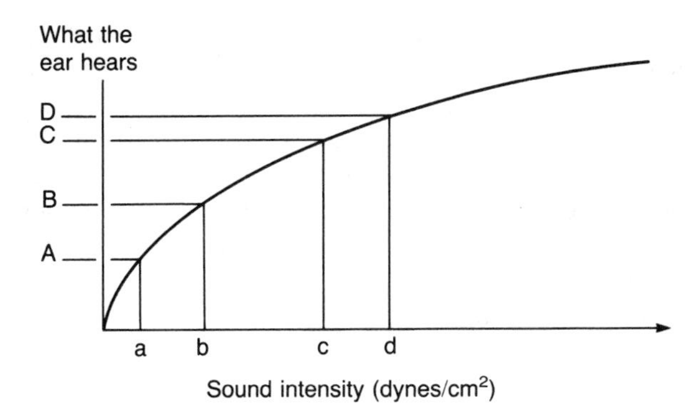

FIGURE D–2
Non-linear audition.

words, very loud sounds will not be as painful or destroy the sense of hearing quite as readily as was the case with linear audition. Therefore, non-linear hearing is a much better idea. In fact, humans do hear in this way because it is biologically adaptive to the survival of the species.

Some History

With the foregoing as background, it is now appropriate to say that, historically, the whole idea of decibels actually got started in the field of audio work because engineers needed a way to relate audio amplifier performance to how humans really hear sounds. It became customary to specify performance characteristics in terms of decibels, and the idea eventually spilled over into other areas of electronics and has remained with us since.

For example, suppose you were listening to a stereo amplifier with 1 watt output from the speaker, and suddenly the output changed to 10 watts. You would say that the sound got louder. But now assume that the output changed again from 10 watts to 100 watts. Again, you would say the sound got louder. Amazingly, though, you would probably report that the amount of *change* was the same in both cases.

Note, however, that the first change was $10 - 1 = 9$ W and the second change was $100 - 10 = 90$ W. In other words, the second change was actually 10 times greater than the first. But your ear heard the changes identically. Obviously, this means that your ear does not hear in a linear way, but in some other way.

If we were to make an actual graph of sound intensities versus sound perception, the locus of points obtained could be expressed mathematically as

$$y = \log_{10} x \qquad \textbf{(D–1)}$$

Equation (D–1) is called a logarithmic equation. More specifically, it is a mathematical model of how the ideal human ear hears sounds. Table D–1 gives just a few important points given by equation (D–1). The table illustrates that for big changes in sound intensity, only slight changes occur in the perception of those changes.

If the student had a background in logarithms and exponents at this point, we could make our last quantum leap to the actual definition of the decibel, and show the beauty and simplicity of its many applications. However, such a background is not assumed, and so we will pause here to discuss some simple ideas. Later, we will pick up where we left off here.

TABLE D–1
Logarithmic scale.

x (sound intensity)	y (what the ear hears)
10	1
100	2
1000	3
10000	4
100000	5
1000000	6

SOME MATH

An *exponent* is a number written to the upper right of another number, called the *base*, that shows how many times that base is to be used as a *factor*. For example, in

$$B^e = N \tag{D–2}$$

the base is B, the *exponent* is e, and the result is some number N. As a more concrete example, in

$$3^5 = 3 \times 3 \times 3 \times 3 \times 3 = 243$$

the base is 3, the exponent is 5, and the result, called the *power*, is 243. Note that the exponent told us to use the number 3 as a factor 5 times.

An expression written in the form of equation (D–2) is said to be in *exponential form*. However, there is an exactly equivalent way to write this same expression using the same variables. It is called the *logarithmic* form, and looks like this

$$e = \log_B N \tag{D–3}$$

This is read as "e is the logarithm of the number N to the base B."

In our last example, $3^5 = 243$, we could have presented this same information in logarithmic form as

$$5 = \log_3 243$$

This is read as "5 is the logarithm of 243 to the base 3."

We now have two ways of conveying the same information. Let us compare these two forms by using a few examples.

Exponential form	**Logarithmic form**
$2^3 = 8$	$3 = \log_2 8$
$4^2 = 16$	$2 = \log_4 16$
$10^4 = 10{,}000$	$4 = \log_{10} 10{,}000$

So, we can say that a logarithm is an exponent. At this point you might ask, "If a logarithm is just another name for an exponent, why bother with it? What's the sense of merely attaching a different label to the same idea?" The answer to that perfectly logical question is that by defining an exponent in this new way, we have opened up novel possibilities for manipulating numbers in ways we couldn't before. To see these new possibilities, we will now develop a few simple properties of logarithms based on the laws of exponents. Laws of exponents may be found in any elementary algebra textbook.

1. The logarithm of a product

   $$\log_a MN = ?$$

   Let $x = \log_a M$ and $y = \log_a N$. Then, by the laws of exponents, $a^x a^y = a^{x+y} = MN$, we conclude that

   $$\log_a MN = \log_a M + \log_a N \qquad \textbf{(D–4)}$$

2. The logarithm of a quotient

   $$\log_a M/N = ?$$

   Let $x = \log_a M$ and $y = \log_a N$. Then, by the laws of exponents, $a^x/a^y = a^{x-y} = M/N$, we conclude that

   $$\log_a M/N = \log_a M - \log_a N \qquad \textbf{(D–5)}$$

3. The logarithm of a power

   $$\log_a N^e = ?$$

Let $x = \log_a N$. Then, $a^x = N$. Raising both sides of this equation to the power of e, we obtain, $a^{xe} = N^e$. Writing this in logarithmic form, we get $\log_a N^e = xe$. Upon substituting for x, we obtain:

$$\log_a N^e = e \log_a N \qquad \textbf{(D–6)}$$

Before leaving our brief (and necessarily incomplete) discussion of logarithmic properties, some mention must be made of what are called antilogs. We make the following definition: An antilog, often called an inverse log, is the number we obtain by raising the base to the power of the logarithm. For example, if $e = \log_b N$, then $b^e = N$. (This should look familiar!) Using antilog notation, antilog $\log_b N = N$. In other words, the antilog simply "undoes" the log in much the same way as the square root "undoes" a square. That is,

$$\sqrt{a^2} = a$$

The antilog is quite often symbolized by $\log^{-1}$ and is read, "the inverse log of . . . ." For example, $\log^{-1}\log N = N$. As a more specific example, $\log^{-1}\log_2 32 = 32$. Note that it was not necessary to actually know the value of the log of 32 to the base 2 (which is 5).

A LITTLE MORE HISTORY

Recall our earlier discussion of the change in loudness of the stereo. In the first case, the sound had changed by 9 watts. In the second case, the change in intensity was 90 watts. Yet, even though the second change was 10 times the first, we concluded that the change in loudness was the same in both cases. How do these results fit with our discussion of logarithms and human audition?

Let's see what happens if we take the common log (log to base 10) of each of these power differences.

$$\text{amount of change} = \log_{10} 10 - \log_{10} 1 = 1 - 0 = 1$$
$$\text{amount of change} = \log_{10} 100 - \log_{10} 10 = 2 - 1 = 1$$

We see that the result is the same for both changes. We are, therefore, forced to conclude that the human ear must hear according to this base-ten *logarithmic model*.

It was this result that provided the first solid link between the logarithmic nature of human audition and electrical power. Since Alexander Graham Bell had done so much pioneering work with human hearing, including the invention of the telephone, his name became associated with the unit of sound intensity, the Bel, or the 10th part of a Bel, the decibel, abbreviated dB. The *only* correct abbreviation for decibel is dB (*not* db, or Db, or DB or anything else).

A Little More Math

In the above analysis of the stereo output power, we calculated the differences between two power levels. Recall that earlier we had discovered that the difference between the logs of two numbers was the same as the log of their quotient (see equation D–5). That is,

$$\log_a M/N = \log_a M - \log_a N$$

This says that subtracting the logs of numbers is the same as taking the log of their *ratio*. You probably already know that, by definition, the ordinary power gain (G) of a device is simply the ratio of power out (P_o) to power in (P_i). That is,

$$G = \frac{P_o}{P_i} \quad \textbf{(D–7)}$$

If we were now to take the common log of each side of this equation, we would obtain

$$\log_{10} G = \log_{10} \frac{P_o}{P_i} \quad \textbf{(D–8)}$$

or, since base 10 is implied

$$\log G = \log \frac{P_o}{P_i} \quad \textbf{(D–9)}$$

and finally

$$\log G = \log \frac{P_o}{P_i} = \log P_o - \log P_i \quad \textbf{(D–10)}$$

We symbolize the log of the ordinary power gain (G) as G' and understand that it has units of Bels. Therefore, the Bel power gain is

$$G' = \log G \quad \textbf{(D–11)}$$

For electronics, the Bel is too big a unit, and so we usually work with 1/10th of a Bel, the decibel. Therefore, equation (D–11) becomes

$$G'(\text{dB}) = 10 \log \frac{P_o}{P_i} \quad \textbf{(D–12)}$$

Note that we had to multiply by 10 because 10 dB = 1 Bel.

Equation (D–12) is the one that most students have thrown at them early in their careers. Since equation (D–12) is only a ratio of two power levels, we say that the dB is a *relative* unit, not an *absolute* unit like the volt, ampere, or ohm. In other words, if we say the power gain is 20 dB, all we know is that the output is 100 times as much as the input power. But we don't know how much power (in watts) such a stated gain actually represents. However, the dB can be made into an absolute unit by specifying a reference level against which other power levels can be compared. We will touch on this subject in more detail later.

From all that has been said, it should now be evident that the change in electrical power produced by the stereo in our previous example was 10 dB in both cases, and that, consequently, the human ear, being a logarithmic device, must have detected the same amount of change both times.

SUMMARY OF FIRST PART

These first few sections should have answered the following questions. You should try answering these questions in your own words without looking back at the text.

1. What is a decibel?
2. Why deal with logs of ratios rather than with the ratios themselves?
3. What is an exponent?
4. What are three common properties of logs?
5. From where did the name for the decibel unit come?
6. What is the abbreviation for the decibel?
7. Why is the decibel often considered a relative unit?
8. What is the logarithmic form of an exponential equation?
9. What is an antilog? An inverse log?
10. What is a Bel?
11. What is the formula for decibel power gain?

THE dB IN PRACTICE

If you have paid careful attention and have understood everything up to this point, you are now ready to delve a little deeper into the dB and its practical applications to electronics. On the other hand, if you are still having problems with the basic concepts, you should not go on, but rather reread the previous sections. Otherwise, you are only going to get very confused and discouraged by what is coming up.

In equation (D–12), we defined the decibel power gain as

$$G'(\text{dB}) = 10 \log \frac{P_o}{P_i}$$

Consider now, that since power may be given by $P = V^2/R$, then the dB power gain (G') may be rewritten as

$$G' = 10 \log \frac{V_o^2/R_o}{V_i^2/R_i} \qquad \textbf{(D–13)}$$

where R_i and R_o are the input and output resistances, respectively, of the device under test (D.U.T.). Or

$$G' = 10 \log \frac{V_o^2}{V_i^2} \times \frac{R_i}{R_o} \qquad \textbf{(D–14)}$$

And finally, from equations (D–4) and (D–6)

$$G' = 20 \log \frac{V_o}{V_i} + 10 \log \frac{R_i}{R_o} \qquad \textbf{(D–15)}$$

Note that if $R_o = R_i$, then $10 \log 1 = 0$ and equation (D–15) becomes

$$G' = A' = 20 \log \frac{V_o}{V_i} \qquad \textbf{(D–16)}$$

This says that the dB power gain (G') is numerically equal to the dB voltage gain (A') *if and only if* the device input and output resistances are equal. This result is of great practical importance since it implies that only a voltmeter is needed to *compute* dB power gains if the resistances across which the voltage measurements were taken are identical.

Equation (D–16) also allows us to calibrate voltmeter scales for direct measurement in units of absolute power, either dBm or watts, *if the readings are always taken across the same reference resistance as used to calibrate the meter scale*. Therefore, if we want to use the voltage-measurement idea to determine absolute power by *measurement,* we must specify

1. the value of P_i to be used as the comparison value.
2. the resistance value across which both V_o and V_i will be taken.

Let us see how this is done in practice.

The Simpson model 260 VOM (shown in figure D–3), for example, has a dB scale which ranges from -20 dB to $+10$ dB, and dB measurements are taken with the selector switch on the 2.5 VAC scale. Moreover, a note in the lower left corner of the instrument face specifies a zero dB power level of 0.001 watt across 600 ohms, thus fulfilling the essential requirements above. But what does all this mean?

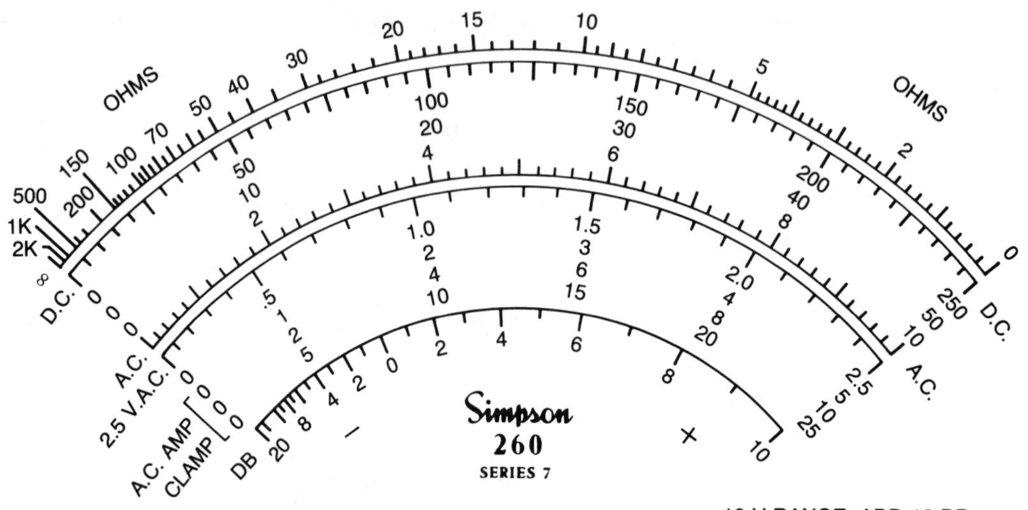

FIGURE D–3
Simpson model 260 VOM. (Courtesy of Simpson Electric Co.)

To begin with, note that from the formula for power, $P = V^2/R$, we may solve for V

$$V = \sqrt{PR} \tag{D–17}$$

or, from the specs given on the instrument face

$$V = \sqrt{0.001 \times 600} = 0.775 \text{ V}$$

So, zero dB will actually correspond to a little over 3/4 volt, which is the RMS voltage required to sustain a power of 1 mW across a 600 ohm load. Thus the Simpson 260 actually measures *absolute* power gains.

Note that from equation (D–16), a dB power gain of +10 requires a voltage of almost 2.5: $A' = 10 \text{ dB} = 20 \log V_o/0.775 \text{ V}$. Solving for V_o,

$$V_o = 0.775(\log^{-1} 0.5)$$
$$= 2.45 \text{ volts}$$

This is why the dB scale may be read *directly* when the selector switch is on the 2.5 VAC range. Of course, we may make dB measurements with the voltage selector switch set on any other range, but we must add a correction factor which is printed on the lower right of the meter scale. These correction factors were derived as follows. Any voltage reading obtained with the selector on the 10 V range is 10/2.5 = 4 times greater than the voltage reading on the 2.5 V range. Therefore, the dB reading is greater by 20 log 10/2.5 = 12 dB. So, 12 dB must be added to any reading. Any other correction factor may be found in a similar way:

0.5 VAC range:	20 log 0.5/2.5 =	−14 dB
50 VAC range:	20 log 50/2.5 =	+26 dB
250 VAC range:	20 log 250/2.5 =	+40 dB

and so forth for any other AC voltage scale.

dB Voltage Gain (A')

By definition, the dB is a power-ratio unit. However, it may also be conveniently used as a voltage-ratio unit. The idea of using voltage measurements to indicate gain results from the ease and convenience of making simple voltage measurements instead of power measurements.

The formula used to compute decibel voltage gain (A') is given by

$$A' = 20 \log \frac{V_o}{V_i} \tag{D–18}$$

Equation (D–18) is derived from equation (D–13), which was used to define dB power gain. The reason for using the coefficient of 20 is to insure that $A' = G'$ whenever the circuit input/output impedances are matched as indicated in equation (D–16). When $R_i \neq R_o$, then equation (D–16) no longer holds, and we must compute power and voltage gain separately using the appropriate formula: equation (D–12) $G' = 10 \log P_o/P_i$, or equation (D–18) $A' = 20 \log V_o/V_i$.

Note that if the correction factor specified by equation (D–15) is applied to equation (D–18), the result is dB power gain (G'). For example, consider the situation shown in figure D–4. However, $A' + 10 \log R_i/R_o = 15\text{ dB} + (-4\text{ dB}) = G' = 11\text{ dB}$.

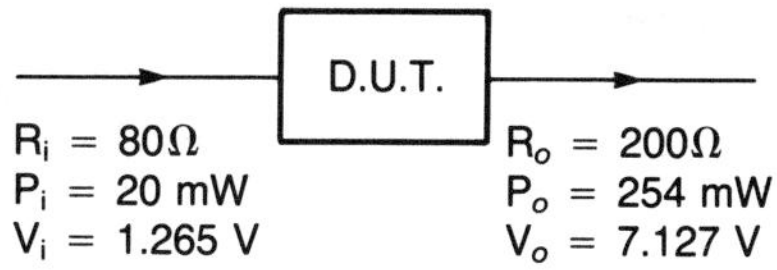

D.U.T.

$R_i = 80\Omega$
$P_i = 20$ mW
$V_i = 1.265$ V

$R_o = 200\Omega$
$P_o = 254$ mW
$V_o = 7.127$ V

$G' = 10 \log P_o/P_i = 11$ dB

$A' = 20 \log V_o/V_i = 15$ dB

FIGURE D–4
Power and voltage gains.

Which dB?

Many different power (voltage) levels and resistance values are used throughout the industry, and it is important for you to always know which one is being applied in your particular test situation. For example, we have just seen that 1 mW across 600 ohms is an important and common reference. However, in microwave work, 1 mW across 50 ohms is the standard. Whenever 1 mW is used as the reference, we use the *absolute* unit of power, the dBm, where m stands for milliwatt. The dB scale of the Simpson 260 is, technically, marked incorrectly in dB rather than in dBm. Similarly, dBmV stands for 1 millivolt as the reference level as in the case of the Sencore model FS73 field strength meter which also specifies this voltage as being measured across 75 ohms, not 600 ohms. The unit of audio frequency signal level is the VU (volume unit), which has the same meaning as the dBm (i.e., 1 mW across 600 ohms). Other dB reference units are the dBv (1 volt) and dBk (1 kilowatt). Any time we have a unit following the dB abbreviation, we know we are dealing with *absolute,* not relative, gains.

SOME PROBLEMS

1. A technician measures 25 mV across the input of an audio amplifier whose input impedance is 2,200 ohms. The audio output voltage is 0.62 V into an 8-ohm speaker.

a. What is the dB power gain (G')?
b. What erroneous dB power gain (G') results using only voltage measurements?
c. What needs to be done to get a true dB power gain (G') using only voltage measurements?

1.

$$P_i = V_i^2/R_i = (0.025)^2/2{,}200 = 2.84 \times 10^{-7}\ \text{W}$$
$$P_o = V_o^2/R_o = (0.62)^2/8 = 48\ \text{mW}$$

By equation (D–12),

$$G' = 10 \log \frac{P_o}{P_i} = 10 \log \frac{(2.84 \times 10^{-7})}{0.048}$$
$$= 52.3\ \text{dB}$$

2. $G' = 20 \log V_o/V_i = 20 \log 0.62/0.025 = 27.9$ dB (this is a gross underestimation of the actual power gain, but is the correct voltage gain).

3. By equation (D–15),

$$G' = 20 \log \frac{V_o}{V_i} + 10 \log \frac{R_i}{R_o}$$
$$= 27.9 + 10 \log \frac{2{,}200}{8} = 27.9 + 24.39 = 52.3\ \text{dB}$$

2. A power amplifier has a gain of 15 dB. *Estimate* (do not calculate) the power ratio.

Knowing some common dB power gain ratios, we write:

$$15\ \text{dB} = 10\ \text{dB} + 3\ \text{dB} + 1\ \text{dB} + 1\ \text{dB}$$

then,

$$10 \times 2 \times 1.26 \times 1.26 = \text{approx } 30{:}1$$

(actual ratio is 31.6:1)

3. An oscilloscope probe has a loss of 3 dB at 100 MHz. If the actual circuit voltage being measured is 12.5 V_p, what voltage will be displayed on the screen?

From $A' = 20 \log V_o/V_i$, we solve for V_o, obtaining:

$$V_o = 12.5 \log^{-1}(-3/20) = 8.85 \text{ V}$$

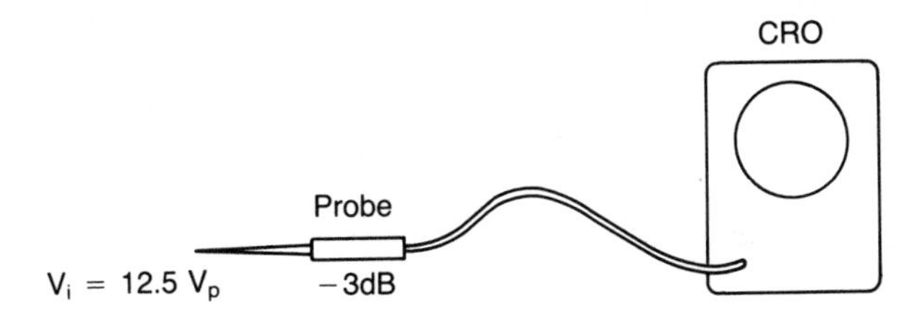

4. A technician measures the stage gain of an amplifier as +9.5 dB (actually dBm) using a Simpson 260. The output impedance of the stage is 1,850 ohms. What is the *actual* dB power gain?

Since the Simpson 260 dB scale is calibrated for use across 600 ohms, the technician's reading is incorrect. We must apply a correction factor as follows. We make use of the relationship from equation (D–15):

$$G' = 20 \log \frac{V_o}{V_i} + 10 \log \frac{R_i}{R_o}$$

Since the output resistance is larger than the 600 ohm reference value, there will be a larger voltage drop which will tend to overestimate the actual gain. Therefore, we must make the ratio R_i/R_o such that the correction factor will lessen the dB meter reading. Consequently, $G' = 9.5 + 10 \log 600/1{,}850 = 9.5 + (-4.9) = 4.6$ dB. Had the output resistance been less than 600 ohms, the ratio R_i/R_o would have been chosen to *increase* the reading obtained on the meter scale.

5. The power output of a transmitter is 20 kW, but at the antenna, it is only 8 kW. What is the attenuation factor of an 18,000 foot transmission line connecting the transmitter with the antenna?

$$G' = 10 \log \frac{P_o}{P_i} = 10 \log \frac{8}{20} = -4 \text{ dB}$$

Therefore, line attenuation is 4 dB/18,000 ft = 0.00022 dB/ft.

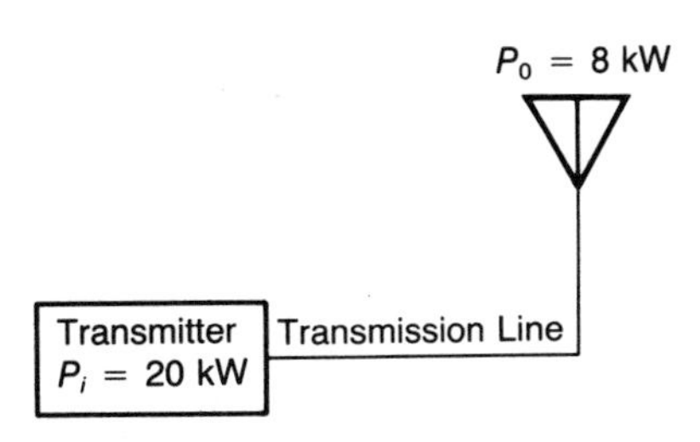

6. The response of a low-pass filter should be -6 dB at 4 kHz. If the output voltage is 0.25 V at 3 kHz and drops to 0.125 V at 4 kHz, is the filter working correctly?

$A' = 20 \log V_o(4 \text{ kHz})/V_o(3 \text{ kHz}) = 20 \log 0.125/0.25 = -6$ dB. Therefore, the filter is working properly. Note that if we had accidentally reversed the ratio as 0.25/0.125, we would have obtained $+6$ dB instead of -6 dB. The absolute dB value obtained by such unintentional reversals is always the same, but the sign *is* important in subsequent calculations. Therefore, be careful when forming ratios.

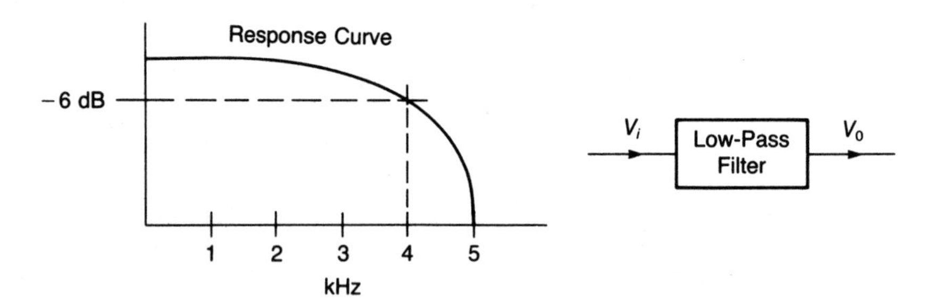

7. A certain stereo amplifier is guaranteed to have an output of 100 watts per channel ± 1 dB. Is this necessarily a good purchase?

From equation (D–12), $G' = 10 \log P_o/P_i$, we know that $+1$ dB corresponds to $P_o/P_i =$ 1.26:1, so $P_{hi} = 1.26P_i = 1.26(100) = 126$ W. Also, -1 dB corresponds to $P_o/P_i =$ 0.79:1, so $P_{lo} = 0.79P_i = 0.79(100) = 79$ W. Therefore, it is quite possible for the maximum outputs between channels to differ by as much as 47 W. This is probably not a good purchase, and points up the fact that although the claims of the seller are not false, they are difficult to interpret for the unwary public. Hence the caveat, "Let the buyer beware!"

8. A multicavity klystron has a power gain of 87 dB. If this represents an output of 12 kW, how much input power is required?

From $G' = 10 \log P_o/P_i$, we solve for P_i to obtain

$$P_i = \frac{P_o}{\text{antilog } \dfrac{G'}{10}} = 2.39 \times 10^{-5} \text{ W} = \text{about } 24\ \mu\text{W}$$

9. The power gain of an amplifier is to be 16 dB when the output power is 3.6 W. If the input power is 5 mW, by how much must it be increased to obtain the desired gain?

Solving for x:

$$x = \frac{P_o}{\text{antilog}\left(\dfrac{G'}{10}\right)} - P_i = 85.43 \text{ mW}$$

Checking, $G' = 10 \log 3.6/89.43 \text{ mW} = 16$ dB.

16 dB Amplifier

$P_i = P_i' + x$
$= 0.005 + x$

$P_o = 3.6$ W

$$G' = 10 \log \frac{P_o}{P_i' + x}$$

10. Solve the following logarithmic equation for x.

$$\log (100 + x) = 5$$

Taking antilogs of both sides, $100 + x = 10^5$. Therefore, $x = 10^5 - 100 = 99{,}900$.

11. Solve the following exponential equation for x.

$$10^{(3-x)} = 856$$

Taking logs of each side, $3 - x = \log 856$, so $x = 3 - \log 856 = 3 - 2.93 \approx 0.0675$.

12. Solve the following logarithmic equation for x.

$$\log \frac{15}{x} = 27.48$$

Taking antilogs, $15/x = \text{antilog } 27.48$. Therefore, $x = 15/\text{antilog } 27.48 = 5 \times 10^{-27}$. Note that we could have rewritten the problem as $\log 15 - \log x = 27.48$, and then solved for x.

13. $\log_3 16 = ?$

The solution to this problem allows us to derive another useful property of logs, namely, change-of-base. Since $\log_a N = e$, then, $a^e = N$. Also, $\log_b N = \log_b a^e$ (because $a^e = N$). Therefore, from equation (D–6), we may write

$$\log_b N = e \log_b a$$

Solving for e,

$$e = \frac{\log_b N}{\log_b a} \qquad \textbf{(D–19)}$$

In our problem, we will let $b = 10$ (the common log base). Then, $\log_3 16 = x = (\log_{10} 16)/(\log_{10} 3) = 1.204/0.477 = 2.524$ Checking, $3^{2.524} = 16$.

Equation (D–19) gives us a convenient method of finding logs to any base using only common logs (or natural logs), which are on most engineering calculators.

SUMMARY OF THE SECOND PART

You should be able to answer the following questions without looking back at the text.

1. What constraints (if any) regarding circuit resistance are placed on the use of the formula $G' = 10 \log P_o/P_i$?
2. What constraints regarding circuit resistance are placed on the use of a VOM type instrument used to measure dBs?
3. Why is the dBm called an absolute unit?
4. Why must the resistance and power levels be specified when dealing with absolute dB power or voltage levels?
5. Why do you think there are so many different reference units in dB measurements?
6. Describe the pattern of entries in the handy dB chart shown in Appendix E.
7. What is the advantage of using voltage measurements to calculate dB power gain?
8. Why do you need correction factors as you switch from one voltage scale to another when measuring dBs on most analog VOMs?
9. Is it necessary for $R_o = R_i$ when computing A' using equation (D–18)?

COMMON VOLTAGE AND POWER RATIOS

Common voltage and power ratios and their dB equivalents are shown in the following table.

P_o/P_i or V_o/V_i		$G' = 10 \log P_o/P_i$	$A' = 20 \log V_o/V_i$
1,000,000		60 dB	120 dB
100,000		50 dB	100 dB
10,000		40 dB	80 dB
1,000		30 dB	60 dB
100		20 dB	40 dB
	64	18 dB	36 dB
	32	15 dB	30 dB
	16	12 dB	24 dB
10		10 dB	20 dB
	8	9 dB	18 dB
	4	6 dB	12 dB
	2	3 dB	6 dB
	1.26	1 dB	2 dB
1		0 dB	0 dB
	1/2	−3 dB	−6 dB
	1/4	−6 dB	−12 dB
	1/8	−9 dB	−18 dB
1/10		−10 dB	−20 dB
	1/16	−12 dB	−24 dB
	1/32	−15 dB	−30 dB
	1/64	−18 dB	−36 dB

1/100	−20 dB	−40 dB
1/1000	−30 dB	−60 dB
1/10000	−40 dB	−80 dB
1/100000	−50 dB	−100 dB
1/1000000	−60 dB	−120 dB

There is a simple pattern contained in the chart which allows the student to easily memorize most of these dB equivalents, and thereby be able to mentally estimate the corresponding output/input ratio of any device. For example,

1. Any time the power is *doubled,* there is a 3 dB *increase*. Similarly, a 3 dB *decrease* is obtained each time the power is *halved*.
2. For *decade* changes in power, the number of zeros in the ratio represents the non-zero integer associated with the dB unit. For example, $P_o/P_i = 1{,}000{,}000$, hence the 6 zeros are associated with 60 dB; for 1/10,000 ratio, 4 zeros correspond to −40 dB; and so forth.

F

BANDWIDTH

Since the concept of bandwidth is so important to our understanding of microwave circuits, we shall now derive some fundamental definitions and formulas. We will use the sample circuit shown in figure F–1 to verify our assertions concerning resonant circuit behavior in general.

The graph shown in figure F–2 is a typical response curve for the series-resonant circuit. From $P = I^2R$, if $I = 1$ (at f_r), $P = R$. But at $I = 0.707$ (at f_1 or f_2), then $P = (0.707)^2R = 0.5R$.

FIGURE F–1
Series-resonant circuit.

L C R → fr = 1 MHz

$L = 8.9 \times 10^{-5}$ H
$C = 2.84 \times 10^{-10}$ F
$R = 56\ \Omega$

$XL = Xc = 560$ ohms

$$fr = \frac{1}{2\pi\sqrt{LC}} = 1 \text{ MHz}$$

FIGURE F–2
Series-resonant response curve.

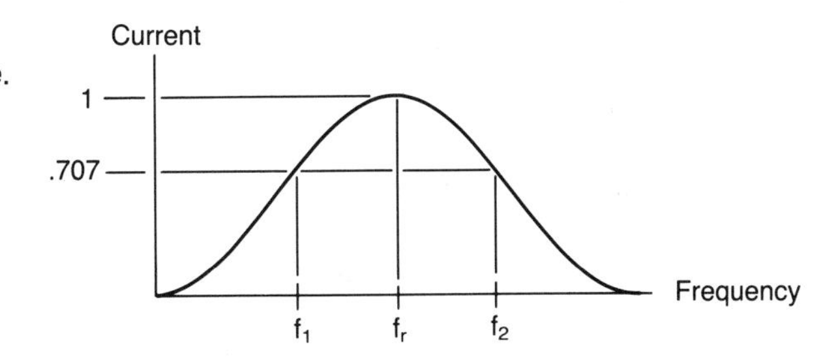

Any two values of I would give the same result provided the value of current at $f_{1,2}$ was 70.7% of the value at f_r. Therefore, when the frequency is off-resonance to the point where the current is 70.7% of its value at resonance, the power is reduced to one-half its former value. Obviously, then, this point is called the half-power point, and is used as a convenience in defining the bandwidth of the circuit—that is, the bandwidth is $f_2 - f_1$ at the half-power point. This half-power point is also called the 3-dB down point since $G' = 10 \log 0.5 = -3$ dB. Any other point might have been chosen for bandwidth measurement, but the half-power point is probably the simplest to use.

At the half-power point, the net reactance equals the resistance. That is, $X_L - X_c = R$. This important conclusion, which will be used later, may be shown as follows. From $I = V/Z$

$$0.707I = \frac{V}{\sqrt{R^2 + (X_L - X_c)^2}}$$

$$0.5I^2 = \frac{V^2}{R^2 + (X_L - X_c)^2}$$

$$(X_L - X_c)^2 = \frac{2V^2}{I^2} - R^2$$

$$= 2R^2 - R^2 = R^2$$

$$X_L - X_c = R \tag{F–1}$$

We may use the above conclusion to derive an equation for the bandwidth at the 3-dB down point:

$$2\pi fL = \frac{1}{2\pi fC} = R$$

from which, $f^2 \times 4\pi^2 LC - f \times 2\pi RC - 1 = 0$

Using the quadratic formula to solve for f

$$f_{2,1} = \frac{| RC \pm \sqrt{R^2C^2 + 4LC} |}{4\pi LC} \tag{F–2}$$

Therefore,

$$\text{bandwidth} = f_2 - f_1. \tag{F–3}$$

Then, from the sample circuit above, $f_1 = 0.95$ MHz and $f_2 = 1.05$ MHz so the bandwidth is 100 kHz.

While the above formula will give the bandwidth in terms of individual component values, there is a much more useful formula: $\text{BW} = f_r/Q$, where Q is a figure of merit defined by $Q = X_L/R$. We will now develop the bandwidth formula in terms of Q.

We may write the general expression for the series resonant circuit as

$$Z = R + j(\omega L - 1/\omega C) \tag{F–4}$$

where $\omega = 2\pi f$.

Now define Q (at resonance) as

$$Q = X_L/R = \omega L/R \tag{F–5}$$

And since, at resonance, $X_L = X_c$, then

$$Q = \frac{\dfrac{1}{\omega C}}{R} = \frac{1}{\omega CR} \tag{F–6}$$

that is,

$$\frac{\omega L}{R} = \frac{1}{\omega CR} \tag{F–7}$$

Now, writing (F–7) for $f \neq f_r$ in terms of Q,

$$Z = R + jR\left(\frac{\omega_o}{\omega} \times \frac{\omega L}{R} - \frac{\omega}{\omega_o} \times \frac{1}{\omega CR}\right) \tag{F–8}$$

where $\omega = 2\pi f_r$ and $\omega_o = 2\pi f_{1,2}$

Note that the second term of (F–8) was so written as to eliminate the R, L, and C values, giving us a parenthetical expression entirely in terms of frequency. On factoring R,

$$Z = R\left[1 + j\left(\frac{\omega_o}{\omega} - \frac{\omega}{\omega_o}\right)Q\right] \tag{F–9}$$

If we label the parenthetical quantity of equation (F–9) as x, we may write

$$Z = R(1 + jxQ) \tag{F–10}$$

We may expand (F–10) and rewrite this as

$$Z = R + jRxQ = \sqrt{R^2 + (xRQ)^2} = R\sqrt{1 + x^2Q^2}$$

And since at $f_{1,2}$ $X_L - X_c = R$, then $Z = R\sqrt{2}$. So,

$$R\sqrt{1 + x^2Q^2} = R\sqrt{2}$$

$$1 + x^2Q^2 = 2$$

$$x^2Q^2 = 1$$

$$x = \frac{1}{Q}$$

Then, x at f_2 becomes

$$\frac{1}{Q} = \frac{f_2}{f_r} - \frac{f_r}{f_2} \qquad \textbf{(F–11)}$$

Note that $1/Q$ must be positive, and will so be since $f_2/f_r > 1$, and $f_r/f_2 < 1$.

On solving (F–11) for f_2 using the quadratic formula

$$f_2 = \frac{f_r}{2Q} \pm \sqrt{\left(\frac{f_r}{2Q}\right)^2 + f_r^2} \qquad \textbf{(F–12)}$$

Also, x at f_1 becomes $1/Q = f_1/f_r - f_r/f_1$. But $1/Q$ must be positive, so we write

$$\frac{1}{Q} = \frac{f_r}{f_1} - \frac{f_1}{f_r} \qquad \textbf{(F–13)}$$

which is now a positive number. Then, solving (F–13) for f_1 we obtain

$$f_1 = -\frac{f_r}{2Q} \pm \sqrt{\left(\frac{f_r}{2Q}\right)^2 + f_r^2} \qquad \textbf{(F–14)}$$

And finally, bandwidth (BW) $= f_2 - f_1$, or equation (F–12) minus (F–14) gives

$$\text{BW} = \frac{f_r}{Q} \qquad \textbf{(F–15)}$$

Equation (F–3) showed the bandwidth of our sample circuit to be 100 kHz. If we now plug the appropriate values into equation (F–15) we obtain

$$\text{BW} = \frac{f_r}{Q} = \frac{1\ \text{MHz}}{10} = 100\ \text{kHz}$$

GLOSSARY

Amplification Factor (μ) A figure of merit for triode vacuum tubes that indicates how much more effective the grid is than the plate (anode) in controlling the flow of electrons from cathode to anode.

$$\mu = \frac{\Delta V_p}{\Delta V_g} \quad \text{(with plate current constant)}$$

where ΔV_p = change in plate voltage
ΔV_g = change in grid voltage

$$\mu = G_m R_p$$

where G_m = transconductance
Rp = plate resistance

Amplitude Modulation A method of encoding amplitude and frequency variations of an RF carrier by means of varying (modulating) the carrier amplitude. (See figure A–2 in *Laboratory Activities* Appendix A.)

Antenna Efficiency (η) The ratio of the radiated power to the power delivered to an antenna. Antenna efficiency is usually calculated indirectly using $\eta = R_r/(R_d + R_r)$ where R_r = radiation resistance and R_d = loss resistance **[Eqn. 3–4]**. In general, $75\% < \eta < 95\%$.

Antenna Impedance (Z_a) A complex AC quantity given by $Z_a = R_a + jX_a$ **[Eqn. 3–1]** which specifies the opposition to signal energy as measured at the feed point of an antenna. R_a is the antenna resistance and X_a is the antenna reactance. For resonant antennas, $X_a = 0$, so $Z_a = R_a = R_d + R_r$, where R_d is the loss resistance due to actual losses in the antenna conductors themselves and R_r is the radiation resistance (a fictitious quantity) calculated as $R_a = P/I^2$ **[Eqn. 3–3]**. For a lossless Hertzian antenna in free space, R_a = 73 ohms. The radiation resistance of an ideal dipole varies with height above ground.

Antenna Resistance (R_a) *See* Antenna Impedance.

Applegate Diagram A diagram showing the relationship between RF signal phase and maximum bunching of the electron beam in velocity-modulated devices such as the klystron.

Aquadag A material used as an absorber of microwave energy consisting of a highly conductive suspension of graphite. The Aquadag is applied to the absorbing structure, which is almost always pointed or tapered to reduce reflections.

Atmospheric Signal Attenuation (A_{v_i}) The attenuation in voltage intensity (v_i) of a signal in free space is given by $A_{v_i} = 20 \log r_2/r_1$ **[Eqn. 2–20]** where r_2 and r_1 are the distances along the signal path. The formula does not apply where there is atmospheric attenuation of the signal. Attenuation due to rain or fog is usually derived from empirical data, and is often presented in chart or graph form.

Attenuation Any reduction in signal power or voltage. Attenuation is most commonly expressed in decibels as follows:
Power attenuation: $G(\text{dB}) = 10 \log P_o/P_i$, where P_o = power output and P_i = power input; Voltage attenuation: $A(\text{dB}) = 20 \log V_o/V_i$, where V_o = output voltage; V_i = input voltage. Care should be exercised to place the smaller value in the numerator; otherwise, the incorrect sign will result even though the absolute value is the same in either case. For example, $10 \log 3/5 = -2.22$, but $10 \log 5/3 = +2.22$. The first result is attenuation; the second, gain.

Automatic Leveling Control (ALC) A feedback loop designed to keep the output power of a BWO constant across its range of frequencies. Used in swept measurements of microwave devices and components.

Azimuth A heading relative to true north.

Backlobe Radiation A form of destructive radiation working against the main beam caused by feeding a parabolic reflector with an isotropic source.

Backward Wave Oscillator (BWO) A microwave oscillator capable of being tuned over a broad range of frequencies. Essentially a TWT with the attenuator removed.

Balanced Line A parallel-wire transmission line whose conductors experience the same electrical capacitance relative to ground.

Bandwidth The range of useable frequencies of a device between the 3 dB down (half-power) points. In a resonant circuit, the ratio of the resonant frequency (f_r) to the circuit Q, giving the range of frequencies between the -3 dB (half-power) points.

Beamwidth The angle measured between the 3 dB down (half-power) points on the major lobe of an antenna's radiation pattern.

Bearing The direction of one object relative to another. Bearings are commonly given relative to true or magnetic north, but may also be specified relative to various quadrants of the boat.

Bearing Resolution The ability of a radar receiver to discern between two targets at the same range, but on slightly different bearings.

Blip The spot on a radar display caused by target echo.

BNC A type of bayonet connector commonly used in microwave work with coaxial cables.

Broadside Antenna An array of simple dipoles fed in phase and having a radiation pattern whose maximum directivity is along the axis normal to the plane of the array.

Cassegrain Antenna A type of paraboloid in which the feed point is located at the vertex of the parabola and is directed against a secondary hyperbolic reflector. Used primarily for deep-space applications where noise must be reduced to an extremely low level.

Cavity Resonator A physical cavity formed by a solid of revolution which, when properly excited, is capable of sustaining oscillations at microwave frequencies. Resonant cavities are functionally equivalent to conventional resonant circuits made from discrete inductors and capacitors, except that the cavities have higher Qs.

Center Feed Antenna A type of parabolic antenna feed arrangement in which a spherical reflector located at the focus of a paraboloid is used to redirect stray radiation back to the main dish.

Characteristic Impedance (Z_0) The frequency-independent opposition offered to the propagation of TEM energy along a transmission line due to the distributed inductance and capacitance along the line. Also called surge impedance.

$$Z_o = \sqrt{\frac{L}{C}} \qquad \text{[Eqn. 1–16]}$$

where L = H/m; C = F/m
For coaxial transmission lines

$$Z_o = \frac{138}{\sqrt{\epsilon_r}} \log_{10} \frac{D}{d}\ \Omega \qquad \text{[Eqn. 1–24]}$$

where D = outer conductor diameter; and
d = inner conductor diameter
ϵ_r = dielectric constant
For parallel-wire transmission lines

$$Z_o = \frac{276}{\sqrt{\epsilon_r}} \log_{10} \frac{2D}{d}\ \Omega \qquad \text{[Eqn. 1–21]}$$

where D = center-to-center distance between conductors
d = diameter of the conductors
ϵ_r = dielectric constant

$$Z_o = \sqrt{\frac{\mu}{\epsilon}} \qquad \text{[Eqn. 1–17]}$$

where μ = absolute permeability of free space (1.26×10^{-6} H/m)
and ϵ = absolute permittivity of free space (8.854×10^{-12} F/m)

Choke Flange A waveguide flange with an L-shaped channel whose total cross-sectional length is $\lambda_p/2$. Used to suppress wave reflections by acting like a short across mating flange discontinuities. Choke flanges mate with flat flanges only.

Circular Waveguide Waveguide having a circular cross-section. Circular guides are capable of handling greater power than comparably sized rectangular shapes, and have less attenuation for a given value of λ_0.
The dominant mode for circular guide is $TE_{1,1}$ and the symmetrical mode is $TM_{0,1}$. The cutoff wavelength for circular guides is given by $\lambda = 2\pi r/(B_{m,n})$ where r = guide radius and $B_{m,n}$ = a Bessel function solution.

Circulator A ferrite device having the property that only rotationally-adjacent ports are coupled. Used extensively in low-power radar duplexers to prevent the SHF burst from damaging the receiver.

Coaxial Cable A type of transmission line consisting of an inner conductor surrounded by, but insulated from, an outer conductor. It is used for the transmission of signals up through the

SHF frequency range (0.3–3.0 GHz). It has a lower dielectric and radiation loss than parallel-lead transmission line. (*See* characteristic impedance.)

Clutter Any echo that obscures real targets on the radar display screen. Sea return and rain clutter are two notable offenders, but desensitization circuits are employed to remove or reduce the effect.

Compass Rose A system of two concentric circles prominently displayed on a navigational chart showing the amount of magnetic variation from true north at that location.

Conduction Loss Power losses due to increased effective resistance of a conductor at microwave frequencies due to "skin effect."

Conical Horn Antenna A type of microwave antenna formed from a circular waveguide. (*See* Horn Antenna.)

Coupling A directional coupler rating specifying the amount of energy coupled from the main arm to the auxiliary arm. Coupling is calculated as dB(coupling) $= -10 \log P_{aux}/P_i$ **[Eqn. 4–26]** where P_{aux} = power output of the auxiliary arm and P_i = power input to main arm.

Crystal Detector Essentially a crystal diode square-law device used for sensing minute amounts of microwave energy. The detector forms an essential part of the slotted line.

Cutoff Frequency (f_o) (*See* Cutoff Wavelength, λ_0.)

Cutoff Wavelength (λ_0) For the $TE_{m,0}$ mode, the longest wavelength (lowest [cutoff] frequency, f_0) which will just barely fail to propagate. The cutoff wavelength is given by $\lambda_0 = 2A/m$ **[Eqn. 4–11]** where A = long dimension of the waveguide and m = an integer representing the number of E-field bunches between (and parallel to) the narrow walls. For the $TE_{m,n}$ modes, λ_0 is calculated by

$$\lambda_0 = \frac{2}{\sqrt{\left(\frac{m}{A}\right)^2 + \left(\frac{n}{B}\right)^2}} \qquad \textbf{[4–22]}$$

where B = short dimension of the waveguide and n = an integer representing the number of E-field bunches between (and parallel to) the long A walls.

Decibel (dB) A logarithmic unit used to express the ratio of two powers or voltages. One decibel is one-tenth of a Bel. For power ratios, $G(\text{dB}) = 10 \log (P_{out}/P_{in})$. For voltage ratios, $A(\text{dB}) = 20 \log (V_{out}/V_{in})$. If the denominator of the ratio (P_{in} or V_{in}) is taken as a reference zero, the decibel may be used as a measure of absolute gain (loss) rather than merely a relative indication of gain (loss). In microwave work, 1 mW is a common reference zero for decibel power gain: $P_{in} = 0.001$ W. In television work, 1 mV across 75 ohms is a common reference zero: $V_{in} = 0.001$ V. Almost any appropriate reference zero may be specified where absolute measurements are desired.

Deviation The inherent compass error due to the magnetic properties of the boat or its contents.

Dielectric Lens Antenna A polystyrene or other dense dielectric material used to collimate spherical wave fronts at microwave frequencies. The lens and feed together function as the antenna system. Various stepped or zoned configurations are often used to minimize weight and attenuation problems.

Dielectric Loss Loss of the energy along a transmission line due to the energy absorbed in heating the dielectric by the passage of the TEM wave. Dielectric loss limits the upper useable frequency of coaxial cables to less than 18 GHz.

Diplexer A bridge circuit used in conjunction with a turnstile antenna for the purpose of feeding two separate signals to a single antenna.

Directional Coupler A waveguide arrangement in which microwave energy moving along the main arm may be sampled through an auxiliary arm. Energy moving in the other direction is not coupled.

Directive Gain (G_d) Gain of an antenna in a particular direction compared with an isotropic source. In the computation of directive gain, the input power to the test antenna and to the isotropic or reference antenna is held constant. Directive gain does not take into account antenna losses.

$$G_d\ (\text{dB}) = 10 \log \frac{P_{d(\text{test})}}{P_{d(\text{iso})}} \qquad \textbf{[Eqn. 2–3]}$$

Directivity (d) A measure of the degree of isolation between the main and auxiliary arms of a directional coupler when the direction of energy flow is reversed. Typical directivities are on the order of 40 dB or more.

$$d(\text{dB}) = 10 \log \frac{P_{\text{aux(fwd)}}}{P_{\text{aux(rev)}}} \qquad \textbf{[Eqn. 4–27]}$$

Dish (*See* Paraboloid Antenna.)

Dominant Mode The mode for which the lowest possible frequency (cutoff frequency, f_0) may be propagated in a waveguide for a given A dimension.

Doppler Radar A type of radar using the Doppler effect to remove clutter and render moving targets visible.

Duplexer A device that allows a single antenna to serve as both transmitter and receiver for a radar set. One type is a circulator, while another uses spark-gap switches.

D.U.T. Abbreviation commonly used for ''Device under test.''

Duty Cycle The ratio between the active (on) time of an event and the total event time expressed as a percent. Small duty cycles allow magnetrons to deliver more peak power for short intervals without burning up. Duty cycle is computed as

$$\text{Duty cycle} = \frac{\text{time on}}{\text{total event time}} \times 100 \qquad \textbf{[Eqn. 5–1]}$$

Duty Ratio Duty cycle expressed as a decimal.

Edison Effect The liberation of electrons by thermionic means. Used as the principal source of electron emission in microwave vacuum tube devices.

Effective Radiated Power (ERP) An expression of the maximum utilization of power into the antenna system which includes losses from all sources between transmitter and antenna. ERP may be calculated as

$$ERP = \log^{-1}\left(\frac{G_p'}{10}\right)P_t \qquad \textbf{[Eqn. 3–7]}$$

ERP may also be defined as ERP = overall power gain ratio × total power.

End-Fire Array An antenna array consisting of a linear arrangement of simple dipoles fed 90 degrees out of phase and having a radiation pattern of maximum directivity in the plane of the array.

Faraday Isolator *See* Isolator.

Far-Field Distance The distance from an aperture antenna that insures that the induction field is not affecting the receiving antenna, and where the transmitting antenna acts essentially as a point source. The far-field formula is given by

$$R \text{ (range)} = \frac{2D^2}{\lambda} \qquad \textbf{[Eqn. 3–21]}$$

where D = largest dimension of the aperture;
λ = wavelength

Feed Antenna One of several different antennas (usually a dipole or horn) used at the focus of a paraboloid antenna for the purpose of projecting (''feeding'') microwave radiation at the parabolic reflector.

Ferrite An insulating material having magnetic properties which may be modified when placed in a magnetic field of appropriate value and orientation. Ferrites are used in microwave work as isolators and circulators.

Field Strength (V_i) A measure of absolute signal strength defined as the voltage induced in a one meter antenna by a radiated TEM wave. In free space, the field strength may be calculated by

$$V_i = \frac{\sqrt{30P_t}}{r} \text{ V/m} \qquad \textbf{[Eqn. 2–18]}$$

where P_t is the total transmitted power in watts; r is the distance in meters. Also called field intensity.

Flexible Waveguide Pliant waveguide is used where lateral displacement or vibrational effects must be taken into account. Flexible guide has considerably more attenuation than rigid rectangular guide of comparable size.

Folded Dipole An antenna formed by connecting two simple dipole sections in parallel, thus increasing the radiation resistance of a simple dipole from 73 ohms to 292 ohms.

Frequency Domain The representation of a complex waveform in such a way that the sinusoidal components are displayed as a function of time.

Frequency Meter An adjustable resonant cavity calibrated in frequency and which undergoes some power absorption at resonance. Used to measure frequency in the microwave region.

Frequency Modulation A method of encoding amplitude and frequency variations of a signal as frequency variations of the RF carrier.

Gain The increase in power or voltage at the output of a device compared with its input or reference standard. For antennas, gain is the increase in signal intensity along the axis of the major lobe of the radiation pattern as compared with a reference source. Often a theoretical source, the isotropic radiator, is used as the comparison device. Gain is often expressed in decibels as follows:

Power gain in general: $G = 10 \log P_o/P_i$
where P_o = power out
P_i = power in

Voltage gain in general: $A = 20 \log P_o/P_i$
where V_o = voltage output
V_i = voltage input

(See Directive Gain and Power Gain.)

Gain Factor The ratio of the output of a device to its input. For antennas, the gain factor is given by

$$\log^{-1}\frac{G(dB)}{10} \text{ or } \log^{-1}\frac{A(dB)}{20}$$

Gallium Arsenide (GaAs) A microwave substrate material that has high ion mobility and so is capable of increasing the frequency limits of JFET devices. Currently, gallium-indium arsenide (GaInAs) is also being investigated as a material for extending frequency limits.

Ground Wave Vertically polarized TEM waves less than 3 MHz that travel close to the ground as they leave the transmitting antenna. Ground waves are one of three identifiable wave types emanating from any terrestrial antenna. The other two types of waves are the sky wave and the space wave. The wave that dominates depends largely on the frequency of the signal.

Group Velocity (v_g) The linear propagation velocity of a wave down a waveguide. The group velocity is calculated using,

$$v_g = c\sqrt{1 - \left(\frac{\lambda}{\lambda_0}\right)^2} \qquad \textbf{[Eqn. 4–21]}$$

where λ = free-space wavelength
λ_0 = cutoff wavelength

Gunn Diode A low-power microwave source using the transferred-electron device (TED) principle to initiate a domain (electron bunch) excursion through the semiconductor at SHF rates. The Gunn mode also exhibits the property of negative resistance.

Heading Heading is the direction in which the bow of the boat is pointed. Headings may be true or magnetic and are referenced clockwise from the corresponding north cardinal point. A heading relative to true north is called azimuth.

Helical Antenna A type of antenna capable of transmitting (receiving) a TEM wave polarized in any direction.

Helix Structure The site of the electron beam-signal interaction in a traveling wave tube.

Hertz Antenna A half-wave dipole antenna complete unto itself. Its radiation pattern is donut shaped.

Horn Antenna A microwave antenna made by flaring a waveguide structure in either or both cross-sectional directions to achieve a closer match to the free-space impedance of 377 ohms. Horn antennas are often referred to as aperture antennas.

Hybrid Ring A rectangular waveguide bent in the E plane and used as a discriminating port coupling device. Its application is much the same as the hybrid tee. Frequently referred to as a rat race.

Hybrid Tee A multiple waveguide tee formed by the intersection of an E plane and H plane tee and used extensively as a discriminating port coupling device for microwave receivers. Sometimes called a magic tee.

IMPATT Diode A plasma-mode device capable of producing a 180 degree phase shift between the domain (current) and RF voltage, and thereby exhibiting a negative-resistance effect. IMPATT diodes are used mostly for microwave amplifiers.

Induction Field The electromagnetic field that surrounds an antenna but does not radiate. Its effect is local, but is responsible for establishing the radiation field. The E-M components are both in time and space quadrature.

Input VSWR A measure of the degree of match offered by the input port of a microwave device or component. Ideally, input VSWR is unity.

Insertion Loss The amount of power lost in a device due to its presence in the path of energy flow, expressed in decibels. For example, an insertion loss of 0.1 dB (−0.1 dB) means that 98% of the energy put into a device comes out. Low insertion losses are always desirable.

Interdigitation A bipolar transistor design technique used to extend the frequency of operation and increase power handling capability.

Interelectrode Capacitance Unintentional capacitance between the various structures within an ordinary vacuum tube which limits the device's performance at UHF and higher frequencies.

Ionosphere A heavily ionized layer of particles in the upper atmosphere capable of reflecting and refracting radio waves launched at the proper angle and having the proper frequency.

Iris A partition located within the E field or H field of a waveguide for the purpose of introducing a capacitance or inductance into the guide structure. (*See* post.)

Isolator *Faraday Isolator:* A ferrite device using the property of Faraday rotation to form a nonreciprocal microwave device.
Resonant Absorption Isolator: A ferrite device utilizing the precessional frequency effect to absorb microwave energy passing in one direction, but not the other.

Isotropic Source A hypothetical point source of TEM waves assumed to be radiating uniformly in all directions. If the power radiated from such a source is P_t, then the power density in free space is given by $P_d = P_t/(4\pi r^2)$ W/m^2 **[Eqn. 2–1]** where P_t is in watts and r is in meters.

***j* Operator** The quantity responsible for the rotation of a positive real number 90 degrees in either direction to the *Y*-axis or axis of imaginary numbers. Since $j = \sqrt{-1}$ and $j^2 = -1$ the *j* operator is used to express complex loads in rectangular form. For example, $Z_L = R - jx$.

Junction The region between the doped elements in a transistor. It is the ion mobility of this region that limits the upper frequency of semiconductor microwave devices.

Klystron A thermionic device for operation in the microwave region utilizing the principle of velocity modulation. Two basic versions of the klystron exist: (1) Multicavity (single transit) klystron used as an SHF amplifier. Capable of delivering many thousands of watts of microwave energy. (2) Reflex klystron used as a low power SHF oscillator.

Lambda (λ) Greek letter used to denote wavelength. Various subscripts are used to denote specific values. For example, λ_0 = cutoff wavelength; λ_p = guide wavelength; and λ_n = wavelength normal to guide walls. Free space wavelength may be computed as $\lambda = c/f$, where $c = 300 \times 10^6$ m/S and f is the frequency in hertz.

Long-Wire Antenna A non-resonant antenna several wavelengths long, but not cut to any particular wavelength.

Loss Resistance *See* Antenna Impedance.

Magnetron A high-power SHF oscillator tube using the principle of sustained interaction and energy exchange between the electrons circulating within the device and the attendant RF field. Magnetrons find extensive use in radar transmitters where high peak pulse power is required.

Marconi Antenna A quarter-wave, vertical antenna utilizing the reflection of the ground plane to provide the other missing quarter-wave section. Marconi antennas are resonant antennas having an omnidirectional radiation pattern in the plane normal to the antenna element.

M-BWO A high-power BWO with outputs in excess of 300 watts and efficiencies greater than 30%.

MESFET A mesa FET using a Schottky barrier gate to extend the frequency range of operation.

Microstrip A parallel-wire transmission line fabricated as part of a MPCB structure.

Microwave Integrated Circuit (MIC) Gallium arsenide (GaAs) and MESFET technology combined to produce integrated devices to function at microwave frequencies.

Microwave Printed Circuit Board (MPCB) Printed circuit boards made from low-loss laminates to work at microwave frequencies.

Microwaves A range of frequencies generally identified as those between 1–100 GHz (30 cm–3 mm). There is no universal agreement as to the exact frequency allocation. At one time, microwaves were defined as waves shorter than one meter (300 MHz).

Mode (1) The manner in which the electric and magnetic fields arrange themselves in a given waveguide operating at a given frequency.
Transverse electric (TE) mode: No E-field component of the wave in the direction of propagation.
Transverse magnetic (TM) mode: No H-field component of the wave in the direction of propagation.
The m,n subscripts always follow the mode designation (e.g., $TE_{m,n}$). m specifies number of bunches of intensity (E for TE; H for TM) between the A walls. n specifies number of bunches of intensity between the B walls.
(2) For a reflex klystron, the region where oscillations will occur within the repeller (reflector) voltage range.

Modulation Index For amplitude modulation (AM), a number (m) between 0 and 1 expressing the degree to which the carrier has been modulated: $m = V_m/V_c$, where V_m is the peak modulating voltage and V_c is the peak carrier voltage.
The value of m may be determined directly from a time-domain display of the modulated carrier on a conventional CRO by $(V_{max} - V_{min})/(V_{max} + V_{min})$.
If the carrier and sideband voltages are given in decibels on a spectrum analyzer, m may be determined by $2 \log^{-1}\{[V_{sb}(dB) - V_c(dB)]/20\}$.
For frequency modulation (FM), the modulation index (m_f) is given by d/F_m, where d = maximum amount by which the carrier deviates from its unmodulated value due to amplitude variations and F_m = the modulating frequency. m_f may vary from 0 and may become infinite.

Monolithic Microwave Integrated Circuit (MMIC) MIC circuits wherein all components, both active and passive, are fabricated directly on the substrate material.

Nautical Mile (nmi) The practical unit of navigational distance equal to 1/21,600th of the earth's circumference, there being that many minutes of arc along any great circle (1 nautical mile = 6075 ft = 1852 m).

Near Field That region within the influence of the induction field of an antenna, usually $< \lambda/2$.

Negative-Resistance Amplifier Any circuit using a negative-resistance device appropriately connected may be shown to exhibit gain. In the conceptual circuit shown below, gain is given by,

$$G = \frac{4g_L^2}{4g_L^2 + g(g - 4g_L)} \qquad \textbf{[Eqn. 6–8]}$$

Gain occurs when $g < 4g_L$.

Non-Resonant Antenna An antenna terminated at the remote end in an appropriate resistance such that there are no standing waves on the antenna elements. Non-resonant antennas are highly directional. For example, a rhombic antenna is said to be non-resonant.

Normalized Value The value of Z, R, or X obtained by dividing the parameter by the characteristic impedance Z_0. Also, the value of Y, G, or B obtained by dividing by the characteristic admittance Y_0, where $Y_0 = 1/Z_0$.
Normalization is required in order to enter points on the Universal Smith Chart. Normalized values are denoted using lower case letters (for example, $Z_L/Z_0 = z_L$).

O-BWO A low-power BWO capable of full-octave tuning. Has poor frequency stability across the tuning range.

Octave Interval The interval between any two frequencies whose ratio is 2:1. Used in specifying the tuning range of a BWO or similar device. The octave interval is given by

$$\text{octave interval} = 3.322 \log_{10}\left(\frac{f_{\text{HI}}}{f_{\text{LO}}}\right) \qquad \textbf{[Eqn. 5-3]}$$

Paraboloid Antenna A type of microwave antenna (or antenna reflector) formed as a surface of revolution from a parabola. The paraboloid forms a very narrow beam and is highly directional.
The gain of an ideal parabola is given by

$$G = \left(\frac{\pi D}{\lambda}\right)^2 \qquad \textbf{[Eqn. 3–17]}$$

where D = maximum diameter
λ = wavelength
In practice, the illumination of the reflector is never perfect, and the gain becomes

$$G = 6\left(\frac{D}{\lambda}\right)^2 \qquad \textbf{[Eqn. 3–19]}$$

The beamwidth of an ideal paraboloid uniformly illuminated is given by

$$\phi = \frac{70\lambda}{D} \qquad \textbf{[Eqn. 3–20]}$$

Parallel-Lead Transmission Line Also called twin-lead, parallel-wire, or balanced line. An inexpensive transmission line often used with VHF television receivers where its characteristic impedance is 300 ohms. It is susceptible to noise and has a higher dielectric loss than coaxial cable.

Parametric Amplifier (PARAMP) A low-noise, low-power, narrow-band microwave amplifier utilizing a varactor to achieve gain.

Parasitic Array An arrangement of passive antenna elements within the induction field of the active element such that a significant modification of the radiation pattern is achieved. Generally, the passive (parasitic) elements are used to make the antenna more directive. The Yagi-Uda is an example of a parasitic array.

Peak Power The output of a magnetron in relation to the average power and the duty cycle of the device. Peak power is given by average power/DCR [**Eqn. 5–2**], where DCR = duty cycle ratio (duty cycle expressed as a decimal, not a percent).

Phased Array An antenna consisting of many elemental antennas (e.g., dipoles), each one of which is fed with a ferrite phase shifter such that the beam may be electronically steered and the antenna may remain stationary.
Used often in radar antenna systems, referred to as an electronically-steered array (ESA).

Phase Velocity (v_p) The velocity with which a point of constant phase on a progressive TEM wave is propagated.

$$v_p = \frac{c}{\sqrt{1 - \frac{\lambda^2}{\lambda_0}}} \qquad \text{[Eqn. 4–14]}$$

λ = free-space wavelength
λ_0 = cutoff wavelength
c = velocity of light (300×10^6 m/S)

$$\text{vp} = \frac{c}{\sqrt{\epsilon_r}} \qquad \text{[Eqn. 1–7]}$$

ϵ_r = dielectric constant
Note that

$$\text{f} = \frac{v_p}{\lambda_p} \qquad \text{[Eqn. 4–6]}$$

where λ_p = guide wavelength

Pillbox A type of radar antenna consisting of a truncated paraboloid resulting in a narrow horizontal beam and wide vertical beam.

Planar Transistor A type of diffused microwave transistor in which the emitter, base, and collector regions are all brought out to the same plane surface.

Plane Wave Any finite, usually small, local area of a spherical wave front. At any appreciable distance from the source, any wave front may be considered as flat without introducing a significant error into the data.

Plan-Position Indicator (PPI) A type of radar display scheme in which a ''map'' of the surrounding targets and land areas is displayed using intensity modulation of the CRT.

Plate Resistance (R_p) A fictitious resistance associated with thermionic devices.

$$R_p = \frac{\Delta V_p}{\Delta I_p}\ \Omega$$

where ΔV_p = change in plate voltage
ΔI_p = change in plate current

Polarization The orientation of the electric field of an antenna relative to the earth's surface and antenna structure.
The E field is always parallel to the antenna wire itself, and so waves are said to be either vertically or horizontally polarized. Circular polarization may also be established using the helical antenna.

Post A screw inserted into the wall of a waveguide whose depth of insertion may be varied, thus providing a variable reactance.

Power Density (P_d) A measure of the distribution of radiated power over a given area. Power from an isotropic source is assumed to be distributed over the area of a sphere and, for free space, is given by $P_t/4\pi r^2 \text{W/m}^2$ **[Eqn. 2–1]**.

Power Gain (G_p) An expression of antenna gain which takes actual losses into account as opposed to directive gain. In power gain, the ratio compares the input powers that are required to achieve a given power density at a fixed distance. Decibel power gain is calculated as

$$10 \log \frac{P_{\text{test}}}{P_{\text{iso}}} \qquad \text{[Eqn. 3–5]}$$

If the efficiency of the test antenna is known, power gain may be calculated by

$$G_p(\text{dB}) = \eta\ G_d(\text{dB}) \qquad \textbf{[Eqn. 3–6]}$$

(*See* Gain Factor.)

Pulse Duration (t) The duration (width) of the SHF energy burst from the magnetron.

Pulse-Forming Network An artificial transmission line capable of providing sharply defined pulses of precise length for the purpose of triggering the magnetron in a radar set.

Pulse Repetition Interval (T) The interval between energy bursts from the magnetron of a radar transmitter.

Pulse Repetition Rate (PRR) The rate at which the magnetron emits SHF bursts of energy in a radar transmitter. It is the reciprocal of the pulse repetition interval (T), and is defined as $1/T$.

Pyramidal Horn Antenna A type of microwave antenna made by flaring a waveguide in both directions. (*See* Horn Antenna.)

Q A figure of merit for a resonant cavity or conventional resonant circuit given by X_L/R. Q is related to bandwidth by the expression $\text{BW} = f_r/Q$, where f_r = resonant frequency.

Quarter-Wave Transformer A $\lambda/4$ section of transmission line used as an impedance-matching transformer placed between line and load. It is based on the relationship

$$Z = \frac{Z_0^2}{Z_L} \qquad \textbf{[Eqn. 1–46]}$$

which is derived from

$$Z = Z_0 \frac{1 + \Gamma}{1 - \Gamma}$$

where $\Gamma = \Gamma_L\, \epsilon^{-j\pi}$

Radar Horizon Essentially the same as radio horizon, but computed in nautical miles instead of kilometers.

$$D_t = D_1 + D_2 = 2.16\sqrt{h_1} + 2.16\sqrt{h_2} \qquad \textbf{[Eqn. 7–16]}$$

where D_1 and D_2 are in nautical miles; h_1 and h_2 are in meters.

Radar Range Equation The basic distance equation delineating the numerous factors affecting the maximum range of a given radar set.

$$r_{\text{max}} = \left[\frac{P_t A_{\text{tgt}} A_r^2}{4\pi\ \lambda^2\ P_{\text{min}}}\right]^{1/4} \qquad \textbf{[Eqn. 7–9]}$$

where P_t = total power

A_{tgt} = radar cross section of the target

A_r = effective or capture area of the receiving antenna

λ = wavelength

P_{min} = minimum power of echo that receiver will pick up

Radiation Field The electromagnetic field that leaves the antenna in time phase and space quadrature. It is created by the induction field that surrounds the antenna. The radiation field propagates at the speed of light in free space, 300×10^6 m/S.

Radiation Loss Loss of microwave energy due to the mechanism of simple radiation from a conductor carrying the signal. Radiation loss is most prevalent in parallel-wire lines and is a minimum in coaxial transmission lines.

Radiation Pattern A polar diagram of field strength measurements made at a fixed distance from an antenna in a given plane.

Radiation Resistance (R_r) *See* Antenna Impedance.

Radio Horizon The point beyond the true (geometrical) horizon at which the reception of a line-of-sight wave is just discernible. The radio horizon is about 1 1/3 times further than the optical horizon. The empirical formula for deriving the distance to the horizon is given as $D = D_t + D_r = 4\sqrt{H_t} + 4\sqrt{H_r}$, **[Eqn. 2–21]**, where D_t and D_r are distances in km from transmitter and receiver and H_t and H_r are the height of the transmitting antenna and receiver antenna in meters.

Radome A structure made of plastic-like material used to enclose the complete antenna assembly to protect the antenna from the weather and to reduce wind or snow loading.

Range Resolution The ability of a radar receiver system to discriminate between two or more closely spaced targets on the same bearing.

Range Rings Circular distance markers generated and displayed on the radar receiver for the purpose of determining distance to the target.

Reciprocity, Principle of The theory that states that any antenna capable of functioning as a radiator may also function, in principle, reciprocally as a receiving device.

Rectangular Waveguide A waveguide having a rectangular cross-section and used for propagating E-M waves by reflection.

Reflection Coefficient A complex number $\Gamma = \rho \angle \theta$ associated with a load of the form $Z_L = R \pm jX$ where ρ is the ratio of the reflected voltage (V_r) to the incident voltage (V_i) as given by V_r/V_i. Note that $0 < \rho < 1$.
The phase angle θ is the angle between V_r and V_i.
Note that $0 < \theta < 180$ degrees.
In terms of the line and load impedances, Γ is given by $(Z_L - Z_0)/(Z_L + Z_0)$ which was derived from equation (1–43):

$$Z = Z_0 \frac{1 + \Gamma}{1 - \Gamma}$$

In terms of VSWR (σ), $\rho = (\sigma - 1)/(\sigma + 1)$ **[Eqn. 1–41b].**

Reflex Klystron *See* Klystron.

Resonant Antenna An antenna whose resonant properties cause standing waves to exist along the antenna elements. Hertzian or Marconi antennas are resonant. Simple resonant antennas have omnidirectional radiation patterns in the plane normal to the radiator and bidirectional patterns in the plane of the radiator.

Return Loss (L_r) Return loss is given by $-20 \log \rho$ **[Eqn. 1–57],** where ρ is the magnitude of the reflection coefficient. The larger L_r, the closer the standing wave ratio is to unity, and the more efficiently the load utilizes the microwave energy. Note the range of ρ, SWR (σ), and L_r:

$$0 < \rho < 1$$
$$1 < \sigma < \infty$$
$$\text{infinity} > L_r\ (\text{dB}) > 0$$

That is, for a perfect load:

$$\rho = 0$$
$$\sigma = 1$$
$$L_r(\text{dB}) \rightarrow \infty$$

Rhombic Antenna A non-resonant antenna named for its rhombic shape.

Ridged Waveguide A waveguide formed with longitudinal ridges such that the cutoff wavelength may be closer to the free-space wavelength with the result that Z_0 becomes very large. Consequently, ridged waveguides find wide application as impedance matching devices.

Rotary Joint A mechanical union of two sections of circular waveguide designed to rotate relative to one another. Used extensively in connecting radar antennas to their transmitters.

S Parameter A ratio of incident and reflected voltage waves representing transmission and reflection coefficients used to characterize a linear microwave device. S parameters may be determined using a network analyzer, which measures transfer and/or impedance functions.

Sectoral Horn Antenna A type of microwave antenna made by flaring a waveguide in one direction only. (*See* Horn Antenna.)

Ship's Heading Marker (SHM) A mark displayed on the radar screen each time the rotating radar antenna lines up with the bow of the ship.

Skin Effect The tendency of electrons to confine their travel to a thin annulus region near the surface of a solid conductor due to induced voltages caused by microwave frequencies. The effect is that of reducing the cross-sectional area and raising the resistance.

Sky Wave One of three types of identifiable waves leaving any terrestrial antenna. Sky waves have frequencies between 3–30 MHz and propagate by means of reflection back and forth between the ionosphere and the earth. The other two types of waves are the ground wave and the space wave.

Slotted Line A microwave laboratory instrument used for measuring the distance between nodes of a standing wave pattern on a transmission line. The distance may then be used to accurately determine the value of a complex load terminating the line.

Slotted Waveguide Antenna A type of radar antenna made by matching precisely located slots into one side of a rectangular waveguide and terminating the remote end in a microwave-absorbing material. The antenna produces a narrow horizontal beam and a wide vertical beam.

Slow-Wave Structure *See* Helix Structure.

Smith Chart Named for Phillip Smith, American engineer. A transmission line admittance or impedance calculator based on the relationship given by $Z_L = Z_0[(1 + \Gamma)/(1 - \Gamma)]$ **[Eqn. 1–43]**. For example, if $Z_0 = 50$ and $\Gamma = 0.623\underline{/65.6^\circ}$ or, $\Gamma = 0.257 + j0.567$ in rectangular form. Therefore,

$$Z_L = 50\,\frac{1 + (.257 + j.567)}{1 - (.257 + j.567)} = 50\,\frac{1.38\underline{/24.3^\circ}}{0.94\underline{/-37.34^\circ}}$$

$$= 50\,(1.47\underline{/61.6^\circ}) = 50\,(.7 + j1.3) = 35 + j65\ \Omega.$$

The reflection coefficient completely describes the load impedance terminating the transmission line.

Space Wave One of three identifiable types of TEM waves leaving any terrestrial antenna. Space waves have frequencies greater than 30 MHz, and pass through the ionosphere relatively unaffected and continue to propagate in space. The other two types of waves are the ground wave and the sky wave.

Spectrum Analyzer A swept superheterodyne receiver capable of resolving and displaying the sinusoidal components of a complex waveform. The instrument performs the Fourier analysis electronically.

Spillover A condition of excessive illumination caused by feeding a parabolic reflector with an omnidirectional source. Spillover has the adverse effect of increasing the beamwidth.

Square-Law Detector A type of crystal detector in which the output current is proportional to the square of the input voltage.

Stripline A coaxial cable formed as a multilayer MPCB structure.

Stub A short ($<\lambda/4$) length of transmission line, shorted at one end and attached at the appropriate distance from the load for the purpose of matching a complex load to the transmission line. The actual length of the stub and its distance from the load are determined by using the Smith chart.

Surface Acoustical Wave (SAW) Device A microwave device using the piezoelectric properties of quartz or lithium-niobate to produce a narrow-band resonant filter.

Sweep Oscillator An RF oscillator capable of producing a continuous spectrum of output frequencies across a given band.

Swept Reflectometry A method of characterizing a microwave load in terms of its return loss by sweeping the device through a range of frequencies. The results are an indication of performance qualities within a complete band of operation.

SWR SWR (standing wave ratio) is the ratio of the maximum to minimum voltage on a transmission line. Also called voltage standing wave ratio (VSWR). VSWR is often symbolized by the Greek letter σ and is computed by

$$\text{VSWR} = \sigma = \frac{1 + \rho}{1 - \rho} \qquad \textbf{[Eqn. 1–40]}$$

Where ρ is the magnitude of the reflection coefficient.

Tee (1) A waveguide structure formed by the intersection of two guide shapes. Tees may be either E plane or H plane. (*See* Hybrid Tee.)
(2) An in-line device used as a temporary splice for coaxial cable transmission lines.

TEM Wave A transverse electromagnetic (TEM) wave is an electrical disturbance in space consisting of an electric field and a magnetic field at right angles to each other and to the direction of propagation.

Time-Domain The representation of a waveform in such a way that its amplitude is displayed as a function of time.

Time-Domain Reflectometry (TDR) TDR, sometimes called closed-loop radar, is a method of isolating line discontinuities or impedance changes by sending a pulse (or pulse train) down the line and analyzing the "signature" of the reflected pulse.

Timer The section of a pulsed radar set that initiates the SHF burst from the magnetron and contains the pulse-forming network.

Top Loading A short horizontal section added to the top of a Marconi antenna to add inductance, thereby increasing its effective length. The loading device often resembles a wagon wheel.

Transconductance (G_m) A measure of how much grid voltage affects control over plate current.

$$G_M = \frac{\Delta I_p}{\Delta V_g} \text{ siemens}$$

where ΔI_p = change in plate current
ΔV_g = change in grid voltage

Transferred-Electron Device (TED) A semiconductor bulk effect producing negative resistance in a variety of devices which, therefore, are capable of providing gain when properly connected in a circuit.

Transmission Line An arrangement of two or more conductors, having a precise geometry, used to convey microwave energy from source to load with a minimum amount of loss. Two common types of line are coaxial cable and parallel-lead line.

Traveling-Wave Tube (TWT) A microwave thermionic device used primarily as an amplifier, and operating on the principle of sustained interaction between an RF signal and electron beam. The heart of the TWT is the slow-wave helix structure where the interaction takes place.

T-R Switch An abbreviation for transmit-receive switch which is a form of duplexer used in radar sets.

Tunnel Diode A microwave diode that exhibits a negative-resistance property, thus allowing it to be used to achieve gain.

Turnstile Antenna An antenna array consisting of one or more stacked sections of two dipoles at right angles, and producing an omnidirectional radiation pattern. Turnstile antennas are frequently used in commercial TV broadcasting.

Unbalanced Line A coaxial cable or similar transmission line whose conductors do not experience the same electrical milieu. More particularly, the conductors do not have the same capacitance relative to ground.

Varactor A microwave device utilizing the properties of a reverse-biased P-N junction to form a voltage-controlled capacitor. Varactors are used as the operational parameter in parametric amplifiers.

Variation The amount of angular displacement east or west of true north of a compass needle for a given location on the globe.

Velocity Factor (v_f) The fraction of the speed of light that a TEM wave propagates along a transmission line as determined by the dielectric constant of the line.

$$v_f = \frac{1}{\sqrt{\epsilon_r}}$$

where ϵ_r is the dielectric constant.

Voltage Reflection Coefficient See Reflection Coefficient.

Voltage Standing-Wave Ratio (σ) The ratio of the maximum to minimum standing-wave voltages present at the load or existing on the line.

$$\text{VSWR} = \sigma = \frac{V_{\text{max}}}{V_{\text{min}}} \qquad \textbf{[Eqn. 1–39]}$$

$$\sigma = \frac{1 + \rho}{1 - \rho} \quad 1 < \sigma < \text{infinity} \qquad \textbf{[Eqn. 1–40]}$$

Waveguide A generally hollow metallic structure through which microwave energy propagates by reflection rather than conduction.

Waveguide Impedance (Z_0) A frequency-dependent impedance offered to the propagation of a wave by a guide structure. For rectangular guides, Z_0 is given by $377/\sqrt{1 - (\lambda/\lambda_0)^2}\ \Omega$ **[Eqn. 4–23]**.

For circular guides, Z_0 is calculated by $377\sqrt{1 - (\lambda/\lambda_0)^2}\ \Omega$. **[Eqn. 4–24]**.

Wavelength (λ) The distance traveled by a point on a periodic TEM wave in the time required to complete one cycle.

$$\lambda = \frac{v_p}{f} \text{ meters}$$

$$v_p = \text{phase velocity} \qquad \textbf{[Eqn. 1–8]}$$

$$f = \text{frequency in hertz}$$

If $v_p = c$, the resultant wavelength is called the free-space wavelength.

Wave Impedance (Z_w) The inherent impedance of free space equal to 377 ohms.

$$Z_w = 377 = \sqrt{\frac{\mu}{\epsilon}}\ \Omega \qquad \textbf{[Eqn. 1–13]}$$

where = absolute permeability of free space
(1.26×10^{-6} H/m)
and ϵ = absolute permittivity of free space
(8.854×10^{-12} F/m)

Wave Propagation Model A mathematical model used to describe or predict certain aspects of TEM wave behavior. The model assumes an isotropic radiator in free space.

X-Band A range of microwave frequencies from 8.2 to 12.4 GHz. As defined by the FCC for radar use, the X-band is between 9.3 and 9.5 GHz.

X-Y Recorder An analog plotter capable of producing a hard-copy trace of a complex waveform.

Yagi-Uda Antenna A parasitic array designed to increase both the gain and directivity of a simple dipole. Yagi-Uda antennas are frequently used for television reception, and may consist of either a simple dipole fed with a 75 ohm coaxial line, or a folded dipole fed with a 300 ohm parallel-wire (twin lead) transmission line. Yagi-Uda antennas are often incorrectly called "Yagis," but the discovery was actually made by Dr. Uda and the English results were published by Yagi in 1928.

Zoned Lens A type of dielectric lens shape used to reduce excess weight and attenuation.

INDEX

WE VALUE YOUR OPINION—PLEASE SHARE IT WITH US

Merrill Publishing and our authors are most interested in your reactions to this textbook. Did it serve you well in the course? If it did, what aspects of the text were most helpful? If not, what didn't you like about it? Your comments will help us to write and develop better textbooks. We value your opinions and thank you for your help.

Text Title ______________________________ Edition __________

Author(s) ______________________________

Your Name (optional) ______________________________

Address ______________________________

City ____________________ State __________ Zip __________

School ______________________________

Course Title ______________________________

Instructor's Name ______________________________

Your Major ______________________________

Your Class Rank _______ Freshman _______ Sophomore _______Junior _______ Senior

_______ Graduate Student

Were you required to take this course? _______ Required _______Elective

Length of Course? _______ Quarter _______ Semester

1. Overall, how does this text compare to other texts you've used?

_______ Superior _______Better Than Most _______ Average _______Poor

2. Please rate the text in the following areas:

	Superior	*Better Than Most*	*Average*	*Poor*
Author's Writing Style	_______	_______	_______	_______
Readability	_______	_______	_______	_______
Organization	_______	_______	_______	_______
Accuracy	_______	_______	_______	_______
Layout and Design	_______	_______	_______	_______
Illustrations/Photos/Tables	_______	_______	_______	_______
Examples	_______	_______	_______	_______
Problems/Exercises	_______	_______	_______	_______
Topic Selection	_______	_______	_______	_______
Currentness of Coverage	_______	_______	_______	_______
Explanation of Difficult Concepts	_______	_______	_______	_______
Match-up with Course Coverage	_______	_______	_______	_______
Applications to Real Life	_______	_______	_______	_______

3. Circle those chapters you especially liked:
   1 2 3 4 5 6 7 8 9 10 11 12 13 14 15 16 17 18 19 20
   What was your favorite chapter? _______
   Comments:

4. Circle those chapters you liked least:
   1 2 3 4 5 6 7 8 9 10 11 12 13 14 15 16 17 18 19 20
   What was your least favorite chapter? _______
   Comments:

5. List any chapters your instructor did not assign. ______________________________

6. What topics did your instructor discuss that were not covered in the text? ______________________________

   ____

7. Were you required to buy this book? _______ Yes _______ No

   Did you buy this book new or used? _______ New _______ Used

   If used, how much did you pay? ___________

   Do you plan to keep or sell this book? _______ Keep _______ Sell

   If you plan to sell the book, how much do you expect to receive? ___________

   Should the instructor continue to assign this book? _______ Yes _______ No

8. Please list any other learning materials you purchased to help you in this course (e.g., study guide, lab manual).

   ____

9. What did you like most about this text? ______________________________

   ____

10. What did you like least about this text? ______________________________

    ____

11. General comments:

May we quote you in our advertising? _______ Yes _______ No

Please mail to: Boyd Lane
College Division, Research Department
Box 508
1300 Alum Creek Drive
Columbus, Ohio 43216

Thank you!